Humankind's Greatest Steps

Second Edition

July 2020

Copyright © by Dubán H. García S.

This text, as a whole, is due to my own creativity and therefore I own the author's rights; consequently no one should publish or distribute it without a true authorization contract. On the other hand, some sections of the book and some images used correspond to works that are framed within the general context of the Public Domain or also of the Public License, as indicated in each particular case; thus, any person may freely use such sections and images.

To all my relatives and my friends

"That's one small step for a man, one giant leap for Mankind."

Neil Alden Armstrong (1930- 2012)

The Moon, July 21, 1969

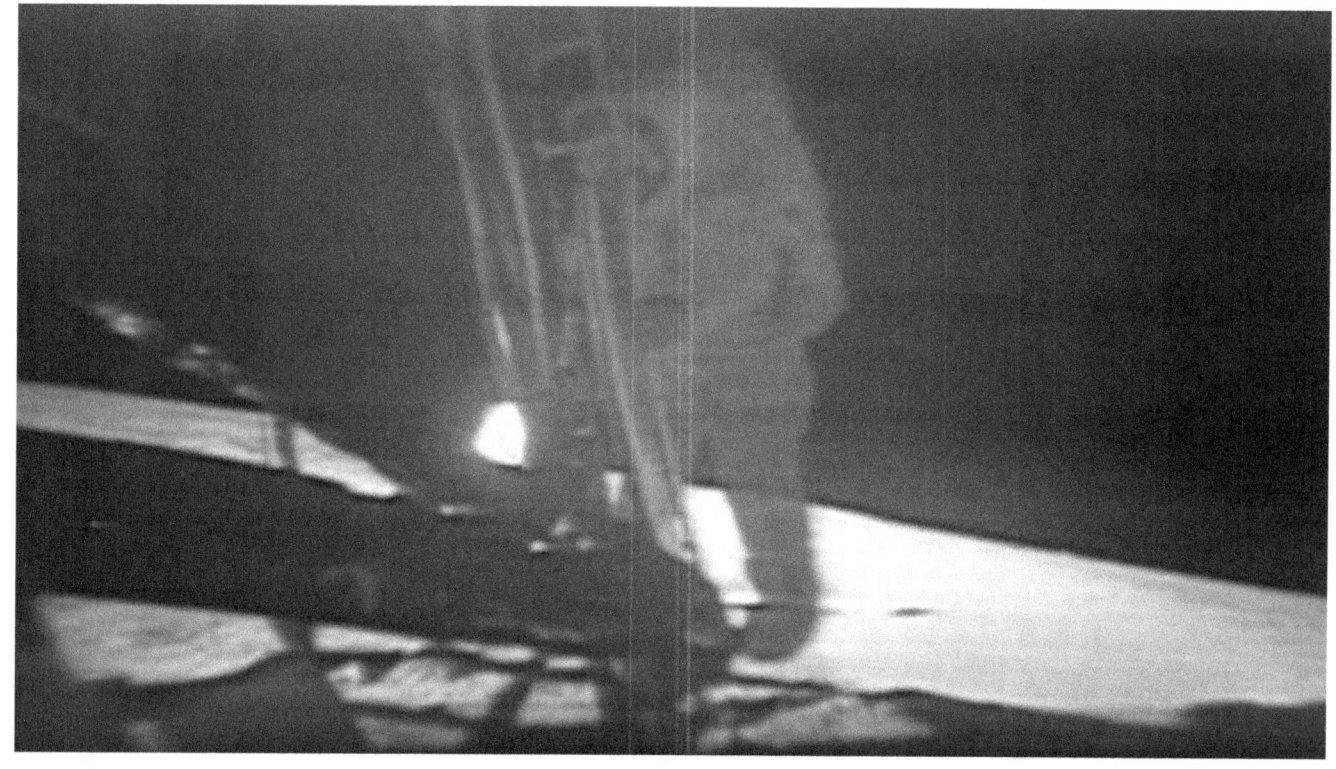

On July 21, 1969, the American astronaut Neil A. Armstrong prepares to take the first step of a man in another celestial body: on the Moon.

Source and credits of the image: NASA.

https://www.nasa.gov/mission_pages/apollo/apollo11.html

Humankind's Greatest Steps

Thrilling History of the Lunar Astronomy and the Fifty Years of the Arrival of Man to the Moon

Dubán H. García S.

July 2020

Acknowledgments

I wish to express my profound gratitude to all those people and institutions that have placed their texts and images in the field of the Public Domain; which have been of great value to illustrate the respective contents in this text.

Special thanks to the National Aeronautics and Space Administration of the United States, NASA, for having their images within the Public Domain; and in particular those contained in its website **NASA's Scientific Visualization Studio:** https://svs.gsfc.nasa.gov/4604

I also want to thank all my family and friends who supported me and encouraged me during the writing of this book.

Content

	Preface	xi
Chapter 1:	The Moon in the Cosmology of Ancient Civilizations	1
Chapter 2:	Lunar Astronomy in Greek Cosmology	23
Chapter 3:	Lunar Astronomy in Greco-Roman Society	51
Chapter 4:	A Medieval Look at Our Satellite	75
Chapter 5:	The Moon in Renaissance Astronomy	95
Chapter 6:	A Telescopic Look at Our Satellite	117
Chapter 7:	Modern Astronomy and the True Lunar Movement	161
Chapter 8:	Twentieth Century: Wars, Space Race and Manned Missions to the Moon	195
	Image Catalog and Their Licenses of Use	239

x

Preface

Surely the fundamental purpose of the study and the dissemination of history is to discover the deepest and most relevant meanings of those human actions that precisely are being considered.

During very immediate or close times to the very development of events, our disposition and ability to fully understand them are seriously influenced by highly subjective elements: strong emotions and animosity generated by the same events, great feelings of euphoria and joy or, on the contrary, of sadness and despair according to the historical circumstances and according to our own expectations and philosophies of life.

Although vivid and valid, the high subjectivity charge contained in such emotions usually hides and eclipses the deeper underlying meanings of those events.

In the present text I refer to the 20th Century as having been bittersweet. And so I consider it, because its first three lustrums brought us the car, the airplane and the most modern and advanced physical and cosmological theories. But then there were two fateful world wars that forever tarnished human history.

The circumstances under which the Second World War culminated led to the development of a Cold War, and within this an Arms Race was unleashed, and within this, a Space Race. This latter had its apex moment on July 21, 1969, when humans put their feet on another celestial body for the first time in history: on the Moon.

According to the above, it is perfectly clear that this ultimate achievement of humanity should have been interpreted and assimilated by the people and societies of that time within an exclusively geopolitical context: the triumph of one nation and the defeat of another. And by its own inertia, this general vision of this event has persisted up to the present.

But fifty years have passed since that historical event. Time has passed, many things have changed; the hustle, the fever, the effervescence and the desperation of that Cold War now no longer exist; and the old rival nations are better devoted in playing enjoyable and popular worldwide football championships.

That is why, after half a century, the pertinent question now is whether to continue treating the arrival of Man to our satellite within an exclusively geopolitical context. My answer is that not. That is my bet and that is the background idea that underlies all this text: Within the scientific and technological field, the arrival of Man on the Moon is the greatest historical achievement of Humanity and must be considered with much broader perspectives. Here I do it from the point of view of the Universal history of knowledge, science and technology; and hence its name: *Humankind's Greatest Steps*.

Thus, this book is written in historical key: Considering the temporal evolution of knowledge in general, of astronomy in particular and especially that of Lunar

Astronomy. To explain the level of understanding that the brightest minds of all times have had about our beloved satellite, knowledge that have made possible travelling to it, is the fundamental purpose of the text. After this, the next goal is to quickly review the history of those Apollo Manned Space Missions that brought the first men to the lunar globe.

To achieve these objectives it is necessary to divide history, as is the normal practice, in different epochs or clearly differentiated periods and that are very obvious. Thus, in the first chapter, entitled *The Moon in the cosmology of ancient civilizations*, the field of prehistory is discussed and it is analyzed how the people of those early civilizations understood and interpreted from the mythological point of view not only the Moon but also the other stars.

In a text on universal history, it cannot be excluded the ancient Greek society, then the *Lunar Astronomy in Greek cosmology* is the subject of the second chapter. Here it can be appreciated how in this ancient civilization the transcendental step from mythology to philosophy took place; and within this philosophical context arose the first great astronomers of humanity, whom were very interested in finding rational explanations for all celestial phenomena. The great Aristarchus of Samos is the most outstanding of them, very famous for being the first person from whom works are preserved that prove that he was the original promoter of a heliocentric theory of the universe, and that additionally developed a scientific methodology to measure the distance of the Moon to the Earth, thus giving rise to the *Lunar Astronomy*.

But this Greek civilization was invaded and almost destroyed by the Roman society of the time; and although some of its material elements disappeared, the Greek culture subsisted and merged with that of its invaders, thus giving rise to a new society. The title of the third chapter is *Lunar Astronomy in Greco-Roman Society*; and there it is studied how the Greek cultural, philosophical and astronomical heritage continued and flourished within the general context of Roman society. Of this period it is worth mentioning Hipparchus of Nicaea, a great continuator of the lunar thought of Aristarchus, and from whom we could say that he is the first really great observational astronomer. Next we have Claudius Ptolemy, who compiled, wrote and transferred, through his work the *Almagest*, the Greek astronomical knowledge to the coming societies. It is also very worth remembering the great historian Plutarch of Chaeronea for his exciting paragraphs about the satellite in his text *On the face that appears in the orb of the Moon*.

And as everything perishes, the colossal Western Roman Empire collapsed definitively in the year 476AD, and from that moment began what in history is recognized as the Middle Ages. And then the magnificent Greek cultural heritage overcame a new crisis, because by the way of the Almagest the Greek astronomical knowledge was transferred to the Byzantine and Arab societies of this period. In *A Medieval Look at Our Satellite*, name of the fourth chapter, we study how medieval Arab astronomers also paid special attention to the Moon and its phenomena within the general framework of the astronomical theory expounded in the Almagest.

Advancing already the time, we arrived at the fifth chapter denominated *The Moon in the Astronomy of the Renaissance*, and we meet with the first authentically modern astronomers. Nicolaus Copernicus captures all our attention for having definitively established the foundations of a modern

heliocentric theory of the Universe; and also for having been the first to place the Moon in its correct place within the Solar System: an earthly satellite, our only natural satellite. Another great character of this era is Tycho Brahe, the best observational astronomer at the time, whose astronomical data were of incalculable value to the next generation of astronomers.

A Telescopic Look at Our Satellite is the name of the sixth chapter, and Johannes Kepler is the first character studied. His *Three laws of planetary movements* perfectly describe the movements of the planets around the Sun, as well as the movement of the Moon around the Earth. The other great astronomer considered is Galileo Galilei, who consecrated within celestial science for having re-designed and introduced the telescope as an astronomical research instrument, for having directed it towards the Moon and for having written in detail which he could contemplate and discover on the satellite surface.

The seventh chapter is called *Modern Astronomy and the True Lunar Movement*. So far all astronomers had managed to alright describe both the position and the movement of the Moon in the sky, but they had not been able to explain them in terms of cause-effect relationships. This could only be possible after the famous English physicist Isaac Newton established modern classical physics, with his Three laws of the movement of bodies and with his Law of universal gravitation. It is this Newtonian physics that makes possible the development of Celestial mechanics, a branch of astronomy that explains the movements of celestial bodies in general, and that of the Moon in particular, within the logic of cause-effect relationships, that is, based on gravitational forces.

Finalizing the text, we have the eighth chapter, *XX Century: Wars, Space Race and manned missions to the Moon*. Here there is a very quick historical synthesis of the beginning of this century, until we reach the point that interests to us: the Space Race with its crewed missions to the satellite. In a more detailed way, the first space programs with their respective achievements are described for both powers of the time, the Union of Soviet Socialist Republics and the United States. The chapter ends with a general description of the manned flights to the Moon, the *Apollo Missions*, and its crown moment with the first men walking over the satellite surface, which is the universal event that was commemorated on last July 20, 2019.

Dubán H. García S.

June 25, 2020

Chapter 1

The Moon in the Cosmology of Ancient Civilizations

At the exact moment of New Moon our satellite is practically invisible from the Earth, this is because its entire illuminated hemisphere is just in front of the Sun, and no light is reflected towards us. The image shows the satellite as it was viewed from Earth on Thursday, January 18, 2018 at 04:00 Universal Coordinated Time, UCT; about 25 hours 43 minutes after New Moon, only the 1,0% of its illuminated hemisphere can be viewed: a tenuous curve of light on the right side. In societies governed by a lunar calendar, this first sighting of the Crescent Moon marks the beginning of a new lunar month.

Courtesy of NASA's Scientific Visualization Studio: https://svs.gsfc.nasa.gov/4604

"And the face of the Sun was bitten. And it became dark and his face went out. And then they were scared up. «¡It has burned! ¡Our god has died! » Said his priests. And they were beginning to think about making a painting of the figure of the Sun, when the Earth trembled and they saw the Moon."

Chilam Balam de Chumayel.[1]
Maya text of the sixteenth century.

"To men it seems that at their sides is that half circle in which it's portrayed how the Sun is bitten. It is here that it is the one in the middle. What bites him is that he paired with the Moon, who walks attracted by him before biting him. She arrives on her way to the north, great, and then they become one and they bite the Sun and the Moon, before arriving at the "Trunk of the Sun". It is explained so that Mayan men know what happens to the Sun and the Moon.

Eclipse of the Moon. It's not that she's bitten. She interposes with the Sun, on one side of the Earth. Eclipse of the Sun. Not that he is bitten. He stands with the Moon, on one side of the Earth.
This is a sign given by God that they are equal; but they do not bite."

Chilam Balam de Chumayel.
Maya text of the sixteenth century.

Prehistory and Archaeoastronomy

The study of the history of astronomy shows us that men of all latitudes and epochs have been assiduous observers of the firmament, and that within the celestial bodies that have most captivated their attention, for their great size and radiance in the sky, are the Moon and the Sun; which were widely studied through the attractive phenomenon of eclipses. Likewise, from this history we can point out two elements common to almost all primitive civilizations, very important both for their recurrence and for their significance. In the first place we have that those ancient civilizations, in addition to their habit of observing the sky and the celestial phenomena, also developed the worthy practice of elaborating written records of their observations, in multiple and lasting formats. Fortunately a good amount of those records has reached the present, and they now allow us to understand that these practices were systematic to a great degree.

The second element concerns the fact that basically all primitive civilizations had well defined periods in which the mythological form of thought prevailed for the interpretation or conception of the respective celestial bodies and their phenomena, as well as of the universe as a whole. Civilizations in their most primitive stages always resorted to mythology, or equivalently to supernaturalism, with the purpose of elaborating explanations for all kinds of natural phenomena. The Moon, the Sun, and their eclipses have such a great notoriety and attraction that they always occupied a central place in the primitive mythological conceptions of the universe; the examples are so abundant that we could say there are so many super naturalist interpretations for these stars, as primitive civilizations existed. In a natural way such stars became supreme deities, and were symbolically represented as a symmetrical brightness or glow that illuminated everything. That being the case, we are going to remember here only some of the most striking cases.

Archaeoastronomy is a relatively new scientific discipline that began to consolidate in the second half of the 20th century. Its objects of study are both the astronomical thought and practices of prehistoric societies; while its methodology resorts to the use of evidences provided by archeology, and the use of ethnological sources such as legends and tribal traditions. It's known that many ancient peoples were fully aware of the celestial phenomena, as indicated by the alignments of its megalithic monuments, including Stonehenge and Avebury in England, which were declared as World Heritage Sites by the UNESCO in 1986 [2]; additionally, Newgrange in Ireland and the megalithic alignments of Carnac in France; all of them from the Middle Neolithic period, about 5500 to 4500 years ago.

Stonehenge is considered the most famous prehistoric monument in the world. The last stage of construction was completed at the end of the Neolithic period around 2500 BC. Currently the most accepted interpretation of Stonehenge is that of a prehistoric temple aligned with the movements of the Sun, since the main axis of the stones corresponds to the line of the solstices: In the middle summer, the sun rises above the horizon to the northeast, near the *Heel Stone*; while in the middle winter, the King star is placed in the southwest, in the gap between the two highest trilithons.[3] These positions of the Sun in the seasonal cycle were obviously important for the prehistoric societies that built and used Stonehenge.

Because the Stonehenge builders had discovered and accurately recorded the positions of the Sun and Moon, and had erected a monument marking their movements and positions accurately, they could have recognized when the Moon was in the process of intercepting the position of the Sun to cause a solar eclipse. Additionally, they could foresee when the Moon was going to a position directly opposite to the King star, which would take it inside the Earth's shadow to cause a lunar eclipse. It is almost certain that they could not predict what kind of eclipse would occur, total or partial one, or where it would be seen; but they might have been able to warn that on a particular day or night the occurrence of an eclipse of the Sun, or of Moon, was possible.

Although there are convincing evidences of alignments of astronomical importance in a variety of archaeological sites, the theses on high-precision primitive astronomy, such as the ability to predict eclipses by using the Aubrey holes in Stonehenge, have not been accepted by the totality of the archaeologists and astronomers. On the other hand, the theses that the Neolithic peoples were recording

the solstices for mythological and ceremonial purposes, and observing the solar cycles to determine the seasonal changes, are much more consistent with the general archaeological and anthropological reconstruction of the cultures of those first sedentary food producers societies of Neolithic times. So, the discussions about the presumed intentions and practices of the prehistoric peoples who built those intriguing sites will continue for a long time.

Perhaps the most decisive characteristic of the Neolithic period is that there is a transition from the nomadic to the sedentary way of life, with the appearance of agriculture and with it the first permanent human settlements, and therefore the first signs of civilization. After centuries of evolution writing and the first manifestations of systematic knowledge arose: geometry, mathematics and cosmology. The Middle Neolithic period was the moment when the geometric idea of space was formulated, and when the cosmos came to be conceived as having a decipherable, understandable pattern. Ancient cosmologies reveal two basic views of the universe: the ancestral cosmologies of circular flat Earth, as in the primitive Babylonian, Egyptian and Greek civilizations; which would later be replaced by spherical cosmologies, so characteristic of the posterior Classical Greek civilization.

Mesopotamia and Babylon

Much of the knowledge about astronomy, the Moon, the Sun, eclipses and calendars that extended from the Middle East to Greece and Rome, and from there to the rest of Europe, arose from apprenticeships developed between the years 3000 and 500 BC. in Mesopotamia, the geographical area of the Middle East located between the Tigris and Euphrates rivers. The Babylonian civilization of Mesopotamia replaced the Sumerian and Acadian around 2000 BC., and had as its headquarters the city of Babylon, that of the mythical *Hanging Gardens*. Sumerians, Acadians, Chaldeans, and Elamites were societies belonging to the Babylonian Empire, whose splendor was reached around 1700 BC.

The Babylonians developed a cuneiform script: characters engraved with a wooden wedge on a clay tablet. In mathematics they developed an advanced numerical system of base 60, instead of base 10 of the modern system. With regard to geometry, they knew the circle that was divided into 360 units, and they had mathematical formulas for calculating its area and its circumference, perimeter or size.

Image 1.1
Archaeoastronomy
Sunset at Stonehenge
Many archaeologists and astronomers argue that Stonehenge had astronomical applications in determining the movements of the Sun and the Moon, and predicting their eclipses.

The Babylonians' cosmology was a compound of the ideas that originally prevailed in the territories around the two ancient sanctuaries, Eridu on the coast of the Persian Gulf and Nippur in northern Babylon. According to the cosmology of Eridu, water was the origin of all things; the inhabited world had emerged from the depths and is still surrounded by *Khubur*, the ocean current; beyond which the sun god grazes his cattle. They assumed that the vault of the sky did not move, but the Sun, the Moon and the stars were living beings or deities that moved along their respective paths. In Babylonian mythology *Sin*, or also *Nanna*, was the lunar God, while *Utu* or *Shamash* was the solar and justice God. As for his abode, the Babylonians conceived the Earth in the form of a flat disk or circle floating in the ocean, with the vault of the sky arched over it.

Image 1.2
Cuneiform Tablet:
Text allusive to the Moon

It is a tablet of clay from Mesopotamia, most probably from Babylon, dated on the fourth century BC. It is preserved in the Metropolitan Museum of Art, New York City.

There are good reasons to study the history of Mesopotamian astronomy; first, it offers a starting point to understand their cultural development. Then we have the theme of astronomical records, which were not only subsequently used by the ancient Greeks, but have also reached the present and are now relevant to modern science. What is usually called applied historical astronomy makes use of old records to derive scientific value data for modern astronomical research. The recent use of Babylonian records related to eclipses to determine if the Moon is actually moving away from Earth due to a process known as *Secular Acceleration*, and also to determine if the deceleration of the Earth's rotation movement is much greater than previously thought, is a good demonstration of the scientific value of such ancient records.

For centuries the positions of the bright stars were established; and the movements of the Sun, the Moon and the planets were traced at the bottom of those fixed stars. Gradually, the Babylonians acquired a remarkably accurate knowledge of the movements of the Sun, the Moon and the planets; so that they were able to predict the positions of these celestial bodies among the other stars and the recurrence of eclipses; without having formulated, as far as it's known, some kind of geometric theory of celestial movements. All these data were recorded in cuneiform writing on clay tablets, many of which are still available today.

Babylonian astronomers made systematic eclipse records for long periods of time, and there is direct evidence that they discovered the periodicity, and therefore the predictability, of such celestial events. They understood well that a lunation, the period between two successive new moons, also known as the lunar month, or also a synodic month, is a little more than 29.53 days, and they were very precise in their predictions of new moons. Additionally, they noted that each lunar eclipse was part of a set of eclipses that took place at equal time intervals; each set generally included five or six eclipses; and the sets were separated by long time intervals of 17 lunations, during which there were no eclipses. Since the Babylonian calendar was based on the lunar months, once a series of eclipses had begun, it was possible to predict lunar eclipses at intervals of six months, when the full moon was visible. Then, and within a primitive superstitious framework,

Babylonian astronomers predicted good or bad events according to the concordance between observation and prediction; in such matters, solar eclipses were the most important for astrological purposes. And although they didn't have an accurate representation of the complicated lunar movements, they had devised some sufficiently reliable prediction means, based on long data sets for lunar eclipses. In the reports of the court of astronomers-astrologers of Nineveh we can read such predictions: *"To the King my lord I have written: an eclipse will occur. This eclipse has taken place; it did not fail. This is a sign of peace for the King my lord."*[4]

From the time of Nabonassar, Babylonian monarch who reigned between 747 and 734 BC, such phenomena were systematically observed; and the first lunar eclipse that they observed with great care from the moment of the beginning until its end, was that of March 19, 721 BC. A fragment of a list of eclipses between 373 and 277 BC has survived; and it is divided into columns that cover a period of 223 synodic months, which is equivalent to 18 solar years or 6585,33 days. This period is now known as *Saros*; during it our satellite returns to the same position with respect to its nodes, its perigee and the Sun; and the eclipses of the previous cycle are repeated in the same order. The discovery of the periodicity of the movements of the celestial bodies constitutes a great advance in knowledge, and we could describe it as the humankind's first great step in the direction of understanding the universe and deciphering its phenomena: knowing precisely the movements of the stars, especially the Moon and the Sun, and being able to predict, anticipate, their positions in the future, will always continue to be the engine that drives the development of astronomical knowledge.

Ancient records prove that the Babylonians and Assyrians knew that lunar eclipses can occur only in Full moon, and solar eclipses only in the New moon. The prediction of solar eclipses in those times was more difficult; although they occur in sets analogous to those of the Moon, there is a greater difficulty in relation to the latter, apart from the cloudy weather and the location of the Moon below the horizon. This difficulty refers to the narrowness of the lunar shadow cone on the surface of the Earth, which makes a solar eclipse visible only in very particular and narrow regions: an observer in a given place can lose up to five out of six solar eclipses.

The Babylonians also initiated the concept of the *Zodiac*, the band on the celestial sphere through which the Sun apparently moves and performs a complete revolution in a year; trajectory known as *Ecliptic*, precisely because it is in the plane that contains it where the Moon must be so that there is an eclipse. Band that they divided into 12 equal zones of 30° corresponding to the 12 lunar months of their calendar, and then assigned each one to the nearest constellation; areas that we know today as *zodiac signs*. The Moon and the planets also move in the same band, but in totally different ways, and we now know that it represents the plane of the ecliptic in which the planets revolve around the Sun, and the Moon around the Earth.

Ancient Egypt

The primitive Egyptian cosmology was deeply intertwined with its mythology: the sky was represented by *Nut*, a goddess in the shape of a woman who arched her body extending her extremities to encompass the whole firmament; the god *Geb*, the Earth, served him as a support, corresponding the points where he supported to the four cardinal points. The arched body of *Nut* gave rise to the celestial vault, identified with *Shu*, a cosmic deity who personifies the atmospheric air and celestial light. *Nut* gave daily birth to the Sun, the god *Ra*, who was usually depicted in human form with a hawk's head, crowned with the solar disk surrounded by the *Uraeus*, a stylized representation of the *Sacred cobra*, associated with the mythical travels of the Sun through the sky and the underworld. *Ra* had two boats to complete his daily trip, one for the day and one for the night; during the day the boat was occasionally attacked by a huge serpent, *Apophis*, personification of chaos and great enemy of *Ra*, for which the Sun was eclipsed for a short time. At the beginning of the Old Kingdom, *Ra* was only one of many existing solar deities, but around 2400 BC, during the V dynasty, he was elevated to a national deity, becoming the official god of the Egyptian pharaohs.

Iah was the goddess of the Moon, a personification of that celestial body; it could be represented as a

crescent moon, an ibis or a hawk. Like the Sun, the Moon had its enemies: a sow attacks it on the 15th of each month, and after an agony of fifteen days and increasing pallor, the Moon dies and is born again; sometimes the sow swallows it for a short time, causing a lunar eclipse.

That the Egyptians were accomplished observers and surveyors is evident by the construction of pyramids and temples normally aligned from north to south with a precision of a few minutes of arc. Between the years 3000 and 2500 BC the Egyptians already systematically observed the sky, and realized that the stars made a complete turn in little more than 365 days; they also verified that the Sun cycle was in agreement with that of the seasons and with the annual growth of the Nile River, which helped them to prepare a solar calendar; the 365-day solar year seems to have been adopted by the Egyptians in this period.

The literature suggests that the Egyptians were less likely than other ancient peoples to perceive bad omens in celestial events; although in some texts there seem to be indications about the interpretation of eclipses as evidence of the titanic struggle between good and evil; there are also descriptions that suggest that they were interpreted only as a "*meeting of the Sun and the Moon*".

A possible Egyptian reference to a solar eclipse is found in the demotic papyrus Berlin 13588, in which a priest named *Amasis* mentions having heard in the Egyptian city *Tjeben* the following: "*The sky swallowed the solar disk when he was taken to the room of embalming, in which the body of the Psamtik king was to be prepared for burial.*"[5] Here reference is made to the King Psamtik I, founder of the XXVI dynasty, and the referred eclipse is the partial one of Sun of September 30, 610 BC.

Ancient Greece

The Greek civilization goes back until the year 2800 BC., and although it took much of the surrounding cultures, like the Egyptian and Babylonian, the Greeks built a society and a culture of their own that is the most impressive of all civilizations. The influence of the Egyptians and the Babylonians was felt especially at Miletus, a city of Ionia in Asia Minor and birthplace of Western philosophy, mathematics and science.

**Image 1.3
Egyptian Papyrus:
Astronomical catalog of planets,
in demotic script**

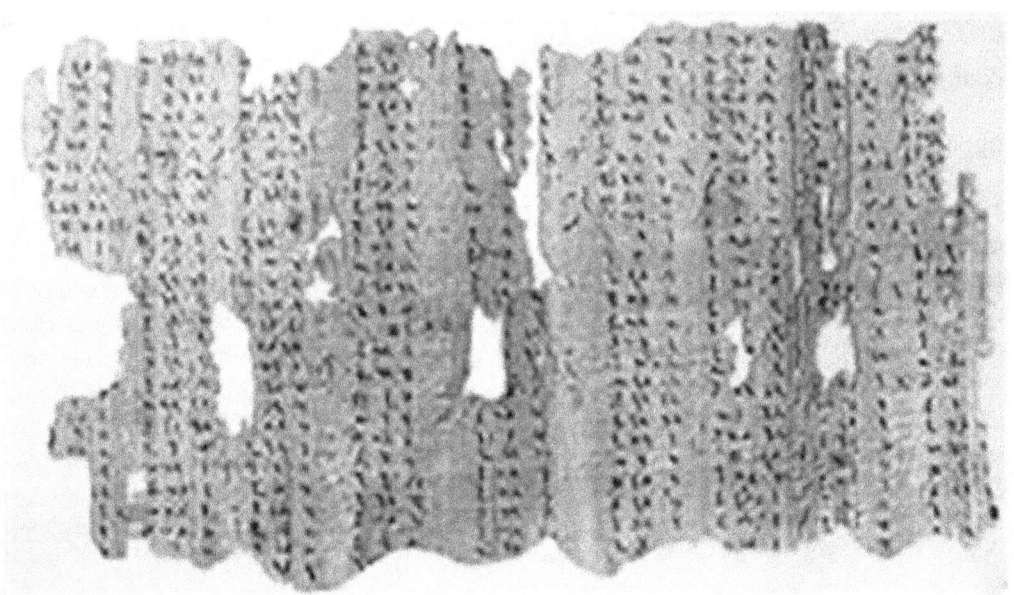

The ancient Egyptians had three types of writing: hieratic, hieroglyphic and demotic. Some of their records have reached the present in the format of papyrus scrolls, support material for writing invented by the Egyptians around the year 3000 BC.

Very early on, as in oriental societies, Greek primitive cosmological thought followed purely mythological lines; it is from the first half of the 6th century BC that Greek philosophy has its roots. The earliest Greek literary creations are materialized in the poems of Homer (700-610 BC), the author of *The Iliad* and *The Odyssey*; very fortunately in such poems there is a valuable source of information about the cosmological ideas that prevailed before the advent of Greek philosophy, since they present us with a striking image of the heavens and of the Earth as they were conceived by the Greeks of such epoch. The firmament was imagined as a great vault or bell that covered everything, and under which move the Sun, the Moon, the planets and the totality of the stars; celestial bodies that rise from the mighty river *Oceans* in the East and sink back into it from the West. The Earth was represented as a flat circular disk surrounded by the Oceans River, where our planet floats like a piece of wood or something. In Greek mythology *Urania* was the muse of astronomy, astrology, poetry and knowledge; while the solar God was *Helios* and *Selene* was the lunar Goddess: just before dawn, by the east and preceded by the *Morning Star* or planet Venus, the *Aurora* rises from the Oceans, after which *Helios* makes its appearance from the *Lake of the Sun*, also in the east.

The word *Selene* comes from ancient Greek and is translated into English as light, or brightness. It constitutes the root of selenology, or lunar geology: the scientific study of the formation and composition of the Moon, as well as its movements in the heavens; of selenography or lunar geography, which includes the study of the geographical characteristics of our satellite. It is also the name of an asteroid discovered in 1905: *580-Selene*; and the Selenium chemical element was baptized in honor of the Greek lunar goddess. The Platonic Dialogues are a set of works written by the Greek philosopher Plato (427 - 347 BC), which have been preserved in their entirety and which constitute his great literary and philosophical legacy. In the dialogue entitled Cratylus, Plato makes reference to the origins of this word, as well as to the thoughts of the Greeks about the Moon and the goddess *Selene*:

"Hermógenes: And the Moon σελήνη? (seleenee ?)
Socrates: That is a word that mortifies Anaxagoras.
Hermógenes: Why?
Socrates: Because it seems to testify the antiquity of the doctrine, recently taught by this philosopher, that the Moon receives sunlight.
Hermógenes: How?
Socrates: The words σελας and φως (selas and foos) have the same meaning (light).
Hermógenes: Without a doubt.
Socrates: All right; the light received by the Moon is always new and old, if the disciples of Anaxagoras say true; because by turning the Sun around the Moon, it sends her an always new light; whereas the one it has received the previous month is already old.
Hermógenes: Agreed.
Socrates: Many call the Moon σελαναία (selanaia).
Hermógenes: Agreed.
Socrates: And since the light is always new and old, no name can agree better than σελαενονεοαεια (selaenoneaeia), where by abbreviation it's say: σελαναία (selanaia, selene)." [6]

In most texts on Greek mythology Selene was the personification of the Moon, and was the daughter of the Titans *Hyperion* and *Tea*; and sister of *Helios*, the god of the Sun, and of *Eos*, the goddess of the dawn. After Helios ends his daily journey through the sky, when night falls on Earth, Selene begins her own. In Athens, Selene was associated with the fecundity of women and other animals, as well as the growth of plants. It was also invoked by lovers and much sought after when practicing magical arts. In art, Selene was represented as a beautiful woman with a pale face, who cross the skies driving a silver carriage pulled by a yoke of white oxen, or also, by a pair of horses. Likewise, she was drawn wearing tunics and riding a horse or a bull, carrying a half moon over her head and carrying a torch.

According to the mythology Selene had love affairs with Endymion, a beautiful shepherd who used to take his flock to the mountain of *Latmos*. One night

the young man fell asleep at the entrance of a cave; the night was clear and in the sky Selene was walking in her carriage; the light of the Moon then entered the cave and thus Selene could see the young man asleep, falling madly in love with him. She descended then from Heaven, leaving the cavern completely illuminated by the silver light of the Moon; Endymion was awakened by the brush of Selene's lips on his ones; when he saw the brilliant goddess, a great passion was born between them. With the passage of time, and wishing to enjoy forever the beauty of the mortal, the visits to Endymion were more frequent and extensive; which caused repeated very dark nights in which only the timid stars shone. Such absences caught the attention of Zeus, who since then forbade Selene from leaving the heavens to visit the pastor; in exchange for this she begged Zeus to grant her lover the gift he wanted. The pastor asked himself two things: eternal youth and sleep; then, Zeus caused the young man to stay deeply asleep forever, but with eternally open eyes. Since then, every night Selene stops in silence to contemplate her beloved, caressing him with her luminous rays; because even while he was asleep, he was always was contemplating her.

The societies of the Far East

In the astrology of ancient China it was considered that celestial events had a mutual influence on human affairs, especially on questions of state and politics. This is based on the correlative thinking of Chinese cosmology: The heavens and the state influence each other in a total system of cosmology; so astronomy was intimately connected with government administration. Abnormal celestial phenomena, such as asteroids and comets, eclipses, supernovas, etc., were interpreted as the result of mismanagement of the state, or as a timely indication of calamities for the monarch or the nation. For Chinese astrologers, the solar God, *XiHe*, represented the *Yang* aspect of nature, the supreme power to give life and the authority to govern, and therefore also represented the Emperor, the same *Son of Heaven*. On the contrary, the Moon Goddess, *ChangXi*, used to associate with the *Yin* aspect of nature, the person and actions of the Empress or the Emperor's concubines.

The Chinese calendar was not so successful with the accurate prediction of solar eclipses, so its occurrence always meant exceptionally disastrous and terrifying omens; it was commonly thought that they were related to great matters of state or of the Emperor himself. The first Chinese word for eclipse, *shih*, means "*to eat*" and it referred to the gradual disappearance of the Sun or the Moon; fact that the Chinese interpreted as if they were eaten by a huge and hungry celestial dragon, which frightened the emperor and all the sudden ones; to scare such a dragon they should throw many arrows into the air, and play the biggest drums and gongs making as much noise as possible.

**Image 1.4
Selene and Endymion**

Painting made in 1713 by the Italian artist Sebastiano Ricci (1659 – 1734)

The Chinese were early in recording eclipses but belated to recognize their cause; it was not until the third or fourth century AD that they understood the solar and lunar eclipses well enough to accurately predict them. Over time the Chinese calendar evolved and provided better methods of predicting eclipses and the positions of the five planets. As the eclipses became easily predictable, they gradually lost their astrological significance.

Chinese astronomy is much appreciated by modern astronomers, especially for providing accurate records of celestial phenomena such as eclipses, novae, comets, meteors, etc., for longer periods of time than any other civilization, and they have found many applications in modern astronomy.

In India astronomy developed a bit later than in China, but in a similar way. A solar calendar was developed; the main celestial objects received the name of gods and goddesses; astrology was the driving force in these developments. In Indian Mythology, *Raju* is one of the *Asuras*, or demons brothers of the gods, who fought against them for the possession of the food of the gods, or *ambrosia*, and for the conquest of *Lakshmi*, the goddess of wealth and beauty.

The Asuras had captured the ambrosia and *Raju* was eating it when the god *Narayana* reached him, released his disc and decapitated him. However, *Raju* had managed to eat something of the gourmet food of immortality, so he did not die: his head and body, separated, float forever in space like two stars invisible to human eyes. His head, also called *Raju*, for revenge periodically attacks and devours temporarily the Moon and the Sun, causing eclipses; and *Ketu*, the rest of the body, has become the constellations.

Pre-Columbian America

The native peoples of Central America, although they did not recognize the shape or movements of the Earth, knew the causes of eclipses, the use of gnomon and the calculation of solstices and equinoxes. In addition to the solar year, these people used the year of Venus, determined by the synodic revolution of that planet.

Image 1.5

Dragons and eclipses:

According to ancient Chinese and Indian mythologies, eclipses were caused by a furious celestial dragon that attacked and temporarily devoured both the Moon and the Sun.

The Mayan people of Central America were the only ones who left inscriptions and records about their culture, in stone and on amate paper in complicated hieroglyphs that are difficult to decipher, although those parts that relate to the calendar can be interpreted. It seems that currently only five Maya manuscript books survive, which fortunately contain lunar and solar calendars, as well as a Venus calendar of great interest; the best of them is the *Dresden Codex*, dating from the 11th or 12th century. From these documents it is clear that the Mayans had a very elaborate calculation of time, as well as tables for the evaluation of the lunar periods, the synodic and sidereal months, and the prediction of lunar and solar eclipses. His year covered 365 days, divided into 18 months of 20 days each and a brief additional month of 5 days.

In their mythology the Mayans believed that the Sun and the Moon were the thrones of the blessed, and thus they deified them, which was part of their religious belief and superstition. During the Classic period, *Itzamna* was the god of the Sun and wisdom, lord of heaven and day; the one who inhabits the celestial world from where he rules the cosmos. He was represented in multiple ways, mainly as an elder; but given its omnipresent faculty, he was represented in Mayan art in animal forms according to the area where it acted: as a bird if he was on a celestial level, or as a crocodile if he was on a terrestrial plane. In the Post classic period *Kinich Ahau* was the god of the Sun and patron of music and poetry; he was married to *Ixchel*, the goddess of the Moon. This last was associated with various elements such as water, fertility and certain trades characteristic of the female gender and motherhood; they used to represent her as a young woman, symbolizing the crescent Moon, or as an elderly woman in the image of the waning Moon.

In their inscriptions it is appreciated that the Mayans tracked the phases of the Moon and also counted the lunations as months of 29 or 30 days, since they did not use fractional numbering. From their calculations it is known that they used different formulations: The *Table of Eclipses* of the *Dresden Codex* is based on 46 multiples of the 260-day calendar, that is, 46 x 260 = 11960 days, which corresponds on average to 405 lunations. In fact, the relationship is very precise since 405 lunations of 29,53 days equals 11959,65 days; this shows how the Maya could be very precise in their arithmetic based only on integers, without the need for fractional numbers.[7]

The native populations of North America had the habit of calling the Full Moon with a different name, but significant for each month. Since this tradition was so rooted, these names have reached the present, like this: in January the Wolf's moon, in February of the Snow, March of the Worm, April Pink moon; in May of the Flowers, June of the Strawberry, July of the Deer, August of the Sturgeon, in September Harvest's Moon; October of the Hunter, November of the Beaver, and finally in December Cold moon, or also Moon of the long nights.

Image 1.6

Ixchel, the Mayan lunar goddess

Detail taken from an original painting by the year 550 AD., preserved in the Museum of Fine Arts in Boston.

The name Wolf's moon in January refers to the frequent presence of wolf packs howling during the cold winters of this month. While the name of the Harvest moon comes from the fact that in the northern hemisphere in September, the Full moon is very close to the autumn equinox, and under these conditions the light of the satellite is more constant at dusk, that allows farmers to extend their working hours a little more during some hours of the night, and pay more attention to the pickup of their respective harvests.

The Celestial Cycles and the Development of the Calendars

The discovery of the precision in the cyclical movements of the stars, and their associated phenomena, facilitated primitive societies to soon develop precise systems of time measurement: the calendars, systematized schemes of measurement and record of the passage of time, extremely useful for the chronological organization of social activities. A first step seems to have been to observe in which positions was the Moon at different times of the agricultural cycle, the planting or the harvest, to then divide the time into *Moon's Stations*, and then to astronomically mark its monthly and annual festivals .

In their historical development, primitive civilizations initially comprised two lunar cycles, normally known as *lunar months*. One of them was the period between two successive full moons, or in more general terms, the time elapsed between two consecutive phases of the Moon, currently known as the *Synodic month*, which is approximately equal to 29,53 days; with which 12 lunar months would give a Lunar year of only 354,36 days, about eleven days less than the typical Solar year of 365,25 days. Because there is not a whole number of full moons in a solar year, the two elements simply cannot be reconciled in a common calendar: while the modern civil calendar is based on the solar year, the dates of religious festivities, such as Christmas, Easter and Pentecost, are still established with reference to the lunar month. The second lunar cycle known in antiquity was the real period of the lunar orbit referred to the fixed stars; now it is known as the *Sidereal month*, and it is the time it takes the Moon to return to the same position between the fixed stars at the bottom of the celestial sphere; it has an average duration of approximately 27,33 days.

Later in time other lunar cycles were detected and understood. The *Anomalistic month* is calculated from perigee to perigee, which is the closest point of the orbit of the Moon to Earth, it has an average duration of 27,554 days. On the other hand, the *Draconitic month* is measured with reference to the nodes of the Moon, the points in the sky where the lunar path crosses the ecliptic, since the lunar orbit is in an inclined plane approximately five degrees with respect to the plane of the ecliptic. The line of intersection of these planes defines two points on the celestial sphere, known as the ascending and descending lunar nodes; which are not static but rotate retrograde with a period of about 18,6 years. The time it takes for the Moon to return to the same node is the Draconitic month, and it has an average duration of approximately 27,2 days. It is very important in astronomy to predict eclipses: since these take place when the Sun, Earth and Moon are in a straight line, which occurs only when the Moon is near the ecliptic, that is, when it is near any of the nodes. Bearing in mind the nature of any kind of eclipse, it is obvious that the Sun and the Moon must coincide in the line of a node so that it can occur. The term draconitic refers to the mythological dragon that inhabited at the nodes and regularly, during eclipses, it ate either the Moon, or the Sun. (The week is not based on any astronomical phenomenon, but is derived from the Jewish and Christian traditions that every seventh day should be a rest day.)

The Babylonians were always based on the lunar cycles for the development of their calendars. At first they had the lunar year that included twelve months of 30 days, that is, it had almost 5,25 days less than a solar year; discrepancy that would soon manifest itself. After a few years, for example, the month of plowing did not fit that agricultural task.

In order to prevent the stations from being out of phase, a thirteenth month was inserted from time to time; although there was no regular system to intersperse this additional month until the fifth century BC, when they began to use the *Metonic cycle*, a time interval of 235 lunar months; which allowed them to establish a much more precise lunisolar calendar. The beginning of the months was established on the day of the New Moon, when the satellite interposes between the Sun and the Earth; in this position the Moon has its dark hemisphere in front of the Earth, so it is practically invisible, appearing to the West immediately after sunset, in the form of a very thin and barely perceptible sickle, with the convexity directed to the West. The name of *Metonic cycle* is due to its discoverer, the Greek astronomer Meton of Athens towards year 432 BC; and it has the peculiarity that it is very close to 19 solar years, a lapse of time in which the Moon returns to go through the same phases in the same days and in the same hours. In a typical lunisolar calendar, most of the years are lunar ones of 12-month, but 7 of the 19 years have an extra month, known as the *intercalary month*.

In Egypt, as in most ancient cultures, the main reason for making astronomical observations was to track the course of long-term cycles, such as weather stations, and short-term periods such as day and night hours. Their agriculture depended on the annual flooding of the *Nile River*, which deposited a layer of silt that fertilized the land, an annual event of vital importance to Egyptian society that had to be predicted. But their original calendar was based on a fixed number of lunar months, and soon it became obsolete with the solar year, the discrepancies in the predictions of the annual flood of the river and the course of the seasons became evident. Through further observations of the heliacal rise of the star Sirius, then called *Sothis* or *Sopdet* too, announcing the flooding of the Nile, the Egyptian calendar finally reconciled with the true solar year for the first time in history towards the third millennium BC. The Egyptians astronomer priests preferred to use the solar calendar for civilian uses, instead of the problematic lunar calendar.

An Egyptian historical source, in which the 365 days of the solar year is mentioned, is the Rhind Mathematical Papyrus, currently in the British Museum. The year was divided into 12 months of 30 days each, organized into three periods of 10 days; at the end of the last month of each year the remaining five days, called epagomenals, were added to complete the solar year.

In 1999 a group of German archaeologists working in a wooded area near the German city of Nebra, southwest of Berlin, discovered a 3600-year-old bronze disk which was hailed as the oldest surviving heavens diagram, and one of the most important archaeological discoveries of the twentieth century.[8]

The *Nebra sky disk* was made in bronze, measures 32 centimeters in diameter, has a blue-green patina and includes golden appliqués representing the Sun, the Moon, and 32 stars. Of the latter, a group of seven points has been interpreted as the constellation of the *Pleiades* as it would have been 3600 years ago. This disk is the oldest visual representation of the cosmos known to date.

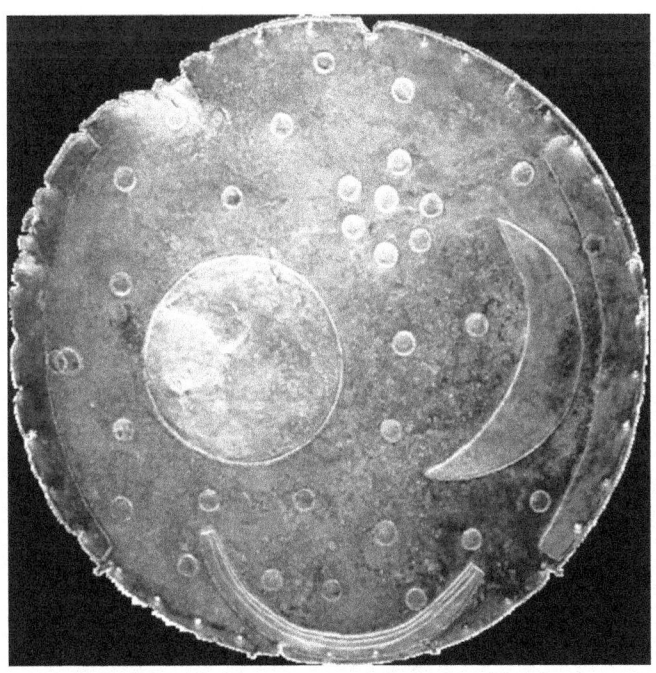

**Image 1.7
Archaeoastronomy
The *Nebra sky disk***

It is considered as the oldest preserved diagram of the heavens. It was discovered in 1999 in the German city of Nebra and it is assigned an age of 3600 years.

The explanation of the disc's purpose sheds new light on the astronomical knowledge and skills of Bronze Age people, who used a combination of solar and lunar calendars as important indicators for agricultural seasons and the passage of time.

A group of German scientists has deciphered the meaning of this great archaeological discovery: they have found evidences that indicate that the disk was used as a complex astronomical marker for the synchronization of solar and lunar calendars. A lunar year is eleven days shorter than the solar year because 12 synodic months, or 12 moon returns to the new phase, take only 354 days. The celestial disk of Nebra was used to determine if a thirteenth month, the so-called intercalated month, should be added to a lunar year to keep the lunar calendar synchronized with the seasons, which are linked to the solar cycle.

According to the ancient Babylonian rule, a thirteenth month should only be added to the lunar calendar when the figure of the Moon and the Pleiades were seen exactly as they appear in the celestial disk of Nebra. Astronomers of the Bronze Age held such an object in their hands and compared its figure with the position of the stars in the firmament. The intercalated month was inserted when what they saw in the sky corresponded to the map on the disk they had in hand. This would happen every two or three years.

Another archaeological finding of great importance for the history of astronomy and knowledge is the *Antikythera mechanism*, which consists of a kind of *mechanical simulator of celestial movements*. It was extracted in 1901 from an ancient shipwreck near the Greek island of Antikythera, and was most likely built by Greek experts between 150 and 100 BC. The instrument should have dimensions of 34 cm x 18 cm x 90 cm in its original state, and consisted of a complex watchmaking mechanism composed of at least 30 bronze gears. Those who have studied it affirm that it was designed and constructed with the purpose of following the movements and predicting the positions of celestial bodies and eclipses for astronomical, calendrical and astrological purposes. It could also serve to predict the exact date of ancient Greek events: the Pythian Games, the Isthmian Games, those of Olympia, those of the island of Rhodes; etc. All the recovered and preserved fragments of the Antikythera mechanism are currently guarded in the *National Archaeological Museum of Athens*.

Image 1.8

**Archaeoastronomy
The Antikythera Mechanism**

Found in 1.901, it was probably used as a mechanical simulator of celestial movements by Greek experts between 150 and 100 BC.

Image 1.9
The Different Cycles of the Moon

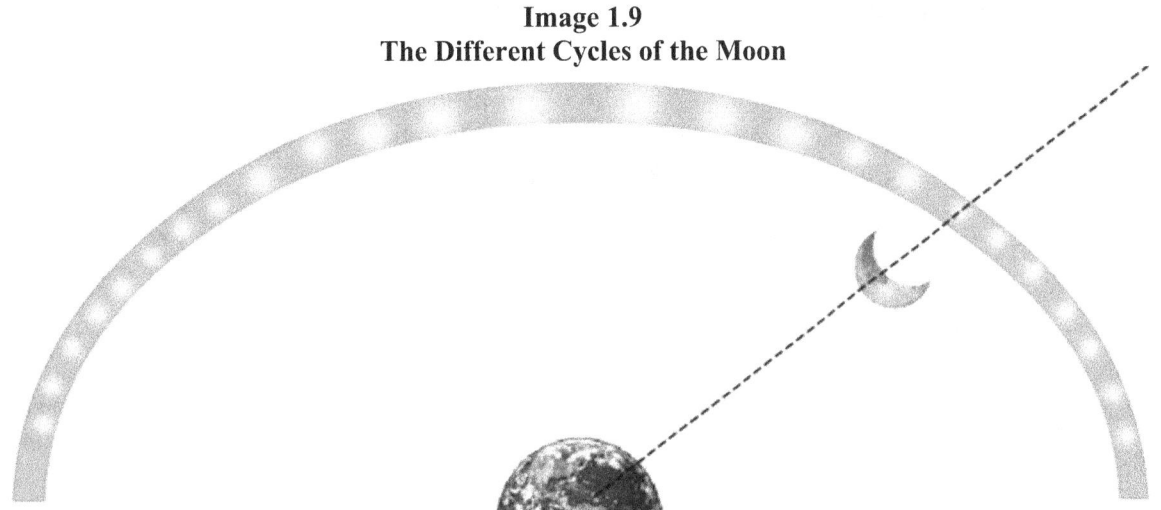

Sidereal Month: It is the time the Moon takes, seen from a same point in the Earth, for returning to the same position with respect to a specific fixed star; its approximated value is 27,33 days.

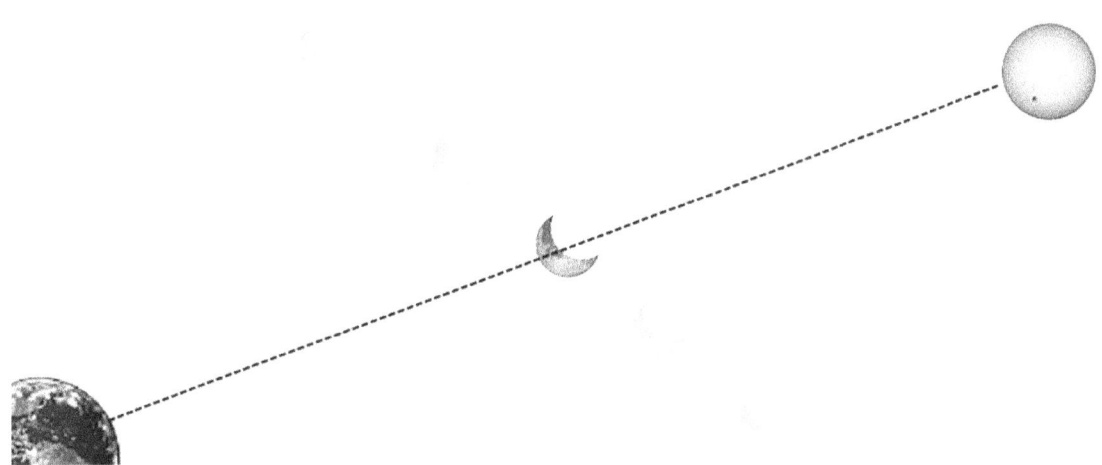

Synodic Month: It is the time our satellite takes to return to the same position between the Sun and the same place on Earth; its average value is 29,53 days.

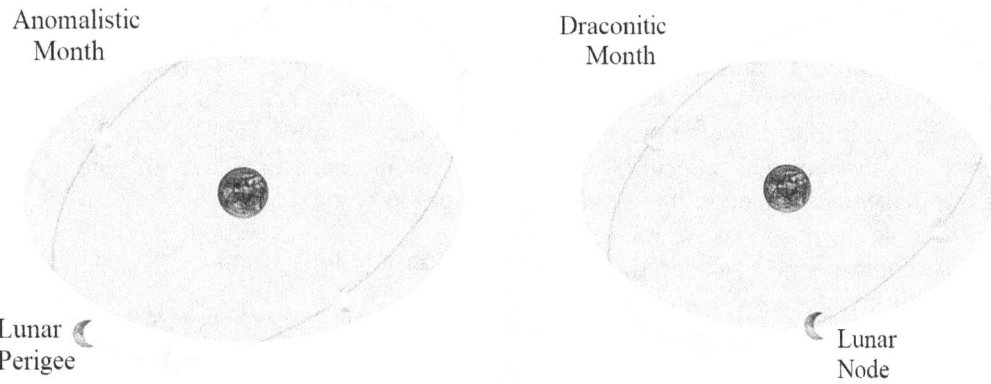

Anomalistic Month: It is the time lapse that the Moon takes to go from one perigee to the next; it has an approximated duration of 27,55 days.
Draconitic Month: The time the Moon takes to return to the same node in the ecliptic, it has an average duration of 27.20 days.

The Synodic Month and the Lunar Phases

This first chapter is the most appropriate to introduce some fundamental concepts of Lunar astronomy that will be widely used throughout the text. As it has been seen, the societies of the old civilizations were very assiduous in the contemplation of the lunar globe and the monitoring of their natural cycles; of which the most notable is the synodic month, or lunation, which occurs in a span of 29,53 Earth days.

During the course of the synodic month, and contemplated from a certain point of our planet, the Moon manifests a series of appearances due to the continuous changes in the relative positions between the Earth, the satellite and the King star; these appearances are normally and widely known as the *Lunar Phases* .

The exposition that now follows is based on images 1.10a and 1.10b, and the numbering and their respective definitions apply equally to both images. For beginning we must make it clear that for all times both the Earth and its satellite always have one of its hemispheres facing the Sun, to the sun's rays, and therefore it looks completely illuminated; while the other hemisphere is opposite to the Sun and does not receive any sunlight, remaining in darkness; and that additionally the Moon does not shine with its own light, but always with borrowed light: reflected sunlight.

Phase 1: New Moon

In this position the three stars, the Sun, the Moon and the Earth, are arranged practically in a straight line with the Moon in the middle; the entire illuminated hemisphere of our satellite looks towards the Sun, and its dark hemisphere looks towards the planet. All this means that no light reflected by the Moon reaches our planet and, therefore, the satellite is practically invisible to us. This astronomical disposition is also known since ancient times as Syzygy, and contemporarily as *Conjunction*: the Moon is on the same side of the Sun, in conjunction with it; and when the alignment is well adjusted, perfect, then eclipses of the Sun take place.

Phase 2: Waxing Crescent Moon

After the New Moon, and with the passage of time, a small portion of the illuminated hemisphere of the satellite begins to be visible from our planet. Depending on the position and weather conditions, the first visibility occurs about 24 hours after the New Moon. Gradually the visible portion of the illuminated hemisphere of our satellite increases, grows, and hence the name Waxing Crescent Moon. The satellite can be drawn with two curves: a convex exterior and a concave interior, and it seems that the Moon is as hollow, or sickle-shaped; it is also normal to express that in this phase the Moon exhibits its *horns*. *Waxing* means that our satellite is growing, while *crescent* refers to the curved shape similar to a banana or a boat.

Phase 3: First Waxing Quarter

At this time the satellite has traveled almost a quarter of its trajectory around the planet, and hence the name for this lunar phase: First quarter, which occurs very approximately 177 hours or 7,4 days after the New Moon. This phase has the remarkable peculiarity that precisely half of the lunar hemisphere illuminated by the Sun is visible from the Earth: the Moon looks like illuminated justly a half. This celestial arrangement occurs when the satellite is located at 90° east of the King star, a point that is also known as the Eastern Quadrature.

Phase 4: Waxing Gibbous Moon

After the First quarter, the visible section of the illuminated satellite's hemisphere continues to grow, now the Moon can be drawn with two convex curves, the satellite looks like a hump or bump, and hence the name Waxing Gibbous Moon. This phase lasts until approximately the end of the second week of the synodic month.

Phase 5: Full Moon

This lunar phase occurs when the satellite has traveled just half of its orbit around our planet, about 354,4 hours, or just over 14 Earth days after the month began with the New Moon. The satellite is now located at the other end, on the side opposite the King star, and that is why this arrangement of the

stars is known as *Opposition*. At this point the entire illuminated hemisphere of the Moon is now visible from Earth, so it looks like fully illuminated, and hence its name of *Full Moon*. Additionally, since the three bodies are practically aligned, this astronomical arrangement is also called Syzygy; and when the alignment is perfect, the lunar eclipses occur there.

Phase 6: Waning Gibbous Moon

Over the time after Full Moon the percentage or illuminated portion of the satellite that can be contemplated from the Earth begins now to decrease, decline or wane; where the waning term comes from. The previous cycle of appearances of the Moon begins to return, but now in its waning modality and changing East-West direction. The first to be seen is the Waning Gibbous Moon phase.

Phase 7: Third Waning Quarter

After having traveled the three first quarters of his trajectory, the Moon is now in its phase known as the Third Waning Quarter, which equals the counterpart of the crescent quarter. The satellite has reached the point known as the Western Quadrature, and it also looks just like it's half illuminated.

Phase 8: Waning Moon

Finally, in its last quarter of the route and the last week of the synodic month, the lunar phase is called the Waning Moon or Waning Crescent. The visible portion of the satellite becomes smaller every day, until disappearing completely, the Moon becomes invisible again marking the beginning of a new cycle, a new lunation and a new synodic month of 29,53 Earth days.

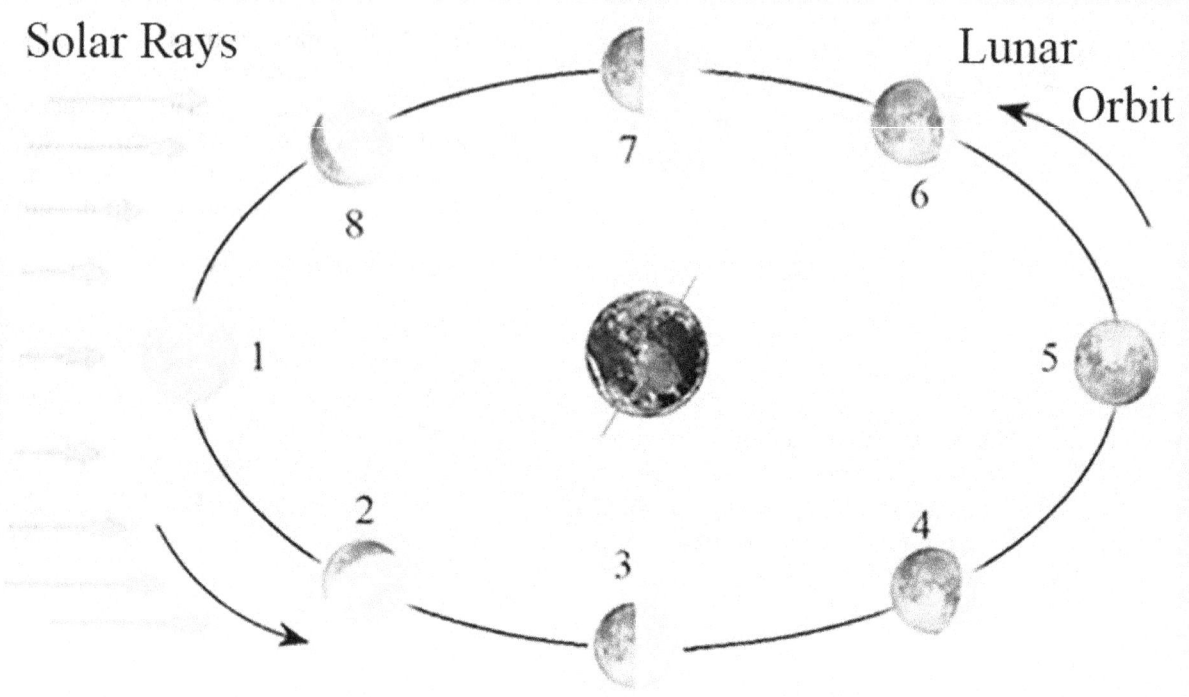

Image 1.10a: The Synodic Month and the Lunar Phases

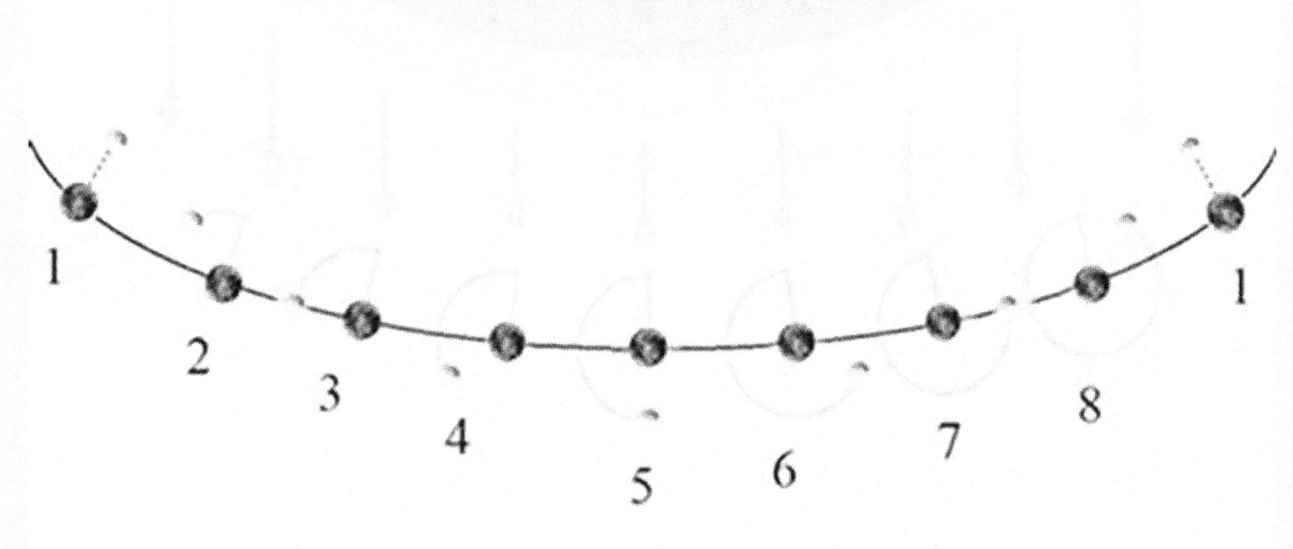

Image 1.10 b: The Synodic Month and the Lunar Phases

Bibliographic Citations

[1] Pueblosindigenaspcn. Literatura Indígena: http://www.pueblosindigenaspcn.net/biblioteca/literatura-indigena/doc_view/135-los-libros-de-chilam-balam-de-chumayel.html

[2] UNESCO: World Heritage List. http://whc.unesco.org/en/list/373

[3] English Heritage: Stonehenge. http://www.english-heritage.org.uk/visit/places/stonehenge/history/

[4] Abetti, Giorgio. *The History of Astronomy*. London: Sidgwick and Jackson, 1954.

[5] Lull, José. *Sobre el eclipse solar del papiro demótico Berlín 13588 y el eclipse lunar de la crónica del Príncipe Osorkón*. Boletín de la Asociación Española de Orientalistas No. 43 (2007):255-266. https://repositorio.uam.es/handle/10486/6572

[6] The Internet Classics Archive. Cratylus By Plato. http://classics.mit.edu/Plato/cratylus.html
Obras completas de Platón. http://www.filosofia.org/cla/pla/azcarate.htm Cratilo. Page 412: http://www.filosofia.org/cla/pla/azf04347.htm

[7] Selin, Helaine. *Mathematics across cultures*. Massachusetts: Springer, 2000

[8] UNESCO: Memory of the world. Nebra Sky Disc. http://www.unesco.org/new/en/communication-and-information/memory-of-the-world/register/full-list-of-registered-heritage/registered-heritage-page-6/nebra-sky-disc/

Chapter 2

Lunar Astronomy in Greek Cosmology

The crescent Moon as seen from our planet on Monday January 22, 2018 at 10: 00 UCT, 5 days 7 hours and 30 minutes after the new Moon, reaching to see a 24,8 % of its illuminated hemisphere. By this time the *horns* of the satellite are already very large and well delineated. This portion illuminated of the Moon that we see has gone continuously and gradually increasing, growing, and hence its name Crescent Moon.

Courtesy of NASA's Scientific Visualization Studio: https://svs.gsfc.nasa.gov/4604

"When we were ready to take up a trip here, the Moon met us and commissioned us, first of all, to greet the Athenians and their allies; She told us later that she was angry, because she has suffered ill-treatment, though she benefits you all, not with words, but in a patent way. Firstly she saves you all no less than one drachma per month in torches; so much so that everyone, when you go out at night, you say: 'Boy, don't buy torches that Selene illuminates enough'."

The Clouds (Comedy)
Aristophanes (444 - 385 BC),
Greek comic playwright.[1]

Classical Greece: From Mythology to Philosophy

Tales of Miletus

Thales of Miletus (624 - 546 BC) was a Greek philosopher, mathematician, geometer, and astronomer; he was born and died in Miletus, a Greek city on the Ionian coast. In his time mythical cosmologies of the world still predominated in Greece, and they had not yet emancipated from the Homeric conception of the Earth as a flat disc; but Thales sought a purely rational explanation of all physical phenomena, what is known as the passage from myth to *logos*, or reason; so he is considered the first philosopher in history. He was also one of the greatest astronomers and mathematicians of his time, and the founder of the *Ionian School of Philosophy*, in whose system questions about the physical constitution of the world occupied a prominent position. His inquisitive approach to the understanding of celestial phenomena marks the beginning of Greek astronomy; his hypotheses were new and audacious, and by freeing the natural phenomena of divine intervention, he paved the way for scientific thought. Important contributions are attributed to him in the field of philosophy, mathematics, astronomy, physics, etc.; it is known that he proposed answers to a series of questions about the earth: its shape, its size, the question of its support or sustenance, and the cause of earthquakes. As for astronomy, he reflected on the dates of the solstices, as well as the size of the Moon and the Sun. No writing of his is preserved, so all that is known about him is by references of other authors.

Given the traditional commercial relationships between the Ionians, Egyptians and Babylonians, it is very likely that Thales visited these regions during his lifetime, and there he could have received cosmological teachings from astronomers, and have acquired knowledge in mathematics, science that the Egyptians had developed at a practical level in order to measure and delimit land parcels, whose boundaries were often erased due to the periodic floods of the Nile River. Many historians argue that Thales acquired information from Near Eastern sources and gained access to the extensive records that dated about the time of Nabonassar, around 747 BC; they also claim that he predicted the solar eclipse of 585 BC by knowing the Saros period, a cycle of 223 lunar months after which the eclipses of both Moon and Sun are repeated with very little change.

A stream of historians argues that he considered the Earth as a flat disk floating in the water, similar to a wooden boat; and that the Sun and the Moon were, like our planet, other flat discs that moved in the sky above it and, at times, aligned; but there is no old testimony that supports that opinion. Conversely, Aristotle, philosopher Aetius and then the historian Plutarch, attributed him the knowledge of the sphericity of the Earth. Aristotle wrote that some thought it was spherical, others considered it circular, and a third group assigned it a drum shape; following the chronological order, the writings of Aristotle seem to indicate that Thales considered the Earth as spherical. So there is no agreement between the historians of philosophy and knowledge, it can be thought that Thales, as well as his disciples in the Ionian School of Philosophy, during the course of their lives could have changed their minds on this subject.

The great philosophical pronouncement of Thales that water is the principle, cause and primitive substance of all what exists, shows that he abandoned and turned his back to the recognition of the gods as provocateurs and determinants of everything that happens. His hypotheses show that he conceived physical phenomena as natural events that have natural causes and, therefore, also have rational explanations. With his new paradigm of observation and reasoning, he devoted himself to study the heavens and to seek logical explanations for the celestial phenomena; which constitutes an obvious step outside the mythology of his time: With Thales of Miletus the long and complex fight of reason against its own mythical past begins.

In the field of the history of astronomy, Thales is famous for being cited as the first Greek to predict a solar eclipse. Herodotus of Halicarnassus (484 - 425 BC), the father or founder of History, was a Greek historian who wrote most of the surviving stories, both those of his time as of earlier periods. He claimed that Thales of Miletus predicted the solar eclipse of 585 BC, which occurred during a battle between the Medes and the Lydians. It seems that Thales assimilated well the theoretical rudiments of solar eclipses, and was sufficiently understood to use the idea coming from the Babylonians about the periodicity of eclipses: the Saros cycle. Undoubtedly, the eclipse seems to have caused the Medes and the Lydians reconsidered their hostile intention and agreed to a peace treaty after five years of war, each seeing it as a bad omen:

> *"As, however, the balance had not been tipped in favor of any nation, another battle occurred in the sixth year, during which, just as the battle was heating up, the day suddenly changed into night. .. The Medes and the Lydians, when they observed the change, stopped fighting and were equally anxious to reach an agreement on the terms of peace."*[2]

However, it is not clear that the Milesian had predicted the exact date and circumstances of such an eclipse, and many historians of science now doubt whether he did more than suggest that an eclipse of the Sun would occur in a certain year. Miletus had cultural relations with Babylon, whose astronomers had discovered that eclipses are repeated in a cycle of approximately nineteen years; and they could predict the lunar eclipses with very good success, but with respect to solar eclipses they were hampered by the fact that such phenomena, unlike those of Moon, not always are visible in all places of the same hemisphere. Thus, they could only know that at some specific date it would be worth waiting for an eclipse of the Sun, and this is surely all that Thales knew. Neither he nor they knew why there is such a Saros cycle. It was a very unusual event, but Herodotus only wrote that Thales guessed the year, so one could wonder if it was a true prediction or just a lucky divination. Today there is no doubt about the occurrence of the eclipse, and this event can be recalculated to show that on the afternoon of May 28, 585 BC, the path of totality extended through the Mediterranean and well centralized from West to East through Asia Minor, where the armed struggle was taking place, and the Sun was hidden by the Moon for just over six minutes.

Either way, if Thales really predicted such a solar eclipse, or if that story were really a myth, the fact that astronomers and writers of that time considered it possible to predict an eclipse implicitly carries the message that these amazing spectacles were no longer considered as a work of the caprice of the gods or, even worse, of the demons; but they had been firmly brought into the realm of predictable natural phenomena.

The diligent record of eclipses could have empowered ancient societies to predict future events, which is much easier for lunar eclipses. Analyzing such data, carefully maintained for generations, ancient scholars would notice that eclipses of the same basic characteristics are repeated to the past and following certain patterns, which would have allowed them to reverse the direction of time and, in an unprecedented event, project the cycles towards the future. In those times these natural phenomena could be predicted, or in best words prognosticate, with a good level of accuracy, without the use of complex mathematical models or elaborate calculations in an electronic computer; and without the need for a full understanding of the complex celestial mechanics involved. The understanding of the characteristics of the celestial cycles allowed the ancient astronomers to foresee their repetitions without understanding that the Earth revolves around the Sun and that the Moon orbits around the Earth. Such an understanding of the pattern of eclipses constituted a powerful "weapon" for astronomers and other scholars who understood them, and for the kings and emperors who used them in their favor.

The accurate prediction and observation of celestial events was necessary for the coordination of different social activities. If an eclipse happened in the terms in what it had been predicted, it meant that the cycles of the Moon and the Sun were well understood; and that therefore the calendar was correct and it determined with precision the seasons, as well as the course of time in general.

Another great merit of Thales in astronomy consists in his determination of the dates of the solstices; which consist of the days of the year in which the Sun reaches its highest or lowest apparent height in the sky, so that the duration of the day or night are the maximum of the year, respectively for each hemisphere, South or North. Equivalently, they are the moments in which the Sun reaches the maximum North (+23° 27') or South (-23° 27')

declination with respect to the terrestrial equator; what happens twice a year: June 21 or 22, and December 21 or 22.

Finally, Thales was interested in the relationship between the size of the Sun and the length of its trajectory in the sky. For which he used the data obtained during his solstice fixation work; with what hi obtained that the diameter of the Sun is to the diameter of its orbit according to the proportion of 1/720. Applying the same methodology to the Moon, he got an identical relationship. These results show highly discrepancy with modern values, which would be due to the precariousness of the water clock that he used for his measurements. In any case, he has the merit of being the first documented astronomer who has thought about quantifying and relating the sizes of the stars and their respective orbits.[3]

Anaximander

The second philosopher in importance of the Ionian school was Anaximander (611 - 545 BC), philosopher, astronomer and geographer, also of Miletus, disciple and follower of Thales. For him the skies were of a fiery nature and spherical in shape, forming a series of layers and enclosing the atmosphere, in the style of a tree bark; in his cosmological model he placed the Sun at a greater distance and the fixed stars nearest our planet. He believed that the Earth was in equilibrium at the center of the world, and that its shape was cylindrical, like a drum; its diameter would be three times its height; the upper part of it, circular and flat, corresponded to the inhabited part of the world. He also boldly affirmed that our planet floated freely in the center of the universe, without the support of water, pillars, or anything; ideas that meant, at the time, a great revolution in the understanding of the universe.

The astronomical speculations of Anaximander can be summarized as follows: (1) the Earth floats free in space and without any support, (2) the celestial bodies lie one after the other, and (3) the trajectories of such bodies describe complete circles and pass also below our planet. These three propositions, notwithstanding their rather primitive nature, constitute the nucleus of Anaximander's astronomy; they meant a tremendous leap forward in the description of the cosmos, and they represent the roots of our western concept of the universe. For him the celestial bodies have no reason to move in any other way than in circles around the Earth, since each point in this trajectory is always as far from our planet as any other; but in his conception of the world he placed the celestial bodies in the wrong order: He located the stars closest to the Earth, then placed the Moon and finally the Sun.

A peculiar feature of Anaximander astronomy is that it asserts that celestial bodies are like rings made of opaque vapor, but internally hollow and filled with fire. These rings would have superficial circular openings through which light and heat escape, and this is what we see as the Sun, the Moon or the stars. According to him, the opening of the ring of the Moon closes and opens rhythmically, which explains the phases of the Moon; sporadically it closes completely causing lunar eclipses. Likewise, when the opening of the solar ring closes for a short period of time, an eclipse of the Sun would occur.

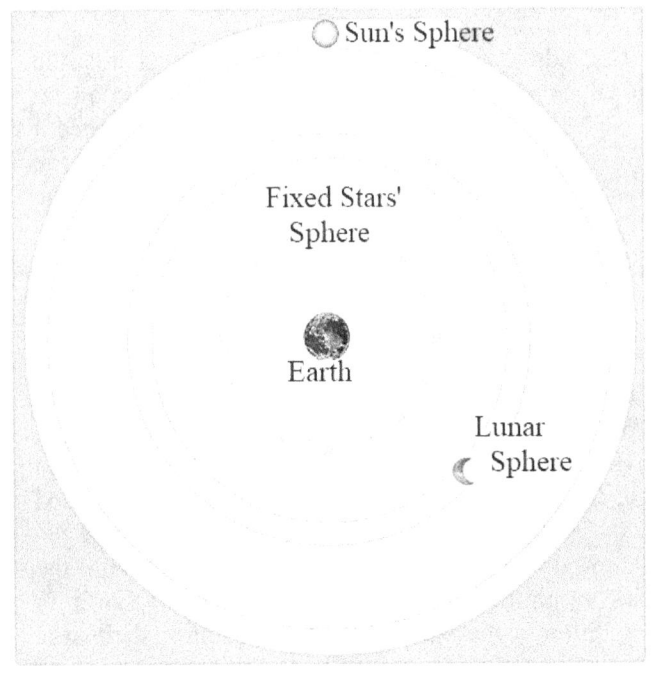

Image 2.1

Anaximander Cosmology

Schematic diagram

Pythagorean cosmology

Pythagoras of Samos (569 - 475 BC) was a Greek philosopher and he is usually considered the first pure mathematician. He was the founder of the *Pythagorean School*, very dedicated to the study of philosophy, mathematics, cosmology, politics, and ethics. Pythagoras was a great mathematician, a fact that is at the same root of his philosophical system; the Pythagoreans based their lifestyle and philosophy on the cult of numbers, taking it almost to the point of frenzy: for them everything was an embodiment of number and everything was ruled by them.

To this Pythagorean school belongs the theory normally known as *Harmony of the Spheres*, or also *Music of the Spheres*, according to which the Universe is based on harmonious numerical proportions. The theory states that, within a geocentric universe, both the movements and the respective distances between the celestial bodies are governed by harmonics musical proportions.

The admiration of the number ten had for the Pythagoreans a transcendental cosmological implication in its conception of the universe, since it was the promoter of the first non-geocentric astronomical system, especially promoted by the Greek mathematician and philosopher Philolaus of Taranto (470 - 380 BC). Nine celestial bodies in this order: the Earth, the Moon, the Sun, Venus, Mercury, Mars, Jupiter, Saturn, and at the end of the heavens the Sphere of the fixed stars, orbit concentrically around the universal Central Fire or *Throne of Zeus*. Lacking a body to reach the tenth value of the Pythagorean Tetractys, Philolaus added to the system the *Counter-Earth*, also called *Antichthon*, located just between our planet and the Central Fire, aligned, in equilibrium and with the same period of daily revolution as the Earth. In addition to the mathematical justification, there was an astronomical argument to accept the existence of the Antichthon, and this was one that accounted for the greater frequency with which the lunar eclipses occurred: sometimes this star was eclipsed by the shadow of our planet, and others ones by the shade of such Counter-Earth.

In this model the Earth always had the same uninhabited hemisphere facing both the Central Fire and the Antichthon, so these bodies were permanently invisible from the hemisphere we inhabit in, which always faces towards the stars. Additionally, the Sun did not occupy the central position of the cosmos, nor was it the creator of its own heat and light, but something like a reflecting crystal that collected these properties from the *Throne of Zeus*, around which it revolved with a period of one year; light and heat that the Sun scatters everywhere after having sifted through its own body; and that it would be the only source of light reflected by the Moon.

The universe proposed by the Pythagoreans is a dynamic one and in which the planets and stars revolve in circular orbits, including the Earth, so that it appears endowed with movement very unlike the Ionian model. The Pythagorean concept of uniform circular movement, which refers to the movements of the celestial bodies, remained unchanged for almost 2000 years, and managed to reach the period of the European Renaissance. By displacing the Earth from the center of the universe, Pythagorean cosmology is a grandiose leap in philosophical thought. In fact the system provided a plausible explanation of the eclipses, for it supposed that those of the Moon were not only caused by the passing of this star through the shadow of our planet, but also occasionally by the shadow of the Counter-Earth, and this would be the reason why there were more lunar eclipses than solar ones. Slowly but surely, the style of Greek rational thought started to develop, which soon began to produce important results too.

For the philosophers of the Pythagorean school of philosophy there is more reliable evidence that they held the doctrine of the Earth's spherical shape, attributed to both Pythagoras of Samos and Parmenides of Elea (530 - 470 BC); and it would be proposed for the first time as a simple hypothesis, not scientifically verified but justified with philosophical arguments. Whether in the first instance to Pythagoras or Parmenides, the doctrine of the spherical figure of our planet must have progressed during the first half of the fifth century BC; and if this doctrine was slow to become acceptable, it is not surprising that the Sun and the Moon were not, even among the Pythagoreans, recognized as spheres for some time.

Anaxagoras of Clazomenae

Likewise of Ionia, the philosopher Anaxagoras (500 - 428 BC) was born in the port of Clazomenae, but spent much of his life in Athens; he was an Ionian that continued with the Ionian rationalist tradition and has a considerable importance in the history of astronomy. He claimed that the Sun and the stars were burning stones, but that we didn't feel such heat of the latter because they were too distant; he also stated that the Moon was of a terrestrial nature, had mountains and, even, inhabitants. Anaxagoras thought that according to the cosmological disposition of the celestial bodies, our satellite would be located under the Sun; and starting from there he developed a correct theory for the luminosity of the Moon and for eclipses. . He was who first explained that this star does not shine with its own light, but that it is due to the light that comes from the Sun and that subsequently reflects it; although other references suggest that Parmenides also affirmed it.

Observing the Moon, the Sun and their respective movements, Anaxagoras was the first, as far as we know, to understand that eclipses occur when a celestial body blocks the light of another. Such rejection of the gods and dragons as causing the eclipses was a thought considerably revolutionary in his time, but he took it even further: If solar eclipses occurred because the Earth is below the shadow generated by the Moon as it passes beneath the Sun, then, Anaxagoras thought, the size of the lunar shadow that covers the Earth should tell us something about the same size of the satellite. Here the great power of philosophical reasoning is: measuring the extent of the lunar shadow on Earth during a solar eclipse, there will be an indicative for the Moon's size. Additionally, since our satellite practically covered the entire Sun by passing under it, giving the impression that they were the same size, the King star should actually be much larger than the Moon the farther it was.

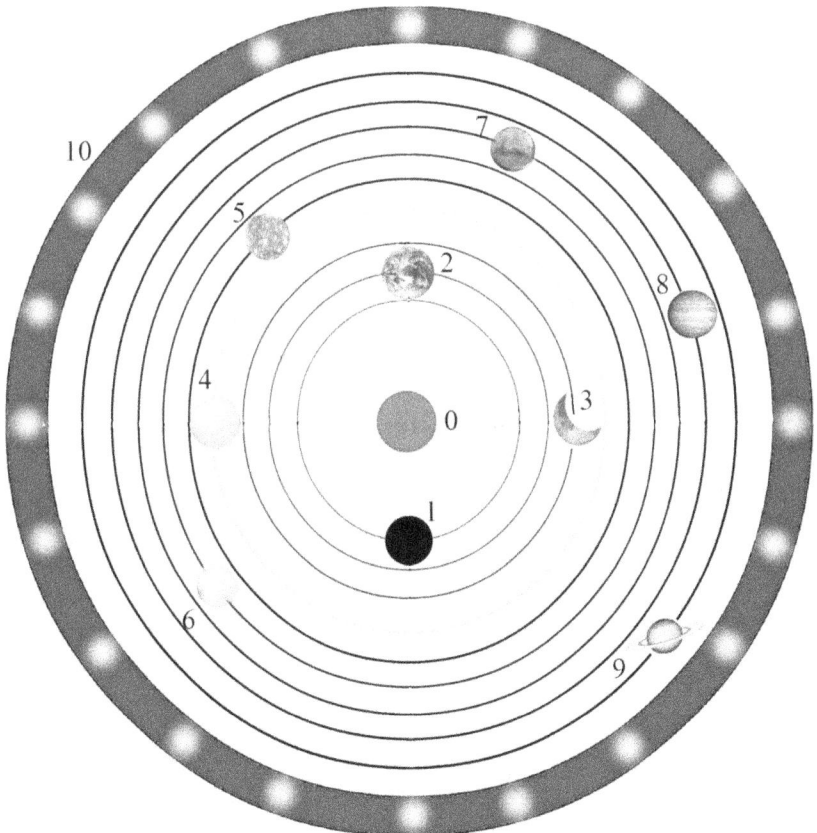

Image 2.2.

Pythagorean cosmology:

Schematic diagram

0 The Central Fire,
1 The *Antichthon*,
2 The Earth,
3 The Moon,
4 The Sun,
5 Mercury Planet,
6 Venus,
7 Mars,
8 Jupiter,
9 Saturn Planet,
10 Sphere of the fixed stars.

Historical literature suggests that Anaxagoras actually made such a measurement by resorting on the occurrence of an annular solar eclipse on February 17, 478 BC; which was visible for almost six minutes in large Greek territories: in its islands, in the Peloponnese peninsula and in Athens, his place of residence. Then, he dedicated to gather information of a good number of sailors and traveling merchants, whom said they had seen, or not, the phenomenon. With such testimonies the philosopher could himself elaborate an idea of the size of the lunar shadow on our planet, and thus, an indicative of the size of our satellite: From these experiences, he concluded that the Moon was bigger than the Peloponnese peninsula.[4] (See Image 3.1)

Anaxagoras was the first to introduce philosophy to the Athenians, who, like other societies in other times and places, maintained certain hostility toward those who tried to introduce a much higher level of culture than they were accustomed to. Under this scheme, they accused and prosecuted him for impiety, for teaching that the Sun was a red-hot stone, and that the Moon had terrestrial characteristics and did not shine with its own light. Disillusioned, he left for Lampsacus in Ionia where he self-exiled and established in; and, probably, he would let himself die of hunger.

The Historian, moralist philosopher, biographer and Greek essayist Plutarch of Chaeronea (50 - 120 AD) in his *Lives of the Noble Greeks and Romans*, commonly called *Parallel Lives*, writes about the astronomical knowledge of Anaxagoras, and also about his sad end:

"For he who the first, and the most plainly of any, and with the greatest assurance committed to writing how the moon is enlightened and overshadowed, was Anaxagoras; and he was as yet but recent, nor was his argument much known, but was rather kept secret, passing only amongst a few, under some kind of caution and confidence. People would not then tolerate natural philosophers, and theorists, as they then called them, about things above; as lessening the divine power, by explaining away its agency into the operation of irrational causes and senseless forces acting by necessity, without anything of Providence, or a free agent. Hence it was that Protagoras was banished, and Anaxagoras cast in prison, so that Pericles had much difficulty to procure his liberty; and Socrates, though he had no concern whatever with this sort of learning, yet was put to death for philosophy."[5]

The Greek Diogenes Laërtius from 3rd century AD was a great historian and doxographer of classical philosophy; in his work *The lives, opinions and remarkable sayings of the most famous ancient philosophers*, which is practically complete, he wrote about the important contribution that Anaxagoras made to the students: *"Finally, having passed to Lampsacus he died there, and asked for the magistrates if he wanted something to be carried out, they said he answered that "every year in the month of my death the boys would be allowed to play", and that nowadays it is observed."*[6] Fact with which school holidays were established in the Greek society of the time.

Like the solar eclipse of Thales, there was a lunar eclipse in ancient Greece that also has its great historical importance, since it happened on the battlefield too; and it is known as the Nicias's eclipse. The Peloponnesian War (431- 404 BC) consisted of a series of military conflicts in which the Greek cities of Athens, at the head of the League of Delos, and Sparta commanding the Peloponnesian League, clashed.

Within this context, the expedition to Sicily was a military campaign of Athens towards that island commanded by Nicias (470 - 413 BC), an Athenian general and statesman who was credited with remarkable victories in those wars; but he was also responsible for his own final defeat due to his highly superstitious character. The expedition extended from the year 415 to 413 BC, and its most important episode was the site of Syracuse. After some initial success, this campaign became an absolute disaster for Athenian forces. After multiple setbacks, Nicias agreed to the withdrawal, which was delayed almost a month, but when the ships were ready to set sail on August 27 BC, it occurred an eclipse of Moon that the Athenians considered as a sign of misfortune sent by the gods; so that both the troops and the sailors refused to embark, a refusal that was approved and backed by the superstitious Nicias. So many delays finally forced the Athenians to a battle in the port of Syracuse for which they were not prepared, and where they were completely defeated. The Syracusan cavalry attacked them without mercy,

killing or enslaving those who remained of the powerful Athenian fleet. Nicias would be one of its many victims, his life lost by obeying the calamitous signs of the Moon Goddess.

Here also Plutarch tells us about the occurrence of this eclipse, as well as the way in which some Greek people, and especially Commander Nicias, thought about the lunar eclipses:

"When everything was ready, without any of the enemies observed it, as they did not expect it, on that same night the Moon was eclipsed; a thing of great terror for Nicias and for all those who, out of ignorance and superstition, were frightened by such events: because as the Sun went dark towards the thirtieth day, almost everyone knew that the darkening was caused by the Moon. But as for this, what was it that opposed it, and being in its fullness suddenly lost its light and changed to different colors; this was not easy to understand, but they had it for something very extraordinary and as an announcement made by the Goddess of great calamities."[7]

Eudoxus of Cnidus and his cosmology

Eudoxus of Cnidus (390 - 337 BC) was a Greek philosopher, mathematician and astronomer. After a sixteen-month stay studying astronomy in Heliopolis, Egypt, he went to Athens to study at Plato's academy; Eudoxus was then 23 years old. Nothing of his work has arrived until our days; all the references we have come from secondary sources, from which it can be concluded that he initiated a great progress in the mapping of the sky and the Earth, and that he was the first to propose a planetary model geometrically and mathematically well founded, for what is considered the introducer of geometry in astronomy.

His great fame in this branch of astronomy is due to his invention of the celestial sphere, divided into degrees of latitude and longitude, in which the positions of the stars, the Sun, the Moon and the five planets known at the time are represented. Also, to his valuable contributions to understand the movement of the celestial bodies, which he explained with his *Geocentric cosmological system*, which he sustained by constructing a *Model of homocentric spheres* that represented the movements the Moon, the Sun, the five planets, and the set of fixed stars; all of them revolving around our planet. In such a model he explained, with geometrical and mathematical arguments, that the Earth was the center of the universe and the rest of the celestial bodies revolved around it; model that would give the Greeks their taste for a mechanical conception of the universe and on which Aristotle would underlie to develop his own cosmological vision.

In his time, the challenge to properly explain the seemingly disorderly movement of the planets was pressing: as seen from Earth they appear to recede in its path, which is known as retrogression; they show moments that seem to be stationary, they also present deviations with respect to the ecliptic and, finally, they show changes in velocity too. Of this erratic character of the movements of these celestial bodies comes the term planet, which in Greek is written πλανήτη, and means precisely *wanderer, vagabond*. The understanding of this complex behavior would lead to Eudoxo to introduce his famous System of homocentric spheres.

The problem was to find the necessary amount of uniform circular movements that, by their combination, would reproduce the movements of the planets as they were actually observed; in particular the variations in their apparent speeds, their stops and retrogradations, and their movements in latitude. Eudoxus was dedicated to finding a combination of movements of several nested or concentric spheres, rotating each one independently around its own axis, which was also transported physically by the revolution of the next sphere that contained it.

Eudoxus explained his system in a book entitled *On Speeds*, which was lost, together with all his other writings; so we depend on the passages in the works of Aristotle and Simplicius of Cilicia (490 - 560 AD) for our knowledge of his system. It does not seem that Eudoxus speculated on the cause of all these rotations of the stars; nor about the material, the thickness or the mutual distances between the spheres. If he simply adopted them as mathematical means of representing the movements of the stars and thereby subjecting them to calculations; or if he really believed in the physical existence of all these spheres, it is something that remains in uncertainty.

But since Eudoxus did not attempt to connect the movements of the various groups of spheres to each other, it seems probable that he only considered them as adequate geometrical constructions to calculate the apparent paths of the celestial bodies. Despite all its complexities, the system of homocentric spheres proposed by Eudoxus demands our admiration as the first serious attempt to deal, mathematically and geometrically, with the seemingly without natural law movements of these stars. It contains the nucleus of all the planetary theories of the next two thousand years, that is, the assumption that irregularities in apparent orbits can be explained as the result of multiple uniform circular motions and superimposed.

In the same way as Plato and the Pythagoreans had held, Eudoxus argued the following order of the stars beginning with the most distant to the Earth and adding for each one the number of moving spheres indicated:

(1) Saturn, 4 spheres   (2) Jupiter, 4
(3) Mars, 4   (4) Mercury, 4   (5) Venus, 4
(6) Sun, 3   (7) Moon, 3 spheres.

For the fixed stars only one sphere was sufficient to represent the daily revolution of the heavens. The total number of spheres was therefore twenty-seven.

For the case of the Moon and the Sun he discovered that it was possible, through an appropriate choice of poles and rotation speeds, to represent their respective movements assuming three spheres for each one of them: each star was supposed to be located at the equator of a first sphere that rotated with uniform velocity around its two poles. To explain the variations observed in their movements, Eudoxus assumed that the poles of this sphere were not immobile but were transported by a larger sphere, concentric with the first, and that rotates with a different speed around two different poles from those of the first one. As this was not enough to represent the observed trajectories, he placed the poles of the second sphere in a third, concentric and larger than the first two and moving with its own speed around two other different poles. But for the most intricate movements of the five planets a set of four spheres were necessary for each of them.

Eudoxus of Cnidus seems to have been the first to try to represent with the help of geometry each of the observed celestial movements: the daily rotation; the monthly revolution, the annual or any other periodic revolution. By means of spheres, whose center coincided with that of the Earth, Eudoxus, also called *the divine,* managed to sustain the movements of the Moon, the Sun, the planets and the stars, resorting exclusively to uniform circular movements; saving the fundamental budget of the geocentric theory. In this way, he came to a general satisfactory explanation of celestial movements, and so inaugurated a new period in the history of astronomy that was marked by attempts to explain the movements of the planetary system through mechanical models.

The Moon in Aristotelian cosmology

Surely the two most universally known philosophers, of whom all people have heard in their lives some comment, are Plato and Aristotle. Plato (427 - 347 BC) very probably was born in Athens, where towards the year 387 BC founded his school of philosophy known as the *Academy of Athens*, in which beside philosophy, rhetoric, mathematics, astronomy and medicine were taught. Among his most notable students or disciples there are the great mathematician and astronomer Eudoxus of Cnidus, whom we have already dealt with, and the famous philosopher Aristotle, who entered to study in the year 367 BC at the age of 17 years and remained there until the death of his teacher in 346. It was at the Academy where Aristotle met the astronomer Eudoxus, who influenced and introduced him in the field of astronomy.

Aristotle (384 - 322 BC), born in Stagira city in the Kingdom of Macedonia, was a true polymath who worked on an enormous variety of topics: political philosophy and philosophy of science, physics and metaphysics, astronomy, ethics, aesthetics, rhetoric, logic, and biology; his ideas have had a profound influence on the intellectual development of the Western world for much more than two millennia. The last great speculative philosopher that figure in the history of ancient astronomy wrote some 200 treaties, which have only survived about 31.

In the year 343 B.C. King Philip II of Macedonia summoned Aristotle to he would be the tutor of his

13-year-old son, who would later be known as Alexander the Great, the future *Conqueror of the world*. Aristotle then traveled to Pella, the capital of the Macedonian Empire, and educated Alexander for about five years. Finally, after the death of Filippo II, the young man occupied the throne and dedicated himself to his territorial conquest. Finished then his functions, the philosopher returned to Athens in 335 BC, which he had not visited since the death of his teacher Plato; and founded there his own school of philosophy, called the *Lyceum* for having as its seat a building called Lyceum, which at the same time was a gymnasium and a temple dedicated to the god *Apollo Lyceus*. This school was also known as the *Peripatetic School*, due to the habit of its members of walking and discussing at the same time.

The central ideas of the Aristotelian cosmology are contained in his work *On the Heavens,* or *De Caelo* in Latin, which is composed of four books. The first two have title *Of celestial bodies,* and in them the author develops the themes relating to simple bodies that make up the universe in general, the nature of the sky, the stars and the Earth.

The shape of the universe as a whole and of the celestial bodies that compose it, are central themes in the astronomy of Aristotle; and he treats them with elements derived from the direct observation of the circular movement of the stars in the sky, and reinforced with philosophical arguments. The system of homocentric spheres and circular movements of Eudoxus was completely accepted by Aristotle. Thus he concludes that the shape of the universe in generally is spherical: *"Therefore, if the sky is moving in a circle and moves faster than anything else, it must necessarily be spherical."*[8] And more still, continuing with his argumentative scheme, he extends this spherical shape to all celestial bodies: *"This is why it seems right that the whole sky and every star should be spherical.*[9] This generalization of the sphericity of the stars is of great importance for the subsequent development of astronomy, because it lays the foundations for the geometric conception not only of the Earth, but also of the Moon and all other celestial bodies.

In astronomy he rejected the Pythagorean approaches of a universe based on a central Fire and an Counter Earth; and in its place he proposed the existence of a spherical and finite Cosmos that would have the immobile Earth, static, as its center; while the Moon, the Sun, the five known planets and all the other celestial bodies revolve around it following circular orbits; which is traditionally known as the *Geocentric Theory of the Universe*. Aristotle dismisses the idea that the Earth has any movement, the rotation on itself or translation around the center of the universe. Like his teacher Plato, he never abandoned the idea that it is the daily rotation of the skies from East to West that explains the movement of the stars.

The evidences that were adduced to prove the sphericity of the Earth varied in different historical periods, but the doctrine by itself was accepted without questions from the times of Aristotle, as well as also the central position of the Earth in the universe. He is the first writer in whose works are clear and definitive statements about the roundness of our planet; of which we will here mention only a few: First, he deduces it from the tendency of all things towards the center, by which, when the planet was in its formation process and the component elements were equally gathered from all directions, the mass thus formed by accretion was constituted so that all circumference should be equidistant from its center. Aristotle sums up the subject well: *"If the earth was generated, then it must have been orchestrated in this way, and thus clearly its generation was spherical; and if it has not been generated and has always been maintained, its character must be that which the initial generation, if it had occurred, would have given it.*[10] Here we could say that the philosopher makes an approximation to modern cosmological theories, because that is how our planet was formed: through a process of accretion of the materials of a protoplanetary disk.

He also infers the roundness of the planet according to the known fact by which, when a person moves in appreciable distances from north to south, the landscape in the sky changes: the stars that were seen from one city, are no longer visible from another. Then, Aristotle states that this can only happen on a curved surface, and uses it as an argument in favor of the sphericity of our planet: *"All of which shows not only that the Earth has a circular shape, but that it is also a sphere without*

large size: otherwise the effect of such a slight change of location would not be so obvious.[11] The fact that he claimed that our planet *is a sphere of no great size* has probably served as a motivation for future astronomers and geographers whom ventured to determine its size.

Finally, Aristotle uses the observational evidence that lunar eclipses show to deduce the sphericity of the Earth, because when it is interposed between the Sun and the Moon the always circular shape of the darkened part of the latter shows that the body that causes the darkening is spherical; for if it was a flat disc there would be some circumstances in which the shadow would be a very deformed circle, similar to an ellipse; or, in an extreme case, a straight line. As the philosopher wrote: *"The evidence of the senses corroborates further this. How else will the lunar eclipses show segments as we see them? Thus, the shapes shown by the Moon each month are of all kinds, straight, gibbous and concave ones; but in eclipses the contour is always curved: and, since it is the interposition of the Earth that creates the eclipse, the shape of this line will be caused by the shape of the surface of the Earth, which is therefore spherical."*[12]

In the Aristotelian astronomy the Moon played a fundamental role, because in it the universe is conceived as divided, precisely by the sphere of the Moon, in two regions or worlds: the *Sublunary* or terrestrial which constitutes the theme of the third and fourth books of his text *On the heavens*; and the *Superlunary* or celestial one. Each of these worlds has well-defined and opposite characteristics. The world or sublunary sphere is the region of the cosmos located below the Moon and consists of the four Empedocles's elements: earth, water, air and fire; but unlike him, Aristotle considered that such elements could transform one another, thus explaining generation and corruption; and for this reason this sublunary world was imperfect and was subject to continuous alterations, and also to linear and random movements.

On the contrary, the superlunary world, which contains the Moon and the remainder higher bodies, is formed by the *Quintessence or Ether*, a special, incorruptible matter. This ether only admits a type of change, the uniform circular movement that, having no beginning or end, he considers it as the perfect, ideal form of movement. With which, in this region dominates the perfection and immutability. Thus, in the Aristotelian vision the Moon, being located in a region of the cosmos constituted by the ether, also exhibits these same ideal characteristics.

After having generalized the sphericity of the celestial bodies and having argued the roundness of our planet, Aristotle then presents his reasoning about the shape of the Moon: *"Once again, what holds for one holds for everything, and the evidence of our eyes shows us that the Moon is spherical. How else should the Moon, as it grows and wanes, show for the most part a figure in the shape of a crescent or humpback, and only in a moment, a half-moon? And the astronomical arguments give a greater confirmation; because no other hypothesis accounts for the gibbous form of eclipses of the Sun. One, then, of the celestial bodies being spherical, clearly the rest will also be spherical."*[13]

As for the movements and the position of the Moon, the philosopher appeals again to the evidence of the observation of the firmament and to the philosophical arguments: *"The movements of the Sun and the Moon are smaller than those of some of the planets. However, these planets are farther from the center and, therefore, closer to the primary body than they, as the observation itself has revealed. Because we have seen the Moon, half full, pass under the planet Mars, which disappears in its dark side and comes out through the luminous and bright part."*[14] Here Aristotle refers to the fact that when the planet Mars intersects with the Moon while it is in its half-full phase, it is hidden by the dark part of the Moon, which shows that our satellite is located in a circular orbit more interior, more near, to Earth; and that Mars is closer to the sphere of fixed stars, like other planets.

Reading *On the heavens* it is concluded that Aristotle accepts the part of the Pythagorean worldview concerning the relative positions of celestial bodies beyond Earth: the Moon, the Sun, the five planets: Venus, Mercury, Mars, Jupiter and Saturn; and finally the sphere of the *Fixed stars*; in that respective order.

But in that geocentric world, in which all the celestial bodies share a same characteristic as it is to

rotate in circular orbits around the Earth, the Moon, apart from its great apparent size in the sky and of being the first in proximity to our planet, it doesn't possess any other special characteristic that allows to differentiate it from other celestial bodies. This means that it has not yet been understood that it is a satellite, our only natural satellite. Because within a geocentric system this is not only unimaginable but also unacceptable.

Hellenistic Greece

The Classical Greek period ended with the rapid rise of the Macedonian kingdom and the conquests of Alexander the Great (356 - 323 BC), who, as already mentioned, was the son of the Greek king Filippo II of Macedonia, and after his decease he was in charge of the government of mentioned kingdom. He is widely known for having formed the Alexandrian Empire, which would extend the Greek rule by the East covering from Turkey to territories of India, and by the South to Egypt, where he founded the city of Alexandria, just where the Nile River savages the Mediterranean Sea. After his early death at 33, in Babylon in 323 BC, began the period of Greek history known as *Alexandrian* or *Hellenistic*, which continued up to the year 146 BC.

During this Hellenistic period the different areas of knowledge became independent of philosophy, a concept that previously covered all of them; therefore, the respective sciences as we today understand them were constituted in autonomous subjects. Which were disciplines studied and taught by great sages like Euclid, Apollonius, Aristarchus, Eratosthenes, Archimedes, etc. Many cities, especially Alexandria, became centers of knowledge and dissemination of science and art.

With the advent of a select group of mathematicians, astronomers, geographers, and other specialists with a notable interest in experimentation, knowledge in this Hellenistic period shows significant progress and achievements. More specifically, the area of astronomy based on previous developments in the philosophical and theoretical planes, began to provide more tangible and practical results of unquestionable importance. The determination of the relative sizes and distances of the celestial bodies, beginning with the Moon, would be the most representative example of such advances.

The Greek geometrical knowledge as a whole is presented to us by the geometer and mathematician Euclid of Alexandria (325 - 265 BC), the father of Geometry, in his work *Elements*; where he collects, organizes and systematizes all geometric and mathematical developments until his era, this being one of the best known academic productions in the world. As summarized by Bertrand Russell: "*Elements of Euclid is undoubtedly one of the best books ever written, and one of the most perfect monuments of the Greek intellect*".[15] The work consists of thirteen books and presents a formal and axiomatic study of the properties of regular forms: lines and planes, triangles, circles and spheres, etc., which we now know as *Euclidean Geometry*; a fundamental work for the formation of astronomers of the time, and for the general development of knowledge.

In the book I Euclid makes a precise definition of the circle, as well as its diameter, and in books VI and XII he deals with the sizes or dimensions of the circles in general. While the sphere is defined in book XI and in XII it is explained the calculation of its dimensions.[16] With which the geometer establish a more consistent basis for quantitative work with the spheres, facilitating the way for astronomers interested in measuring the size of the celestial bodies.

Aristarchus of Samos

Aristarchus (310 - 230 BC) was born on the Greek island of Samos, was a student of the Peripatetic school founded by Aristotle, and became a leading mathematician, geometer and astronomer. Since only one of his writings has survived, his cosmological thoughts have been known mostly from references made by later authors. Ptolemy in the *Almagest* names him as a conscientious observer of the solstices and equinoxes; and it seems that Aristarchus interpreted his observations correctly, attributing these phenomena to movement of the Earth around the Sun, and stating that it was necessary that the Earth's orbit was inclined to explain the cyclic climatic changes , or *Seasons*. Referring to him, Bertrand Russell states that: "*... He is the most interesting of all the astronomers of his time, because he advanced the complete Copernican hypothesis, ...*"[17]

Most likely due to the work of Euclid, from the third century BC the Greek astronomy, which had been fundamentally qualitative, entered a quantitative phase in large part represented by the works of Aristarchus, Eratosthenes, Hipparchus and, finally, the Egyptian Ptolemy. If Thales introduced the rational form of thought, or philosophy, in ancient Greece, Aristarchus of Samos annexed it, during the Hellenistic period, the mathematical and geometric Euclidean supports. He promoted one of the most brilliant and audacious ideas in the history of knowledge, such as extending the usefulness and validity of geometry and mathematics to studying the whole universe; applying its methodologies to the analysis and description of both the stars as of the involved cosmological phenomena, and pointing then the way for future generations of astronomers An apparently simple idea, but of incalculable value for the development of science and knowledge in general, and it is for this reason that we can call him the *founder of mathematical astronomy*.

For the Greeks it was obvious that the heavens should exemplify geometric beauty, which would only be the case if all the stars moved in circles in the sky; but the apparent movements of the planets, which have been very deeply analyzed, seem to be irregular, complicated and incomprehensible, nothing like a divine creation. The problem, then, was raised as follows: Is there any hypothesis that eliminates the apparent chaos in planetary movements and establishes order, simplicity and beauty instead? Aristarchus of Samos, contradicting his generation, dared to formulate such a hypothesis: that all the stars and planets, including ours, turn in circles around the Sun as their center; what amounts to the first recorded approach of a *Heliocentric theory*.

But this worldview was rejected for more than two thousand years, largely due to the great philosophical weight of Aristotle; and very much in spite of the fact that the geocentric theory did not satisfy some astronomers of this epoch, who lamented that it did not explain such an anomalous movement, of recoil, of the planets. Among them was Heraclides Ponticus (390 - 310 BC), disciple of Plato who, studying the movements of Mercury and Venus, clearly understood that their centers of revolution should be the Sun. That Aristotelian geocentric system, based on two erroneous assumptions, that the planets orbit around the Earth and that their orbits are circular instead of elliptical, was already very questionable at that time.

The most daring astronomer of this era was Aristarchus, who proposed the first true formulation of a *Heliocentric theory*; in which he argued that the set of fixed stars, the *Celestial Vault*, was at a practically infinite distance from the Sun; which served him to raise the hypothesis that in the center of the universe was not the Earth but the Sun, and that our planet not only revolved annually around the King star with an inclined orbit, but also on its own axis in a period twenty-four hours. Additionally, he placed our planet in its correct place, between the planets Venus and Mars. In his works, Archimedes, Plutarch and Simplicius relate that Aristarchus formulated a theory according to which the Earth revolves annually around the Sun and at the same time rotates daily on an axis inclined with respect to the plane of its own solar orbit. From the astronomical point of view, it was undoubtedly a valuable suggestion that could explain both the regular daily movement of the fixed stars, as well as that irregular movement so characteristic of the planets. There is no reason to doubt that, in his heliocentric system, this great philosopher would locate the Moon as a satellite of the Earth and orbiting around it.

Archimedes of Syracuse (287- 212 BC.), a contemporary 23 years younger than Aristarchus, in his work *The Sand Reckoner* explains that the latter published a book based on certain assumptions on which it appears that the universe was much larger than was expected. These hypotheses were that the sphere of the fixed stars and of the Sun remain motionless and have the same point as their center, while the Earth travels around the Sun in a circular path, with the Sun at the center of its orbit.[18] The historian Plutarch also made reference to Aristarchus summarizing his heliocentric idea in which the sky is immobile and the Earth moves on an inclined orbit around the Sun, and at the same time it rotates on its own axis. Likewise, he reports that in this last aspect Aristarchus followed Heraclides Ponticus by believing that the apparent daily rotation of the fixed stars was due, in reality, to the rotation of the Earth. Finally, Plutarco relates that the Stoic philosopher Cleanthes of Assos (330 – c. 230 BC) stated that it was the duty of the Greeks to

accuse Aristarchus for impious, for having displaced Earth from the center of the universe.[19]

Indeed, since what is observed is the same in both cases, in theory it seems much easier a single body, the Earth, rotates on itself in 24 hours, than all the celestial bodies rotate in such a period around it. But the double terrestrial movement, translation and rotation, implied consequences practically unacceptable at that time, both from the astronomical point of view and, most importantly, from the philosophical point of view. In short, it was the respect for apparent phenomena and rational argumentation which led the Greeks to reject the terrestrial movement and heliocentric hypotheses of Aristarchus. Therefore, this theory did not convince the wise men of their time and was harshly fought, and finally abandoned; only many centuries later it was retaken and revalued by the priest astronomer Nicolaus Copernicus in the sixteenth century.

It is very disappointing to discover that the only existing work of Aristarchus, "*On the Sizes and Distances of the Sun and the Moon*"[20], is based on the geocentric worldview; but the evidences that he posed the heliocentric point of view are, in any case, quite conclusive; and many historians are inclined to the opinion that he adopted the heliocentric worldview long after writing such a work. In it the author exposes the way to determine such magnitudes by the geometry contained in the *Euclid's Elements*; taking into account the Aristotelian theories of the spherical celestial bodies and the geocentric universe, in which the Earth is at rest and the Moon revolves around it with a period of 29,5 days, while the Sun does so in a year of 365,25 days period; and additionally using a method that allows him to find the relative sizes of the Moon and the Sun with respect to the diameter of our planet.

About the Sizes is structured like other applied mathematical texts of the early Hellenistic period, and it is a work of computational deductive mathematics. Therefore, the objective of the text is to develop geometric schemes for the Earth-Moon-Sun system, and then to introduce numerical parameters in them to derive the lower and upper limits for the sizes and distances of the Sun and the Moon.

Significant, although approximate and erroneous, were the first investigations of this extraordinary scientist of antiquity about the distances between the celestial bodies. Aristarchus achieved such a feat by making detailed observations of the movements of the Moon and the Sun, and their eclipses too; understanding that those of the Moon are produced because the Earth interposes between the Sun and such star, with which the shadow of our planet projected on the surface of the Moon advanced until it completely covers it. In *On the Sizes and Distances of the Sun and the Moon* he provides the details of its remarkable geometric arguments, based on observations, with he determined that the Sun was about 20 times more distant from Earth than the Moon, and that it was also 20 times the size of our satellite. His method was mathematically and geometrically correct, but the precision required was extremely high due to the great distance of the Sun compared to the Moon.

Sir Thomas Little Heath (1861 - 1940) was a classic British scholar, mathematician, historian of mathematics and translator of ancient Greek works. Heath translated into English the works of Euclid of Alexandria, of Apollonius of Perga, Aristarchus of Samos and of Archimedes of Syracuse. For the discussion that follows, we will circumscribe to his text "*Aristarchus of Samos: the ancient Copernicus*"[21]; in which he presents both the original Greek version of *On the Sizes and Distances of the Sun and Moon*, as its translation into English.

For the determination of such magnitudes, the Moon evidently played a major role in the astronomical theory of Aristarchus; his work is based on a set of six assumptions, which himself denominates as *Hypothesis*, all of them referred to our satellite. The first three are geometric and somewhat obvious, and refer to 1) the Moon receives its light from the Sun, 2) that it moves in a sphere that has the Earth as a center, and that this last can be assumed geometrically as it was one point; and 3) that when the Moon is in its *First quarter* or *Third quarter* phases, appearing as half illuminated, the maximum circle that divides the dark and bright regions of the satellite is just in the direction of our eyes, so that we see it as a straight line. The other three hypotheses are quantitative, computational, they are not so obvious and demand

a greater attention on the way Aristarchus elaborated them: 4) that the amplitude of the shadow of the Earth is twice that of Moon; 5) that the apparent angular size our satellite is 2°, and finally, 6) that at the moment when the satellite is half illuminated, the Sun-Moon-Earth angle is 90°, that is, it is a straight one; while the Sun-Earth-Moon angle is 87°. The latter are assumptions about the physical world that allow the application of numerical parameters to the geometric model, and serve to derive numerical solutions to the problems in question.

To determine the distance and size of the Moon relative to the diameter of the Earth, Aristarchus had to make observations tending to evaluate the angular velocity of the Moon against the background of the fixed stars, the apparent angular size of the same and the angular velocity with which it traverses the shadow of our planet during a total lunar eclipse. While to calculate the distance and size of the Sun relative to the diameter of the Earth, he should have used the theory of the lunar phases and the fact that the apparent angular sizes of the Sun and the Moon are practically the same, as shown by the total eclipses of Sun.

Aristarchus correctly uses the physic evidences of the eclipses to argue that the Sun and Moon subtend the same angle in the sky, equivalent to their apparent sizes, but strangely he uses values for such angle as 2°, which is quite inexact, since it is actually four times smaller. Later, Archimedes cites a more accurate value of 1/2° for the angle subtended by the Sun and attributes this figure to Aristarchus; therefore, it must be assumed that the astronomer of Samos wrote his text at the beginning of his career, and that much later he was able to develop better instruments with which to make more precise astronomical measurements and be able to calculate more accurate values for the angular sizes of the Moon and the Sun; and then he would adopt his hypothesis of a heliocentric universe.

Taking the aforementioned six hypotheses as a foundation and support for his work, Aristarchus proceeds to enunciate and demonstrate, always within a geometric and mathematical context, a set of eighteen *propositions* related to the geometry of the Moon-Earth-Sun system, and within which the special cases of the eclipses of the Moon, the eclipses of the Sun and the situation of the First quarter in which the Moon is justly half illuminated, play a crucial role.

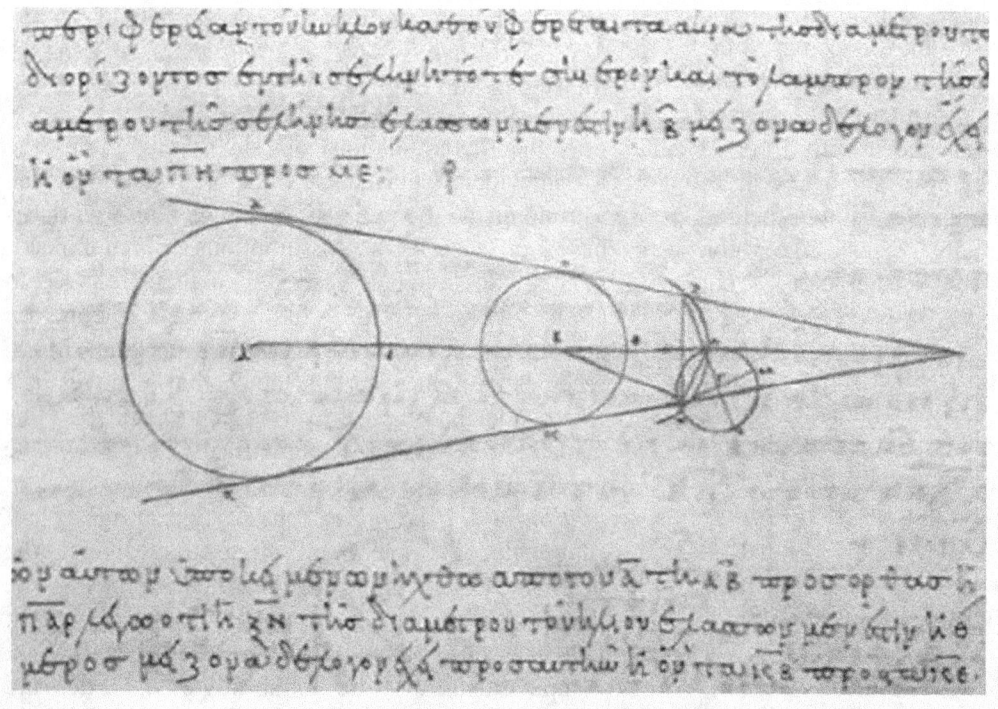

Image 2.3

Geometry of the Sun-Earth-Moon system according to Aristarchus.

Detail of a copy in Greek of the tenth century AD of the text *"On the Sizes and Distances of the Sun and the Moon"*, originally written by Aristarchus of Samos in the 2nd century BC.

Of these propositions we will here consider only six, which are those that make specific reference to the sizes and distances of the Moon and the Sun. As mentioned, the propositions are already demonstrated in the work of Aristarchus, so here we will take them as true, and we will only give some general indications about the way he did it. We will also present them in the same order in which Aristarchus considered them.

In proposition six he establishes that the Moon moves in a lower orbit than the Sun, and shows that when it is illuminated justly in half, its angular distance to the Sun measured from the Earth-Sun line is lower to a quadrant, or less than 90°. This result acquires greater significance when combined with hypothesis number six, according to which at the moment when the satellite is half illuminated, its angular distance to the Sun, which corresponds to the Sun-Earth-Moon angle, is 87°, while the Sun-Moon-Earth angle is 90°, that is, it is straight; All these factors justify that the Moon is precisely illuminated in half. In his text, Aristarchus does not refer to the way in which he obtained this value of 87°, so here we are going to propose a possible way of doing it.

In a first moment, the astronomer must have considered some central elements of the geocentric conception of the universe, according to which our satellite describes a 360° complete circular orbit around the Earth, East-West direction with respect to the fixed stars and in a period known as sidereal month, equivalent to 27,33days, or also equal to 655,73 hours; which gives an angular velocity of the Moon of 13,18 ° per day, or also 0,55 ° per hour. On the other hand, the Sun also describes a geocentric orbit in the same direction with respect to the fixed stars, in a period known as sidereal year and equivalent to 365,25 days; with which the angular velocity of the King star is 0,986° per day. Additionally, Aristarchus had to use old astronomical records or data, as well as those obtained by his own direct observations. Finally, he could have used a *Mechanical Simulator of the celestial movements,* something similar to the *Antikythera* mechanism already mentioned in the previous chapter, and which were probably already known in his time, both to derive and obtain data, such as to test the results of their theories.

Image 2.4
General geometry of a Sun Eclipse

Geometric disposition of the Earth-Moon-Sun system, in the coordinate plane $X_i Y_i$, for the initial moment
of a Total Solar Eclipse.
It is not to scale.

Now let us consider as the starting point the exact moment of a total solar eclipse, in which the three stars are perfectly aligned and our satellite is in its New Moon phase and stands between the Sun and the Earth, as shown in Figure 2.4. This situation would allow us to select an initial system of rectangular coordinates X_i and Y_i with center in our planet, and in which the X_i axis corresponds to the line that passes through the centers of the three bodies, while the Y_i axis would be a perpendicular to it and passing through the center of the Earth. Taking this as the initial moment, Aristarchus, probably using *water clocks* or *hourglasses*, could have measured the time it takes for the Moon to strike its First Quarter phase, in which it looks like exactly half illuminated.

Had he proceeded this way, the astronomer should have obtained a value very close to 171,3 hours, or 7,137 days to visualize our satellite in its First quarter. Next, and considering the angular velocities of the stars, he could determine that in such period of time the Moon should have traveled an angle of 94,07° with respect to the initial coordinate system; while the Sun should have traveled only an angle of 7,07°; with which the difference in the angular displacement of these two stars should be 87 °.

Now we must elaborate a final system of coordinates X_f and Y_f also with center in the Earth, and in which the X_f axis corresponds to the Earth-Sun line, and the Y_f axis is a perpendicular to it. According to what has been exposed, this final system of coordinates would be shifted angularly with respect to the initial system in an angle of 7,07°, which is the same value of the displacement of the Sun.

In this final situation, schematized in Image 2.5, it can be seen how the separation, or angular distance measured from the X_f axis, between the Moon and the Sun is 87°, which corresponds to the data provided by Aristarchus in his number six hypothesis.

This data is of supreme importance for the calculations and subsequent demonstrations of the astronomer. As in proposition seven, in which Aristarchus uses an elegant geometric argument to establish that the distance from the Sun to the Earth is eighteen times greater, but twenty times less than the distance from the Moon to our planet.

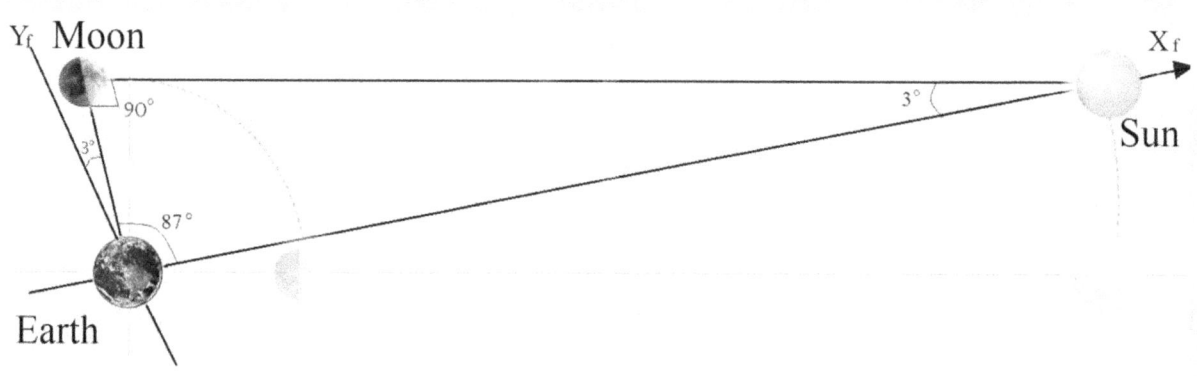

Image 2.5
First Quarter: The Moon justly half illuminated.
General geometry about the spatial arrangement between the Earth, the Sun and the Moon, so that the satellite can be seen from our planet as justly half illuminated: it occur in the First and Third quarter phases. According to the work of Aristarchus, taking as center the Earth, the angle between the Sun and the Moon is 87°. It is not to scale.

Although according to Image 2.5 this is a case of basic trigonometry and we would now express the relationship between the distance of the Moon and the Sun as equal to *cosine* (87°). But it must be taken into account in such time this branch of mathematics had not yet been invented; which we now call trigonometric functions, *sine*, *cosine*, *tangent*, etc., were unknown, so he used instead the *Table of chords* created by Ptolemy; so the philosopher had to approach a different method of calculation to obtain the lower and upper limits for such distances.

In proposition nine Aristarchus also resorts to the basic geometry of a solar eclipse to show that the size of this star is also greater than eighteen times, but less than twenty times the size of the satellite. The last results mean that the size and distance from the Sun have the same proportion to the size and distance of the Moon: Considering the average values, we have that according to Aristarchus's text, the Sun is nineteen times farther from the Earth than the Moon, and it is also nineteen times larger than our satellite.

The distance from our satellite to Earth is addressed by Aristarchus in his proposition number eleven, where he shows that the distance from the Moon to the planet is greater than 22,5 but less than 30 times the lunar diameter; or taking the average value we have that the satellite is retired from Earth 26,25 times its own diameter.

Later, in proposition fifteen, the astronomer is dedicated to finding a relationship between the size of the Sun and that of our planet. This time he resorts to the general disposition or geometry of a total eclipse of the Moon to show that the diameter of the King star has a relationship with the diameter of the Earth that is greater than 19/3, but less than 43/6. What if expressed in terms of the average value means that the Sun's diameter is 6,75 times larger than that of our planet. This last value would make the volume of the King star was about 300 times greater than that of our planet; and it could have been this great size that led Aristarchus to place the Sun at the center of the universe instead of the Earth, since even in his time it might seem absurd to make the much larger body revolves around the smaller one.

Finally, in his proposition seventeen Aristarchus uses the previous results to deduce a relationship between the diameter of the Moon and that of our planet: he states that the size or diameter of the Earth is 2,5 times greater but 3,16 times smaller than the one of the Moon. Then, on average the planet has a diameter 2,83 times greater than that of the satellite; or the inverse relation expresses that the diameter of the Moon is only 0,35 times that of the Earth.

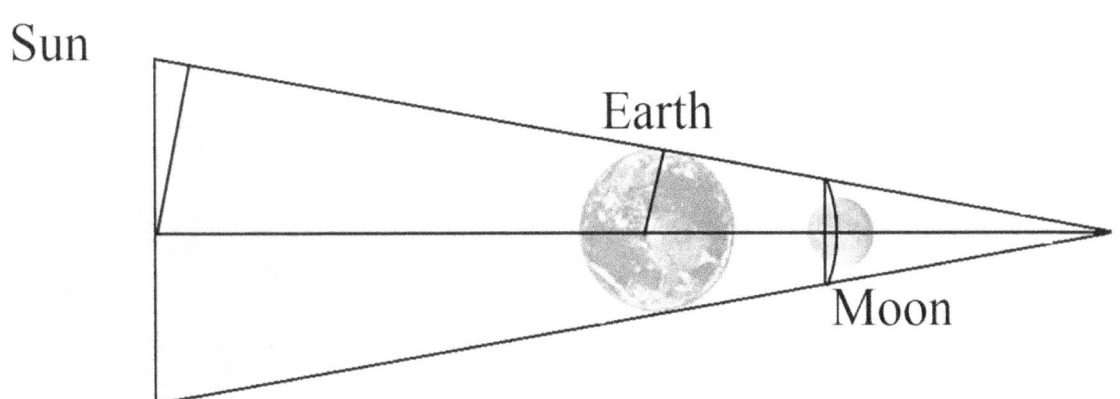

Image 2.6 General geometry for a Moon Eclipse
This general geometric scheme of a lunar eclipse was used by the astronomer of Samos to derive the diameters and, therefore, the relative sizes between the three stars: Earth, Moon and Sun. Schematic diagram, not to scale.

All the previous work of the astronomer of Samos we can synthesize looking for simple relationships that express the distance and sizes of the Moon and the Sun as functions of the diameter of the Earth, D_T; the final results are presented in Table 2.1.

Table 2.2 shows the values corresponding to the modern data for these magnitudes. It can be seen that the estimates were of an order of magnitude too small, but the errors were due to the lack of sufficiently precise instruments, rather than Aristarchus's correct form of reasoning. His method was mathematically and geometrically correct, but the precision required was extremely high due to the great distance of the Sun compared to the Moon.

Following his methodology, Aristarchus obtained numerical values far from reality, with errors of up to twenty percent for the case of the Moon and considerably higher for the Sun; although they are reasonable for that moment and that method. These numerical results now do not matter anymore, but it should be highlighted his audacity and the elegance of both his thinking and methodology. Anyway, using data provided by the observations of the Moon, the Sun and the eclipses, he had shown that the Sun was much more distant than the Moon and that it was much larger than the Earth; and also that this last in turn was larger than our satellite, a considerable achievement for the time.

ABLA 2.1
Sizes and distances of the Moon and the Sun depending on the Earth's Diameter, D_T, According to the work of Aristarchus of Samos.

	Moon	Sun
Diameter	$0{,}353 * D_T$	$19 * 0{,}353 * D_T = 6{,}7 * D_T$
Distance	$9{,}266 * D_T$	$19 * 9{,}266 * D_T = 176{,}1 * D_T$

TABLE 2.2
Sizes and distances of the Moon and the Sun depending on the Earth's Diameter, D_T, according to modern values.

	Moon	Sun
Diameter	$0{,}273 * D_T$	$109{,}3 * D_T$
Distance	$30{,}17 * D_T$	$11741 * D_T$

Aristarchus's work has always been subjected to scrutiny and to criticism both constructive and destructive. But regardless of the validity of his theoretical schemes and his numerical results, what is really important is that he proposes the universe has a certain mathematical structure that is decipherable, and that it can be interpreted and described in a quantitative and precise way. He shows that, from a few simple statements, which are easy to obtain through proper observation, and employing the respective geometric and mathematical methodology, a new knowledge of the universe can be produced, which otherwise would be beyond our reach. In this way the astronomer of Samos left well laid the foundations for the works of future astronomers.

All above makes Aristarchus worthy of a permanent pedestal of honor in the field of the history of knowledge in general, of astronomy in particular and very especially in the study and understanding of our satellite. But the results provided by Aristarchus for the sizes and distances of the Moon and the Sun were relative to the size of our planet, unknown in his time, but that Eratosthenes would be in charge of evaluating.

The Geometry of Apollonius

Apollonius of Perga (262 - 190 BC) was a famous Greek geometer and mathematician, usually known in his time as the *Great Geometer*; as a young man he studied in Alexandria with the successors of Euclid, and later he settled in Ephesus and also in Pergamum. His extensive works in geometry deal with the three-dimensional conical figures, as well as the flat curves. Writings that he compiled in his work *Conics*, in which he coined the terms *ellipse*, *hyperbola* and *parabola* to indicate those geometric figures that correspond to the respective properties of these three mathematical functions.

In his work, Apollonius created the foundations of *Conical geometry* through a compendium of eight books, in which he made respect to the conical figures what Euclid had done previously regarding circular geometry. It is called a conic section to all the curves resulting from the different intersections between a cone and a plane that doesn't pass through the cone's vertex. Such curves are defined more exactly as the geometric places of the points in the plane for which the distances to a line called *directrix* and to a fixed point called *focus* are in a certain ratio or *eccentricity*. It was Apollonius in *Conics* who showed that from a single cone the three types of sections can be obtained varying the angle of inclination of the plane that cuts it. The great scientific value of his work took a long time to be understood, because the importance of conical shapes in the universal system only came to be appreciated with the discovery of the astronomer Johannes Kepler (1571 - 1630) according to which the planetary orbits are elliptical, occupying the Sun one of the focus of such ellipses.

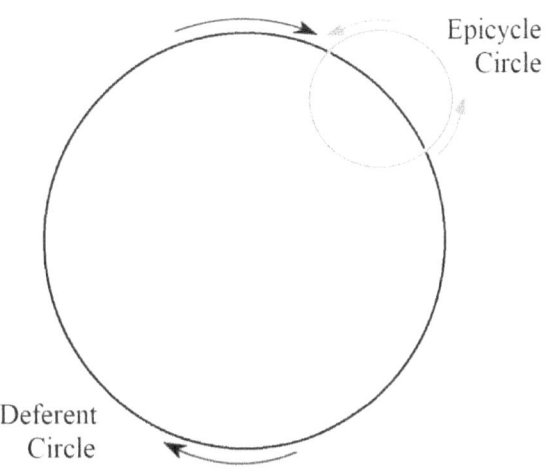

**Image 2.7
Uniform Circular Celestial Movement:**

The geometrical concepts of the Deferent circle and Epicycle circle were initially introduced by the astronomer and geometer Apollonius of Perga, who had the purpose of representing the uniform circular movements of the stars according to geocentric hypotheses.

Equally, it is attributed Apollonius to having originated the hypothesis of the *Deferents* and *Epicycles*, to try explaining the apparently disordered movement of the planets and the variable speed of the Moon. The time was right for the development of a geometric pattern that could better represent planetary movements and still keep Earth at the center of the universe. Particularly the movements of Mercury and Venus, which are always close to the Sun, should have originated the idea of an *epicycle* moving on a concentric *deferent*: the epicycle would be the secondary orbit in which said star revolves around some point of the deferent, and the latter would be a concentric circle with the Sun. Even ignoring the King star, a system of epicycles and deferents would be very useful to find an explanation to the direct and retrograde movements of the planets, and for their apparent stops.

In the celestial movements two aspects were very difficult to explain for the ancients. The first concerns its variable speed, which has the sidereal revolution as its period; the second is the apparent change in the direction of its movement, going from direct to retrograde and vice versa, including a moment of detention in this movement. The first difficulty was explained for the Moon and the Sun by means of an *Eccentric* scheme; but the second, relative to the planets, required the development of the theory of epicycles movement: The star moves first with a uniform speed along a circle, which is the epicycle, whose center moves at the same time along the circumference of the eccentric circle, the deferent, which has as its center our planet.

This was valid not only for the lower planets Mercury and Venus, but also for the higher ones, Mars, Jupiter and Saturn, which could move in the trajectory of an epicycle around an ideal center that, like the Sun, had a primary revolution around the Earth. It might seem strange that a celestial body had a revolution around an imaginary point and without a respective accessory in the sky. Nevertheless, this theory found the support of the mathematicians, since it seems to have been proposed and illustrated initially by Apollonius; it was also adopted by Hipparchus of Nicaea, and then generally accepted in place of the theory of the homocentric spheres of Eudoxus, which was insufficient to explain the observed phenomena.

Apollonius in 225 BC explained how it was possible to determine the ratio between the radiuses corresponding to the two circles; and how the trajectory along the epicycle corresponds to the synodic period, while that along the deferent was related to the sidereal period. Starting from this fundamental theory, first Hipparchus and then Ptolemy could develop a planetary theory that would complete the astronomy of the Greeks of the school of Alexandria.

Eratosthenes of Cyrene

The idea of measuring the circumference, diameter, or size of our planet seems to have been considered by the Greek philosophers at a relatively early time. Thus, the Greek comedian Aristophanes (444 - 385 BC) in his work *The Clouds* states that a disciple of Socrates says that the object of geometry was the measure of the entire Earth; which implies that the solution of such problem was contemplated by the academics of the time. We must also remember that Aristarchus had determined the sizes and distances of the Moon and the Sun refered to the diameter of our planet. In such a way that many astronomers and geographers of the time should have been very anxious to find an effective method to measure the Earth, and in this way unlock the calculations for the sizes and distances of the Moon and the Sun. Eratosthenes would be the most successful in that purpose.

The first great Alexandrian geographer was Eratosthenes of Cyrene (276 - 194 BC), who was also a mathematician, poet, philosopher, and historian. Between 230 and 195 BC he was director of the *Great Library of Alexandria*, the highest academic institution of the time. There Eratosthenes collected the available geographical knowledge and made numerous calculations of distances between significant places on Earth; and he organized the data of the registers for the East provided by the expeditions of Alexander the Great and that of the Mediterranean elaborated by the navigator Pytheas of Massalia (4th century BC). With this information he prepared the first map of the world known to date and based on scientifically proven facts. Both theoretical and practical sense, the Hellenistic geography reached its apogee with the work of polygraph Eratosthenes, who has been assigned a

founding role in the geography, the mapping and the geodesy.

Eratosthenes was 34 years younger than Aristarchus and both were in the Library of Alexandria; for this reason the Cyrenian must have been very aware of the work of Aristarchus in which he reported the sizes and distances of the Moon and the Sun dependent on the diameter of our planet. So Eratosthenes, most likely with the purpose of deepening and advancing the calculations of the astronomer, undertook to measure the Earth by experimentally determining its diameter, this calculation being the one that would lead him to immortality.

Eratosthenes' method of measuring the size of our planet involves understanding geocentric theory of the universe and spherical geometry, understanding proportionality between arcs of circumference, and calculating adjacent angles between parallel lines; also the knowledge on the calculation of the distance between two sites located in the same terrestrial meridian, and the use of the gnomon to measure the angle of incidence of the solar rays.

One of the sites chosen by Eratosthenes was the city of Syene, present-day Aswan in Egypt, where there was a well in which the sun's rays exactly penetrated at midday in the summer solstice on June 20-22, which meant in that city and date sunlight came exactly perpendicular to the ground. The other selected city was Alexandria, which Eratosthenes estimated was just to the north on the same meridian, and whose distance to Syene was estimated at 5000 *Stadia*, (*Stadion* in singular). Thus, all that Eratosthenes had to do was to measure in Alexandria, with the help of the *gnomon*, the angle of incidence of the Sun's rays at noon on June 21; which would be identical to the angle subtended by the two cities in the terrestrial sphere, and that therefore would determine the respective arc of circumference between them. The value he obtained was 7,2°, that is to say, a fiftieth part of the circle; in such a way that the circumference or size of the Earth must be 50 times 5000, which gives 250000 Stadia. For the people of the time this meant an extraordinary achievement of science. The *Stadion* was a unit of measure of length from that period, for which historians have found several modern equivalents, and in this text we will use the value of 185 meters; with which we have that the measurement of Eratosthenes was equivalent to 46250 Km of circumference for the Earth, which expressed in terms of its diameter would give 14722 Km, that is fifteen percent in excess with respect to the accepted modern value for such diameter: 12740 km.

The great contribution of Eratosthenes consisted in having formulated a reliable method for the experimental determination of the size of our planet; and so he cleared too the way to calculate the sizes of the Moon and the Sun in their absolute values : he also determined that the diameter of the Sun was twenty-seven times greater than that of the Earth, or equivalently 397494 km in diameter according to his data; also, he calculated the distance of the Sun from our planet in 148740000 Km., and that of the Moon in 144300 Km. Other contributions of Eratosthenes were to have measured, with great precision too, the angle of inclination of the Earth's rotation axis, or equivalently, the obliquity of the ecliptic in 23°51'15'', and to have elaborated a compilation in a catalog of near of 675 stars.

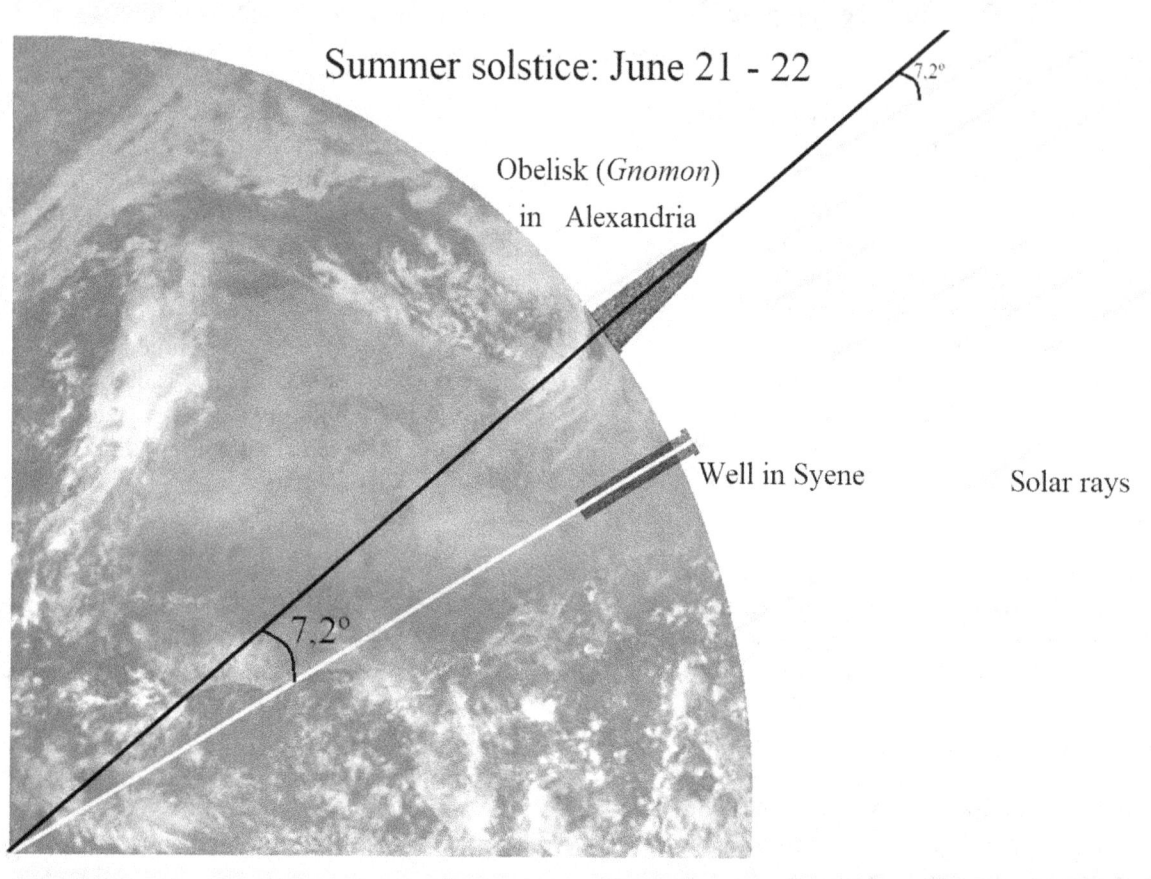

Image 2.8
The Eratosthenes experiment
Having determined the size of the Earth, Eratosthenes cleared the way for the subsequent measurement of the distance of the Moon, of the other stars, of the size of the Solar System as a whole and, finally, of the Universe as a whole.

Bibliographic Citations

[1] Ian Johnston. *Aristophanes: Clouds*. http://johnstoniatexts.x10host.com/aristophanes/cloudshtml.html

[2] The Internet Classics Archive. The History of Herodotus. Book I. http://classics.mit.edu/Herodotus/history.html

[3] Internet Encyclopedia of Philosophy. Thales of Miletus: http://www.iep.utm.edu/thales/

[4] Nordgren, Tyler. *Sun, Moon, Earth: the history of solar eclipses, from omens of doom to Einstein and exoplanets*. New York: Basic Books. 2016

[5] Wikisource. Parallel lives: Nicias. https://en.wikisource.org/wiki/Plutarch%27s_Lives_(Clough)/Nicias

[6] Early English Books Online. Diogenes Laertius. *The lives, opinions, and remarkable sayings of the most famous ancient philosophers*. https://quod.lib.umich.edu/e/eebo/A36037.0001.001/1:5?rgn=div1;view=fulltext

[7] Wikisource. Parallel lives: Nicias. https://en.wikisource.org/wiki/Plutarch%27s_Lives_(Clough)/Nicias

[8] Internet Archive. On the Heavens. 287-a-30. https://archive.org/details/decaeloleofric00arisuoft

[9] Internet Archive. On the Heavens. 290-b-5. https://archive.org/details/decaeloleofric00arisuoft

[10] On the Heavens. 297-b-15.

[11] On the Heavens. 298-a-10.

[12] On the Heavens. 297-b-30.

[13] On the Heavens. 291-b-20.

[14] On the Heavens. 292-a-5.

[15] Russell, Bertrand A, W. *A History of Western Philosophy*. New York: Simon and Schuster Inc. 1945.

[16] Clark University. Euclid's Elements. https://mathcs.clarku.edu/~djoyce/java/elements/toc.html

[17] Russell, Bertrand A, W. *A History of Western Philosophy*. New York: Simon and Schuster, Inc. 1945.

[18] Internet Archive. The works of Archimedes. *The Sand-reckoner*. https://archive.org/details/abw0362.0001.001.umich.edu

[19] Bill Thayer's Web Site. Plutarch: *On the Face in the Moon*. http://penelope.uchicago.edu/Thayer/E/Roman/Texts/Plutarch/Moralia/The_Face_in_the_Moon*/home.html

[20] Library of Congress. Rome Reborn. http://www.loc.gov/exhibits/vatican/math.html#obj6

[21] Internet Archive. Heath, Thomas Little. *Aristarchus of Samos, the ancient Copernicus*. https://archive.org/details/aristarchusofsam00heatuoft

Chapter 3

Lunar Astronomy in Greco-Roman Society

Our satellite as seen from Earth on Wednesday, January 24, 2018, at 22:00 UCT, 7 days and 19 hours after the New Moon, seeing 50% of its illuminated part. At this time the satellite has traveled a quarter of its elliptical trajectory around the planet, so it is said to be in its *First Quarter*; and this is a very special moment in lunar astronomy because from the Earth we see the Moon as justly illuminated in half. Recall that this was the point selected by Aristarchus of Samos to make its measurements of satellite's size and distance.

Courtesy of NASA's Scientific Visualization Studio: https://svs.gsfc.nasa.gov/4604

"Below this is the Moon, closest to the Earth of all the heavenly bodies, in that accepted theory places it at the junction of the air and the ether, which is why its own body is also visibly murky. The illuminated part of it has its luminance from the Sun, since the hemisphere of the Moon that is turned toward the Sun always gets illuminated. The Moon completes its own circuit in $27^1/2$ days, and is in conjunction with the Sun in 30 days."

On the Circular Motions of the Celestial Bodies.
Cleomedes, Greek astronomer of the 1st century AD[1]

"Clearchus says that the face, as we call it, is made up of images of the great ocean mirrored in the Moon. For our sight being reflected back from many points, is able to touch objects which are not in its direct line; and the full moon is of all mirrors the most beautiful and the purest in uniformity and luster."

The face which appears on the orb of the Moon
Plutarch of Chaeronea (50 - 120 AD), Greek historian.[2]

The Greco Roman World

The foundation of the millenarian city of Rome can be approached, in a first view, from the mythology and the legend perspectives, according to which this city would be founded on the banks of the Tiber River around 753 BC by the legendary twin brothers Romulus and Remus. But a more solid and credible perspective is the historical one, according to which Rome arose in progressive form by the installation of different Latin tribes in the area of the traditional *Seven Hills*, and by means of the creation of small villages in their tops, which ended up merging towards century VIII BC. In this scenario, the Etruscan king Lucius Tarquinius Priscus played a leading role, since he was who gave Rome an authentic civic appearance, thanks to his great urbanizing work in the late seventh century BC.

The history of Rome is divided into three major periods: the Roman Monarchy comprises from the founding period until the year 510 BC; the Roman Republic from 509 to the year 27 BC, and finally the colossus Roman Empire that goes from this last date to 476 AD, year in which the Western Roman Empire definitively collapses.

The Roman Republic was basically a warrior state. Between the beginning of the III and the middle of the II century BC the Republic began to expand its zone of influence beyond the Italian peninsula through the Mediterranean, giving rise to the Punic Wars that would lead to the establishment of Rome as the greatest power in the region. The necessities to conquer new lands to install their citizens and dedicate them to agriculture, to secure their borders, defend their allies, expand their trade, or simple military glory, impelled the Romans to a strong and accelerated geographical expansion.

All the great states of the Mediterranean were subdued by Rome in a short time. The final defeat of the Alexandrine Kingdom of Macedonia in the battle of Corinth in 146 BC, within the general framework of Macedonian Wars, triggered in the total subjugation of Greece to the Roman power, leaving Rome as absolute owner of the Mediterranean.

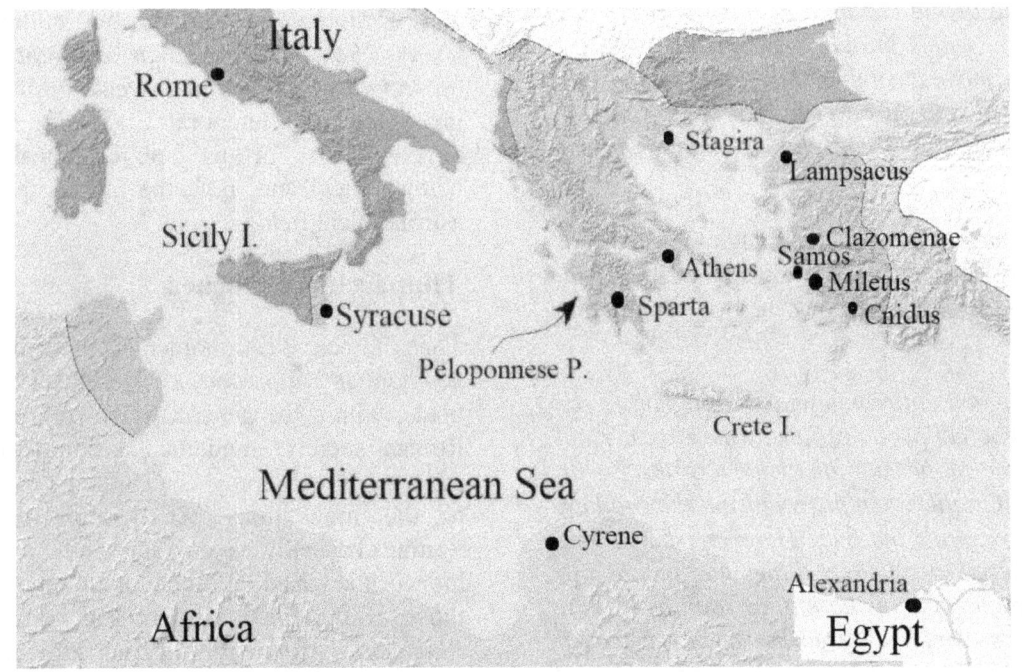

Image 3.1
Some geographical areas of the ancient Greco-Roman world
Recall that in the fifth century BC Anaxagoras, using information on a solar eclipse, determined that the Moon was larger than the Peloponnese peninsula.

The texts on history suggest that the coincidence between lunar eclipses and military battles was very frequent. The *Macedonian Wars* were a series of military conflicts between the Greek kingdom of Macedonia and the Roman Republic, which took place between the third and second centuries BC and that ended with the defeat of the first. The Battle of Pydna took place on June 22, 168 BC near the town of Pydna in the Gulf of Thessaloniki, in the northeast of Greece; and put an end to the Third Macedonian War that had begun three years earlier, showing the supremacy of the Roman legion over the rigid Macedonian phalanx. The Roman army was under the command of the general and politician Lucius Aemilius Paullus Macedonicus (230 - 160 BC); and that of Macedonia was directed by Perseus of Macedon (21 - 165 BC), the last of his kings.

The story of the lunar eclipse that occurred in the course of this battle is presented again by the great historian Plutarch in his work *Parallel Lives*:[3]

"When it was night, and when after the ranch they went to sleep and rest, the Moon, which was full and well discovered, suddenly began to blacken; and collapsing his light, having changed different colors, disappeared. The Romans, as it is of ceremony, implored it so that their light would return to them, with the noise of the metals and raising to the sky many lights with brands and axes; but the Macedonians moved nothing, but terror and fright seized the field and among many secretly ran the voice that this prodigy meant the destruction of their king."

Later Plutarch describes how, although the causes of these natural phenomena were already well known, the superstitious character continued on dominating people:

"Aemilius was not entirely new and a pilgrim in the anomalies that eclipses produce; which at certain times bring the Moon into the shadow of the Earth, and conceal it, until, when passing from the shadow, it again shines with the Sun. However, being him very surrendered to religious things and inclined to sacrifices and fortune divination, as soon as he saw the Moon was completely free he sacrificed eleven bulls; and as sooner as the day came then he offered a new sacrifice of the same kind to Hercules, not stopping until twenty; and at the first and twentieth prodigies were observed, which he said they award victory to those who defended themselves. He therefore made a vow to the same God of another hundred oxen and sacred games, commanding the warlords to order the army for battle; but he waited with all to the inclination and deviation of the brightness so that the Sun from the East did not dazzle them in the fight by giving them in front their faces; so he was giving time, sitting in his tent, which was open for the part of the plain and the field of the enemies."

But, fortunately for Occident, such a military defeat of the Greek state did not mean the destruction or annihilation of its cultural traditions and developments. Fortunately the Romans decided to take cultural relations in more friendly terms: during this period there was the phenomenon of *Hellenization* of the primitive Roman-Latin culture. The contact with the defeated Greeks and Macedonians, whose territories had passed to the Roman administration , brought as a consequence the transfer of Greek and Hellenistic customs and cultural practices to Roman society; Greek professors and philosophers came to the empire to spread by their s territories the Greek and Hellenistic culture. In this sense, the militarily defeated nation becomes an invader in cultural terms; the Greek-Roman society of this era was politically, militarily and economically Roman, but culturally it was Greek: *"Rome is a town that has had by culture that of another town, the Greek."*[4] All the above guaranteed the temporal continuity of the Greek developments in the philosophical, geometric, mathematical and, now the most important for us, astronomical fields.

Hipparchus of Nicaea

The Greek astronomer, mathematician and geographer Hipparchus of Nicaea (190 – 120 BC) made, within the general framework of that Greek-Roman society, fundamental contributions to the advance of astronomy as a mathematical science and to the foundations of trigonometry. Since his writings basically haven't survived the knowledge of his work is based on second hand reports, especially those of the great astronomical compendium *The Almagest*, written by Claudius Ptolemy in the 2nd century AD.

A native of Nicaea city, Hipparchus spent much of his life abroad, mainly on the Rhodes island, which during the last century and a half before the Christian Era was a great rival of Alexandria as a center of cultural, intellectual and literary life of that time. Among the men whose work was most fruitful on the island, the most important place is occupied by Hipparchus; for it seems most of his adult life he occupied there carrying out an observation and astronomical research program.

Nicaea's astronomer transcends all his predecessors and contemporaries reputation, as the relative importance of their work for the advancement of astronomy at the time, is comparable with that of Galileo Galilei and Isaac Newton in their respective moments: He improved Aristarchus's estimates of the sizes and distances of the Sun and the Moon; he devoted himself tenaciously to the problem of representing the observed trajectories of such stars and planets by combinations of uniform circular movements; he discovered the *Precession of the equinoxes*; he estimated the duration of the lunar month with very acceptable errors; and he also made a catalog of eight hundred and fifty fixed stars, giving his celestial *Latitude* and *Longitude*. Finally, he was the first to write systematically on trigonometry and thus he established the foundations of this branch of mathematics, which is why he is considered the founder of it.

Hipparchus also wrote critical commentaries on the works of some of his predecessors and contemporaries. In *Commentaries on the Phenomena of Aratus and Eudoxus*, his only surviving book, he harshly exposed the errors in *Phaenomena*, a popular poem by the Greek writer and poet Aratus (310 - 240 BC), which was based on a homonymous treatise, now lost, of Eudoxus of Cnidus in which the constellations known to date were named and described.

A good part of the astronomical studies carried out during the Greek period were based on data from previous observations made by the Babylonians, which were very extensive, but also quite crude. This situation would change considerably by the great dedication of Hipparchus to the experimental field, of whom we could say that he is the first really great observational astronomer. He had available a lot of information about eclipses from previous Babylonian observations, as well as those made in Alexandria during the last 150 years and further data reported by his own observations. From the combined Babylonian and Alexandrian data, he developed the theories of the Sun and the Moon, and from the Alexandrians alone he made his brilliant discovery of the *Precession of the equinoxes*.

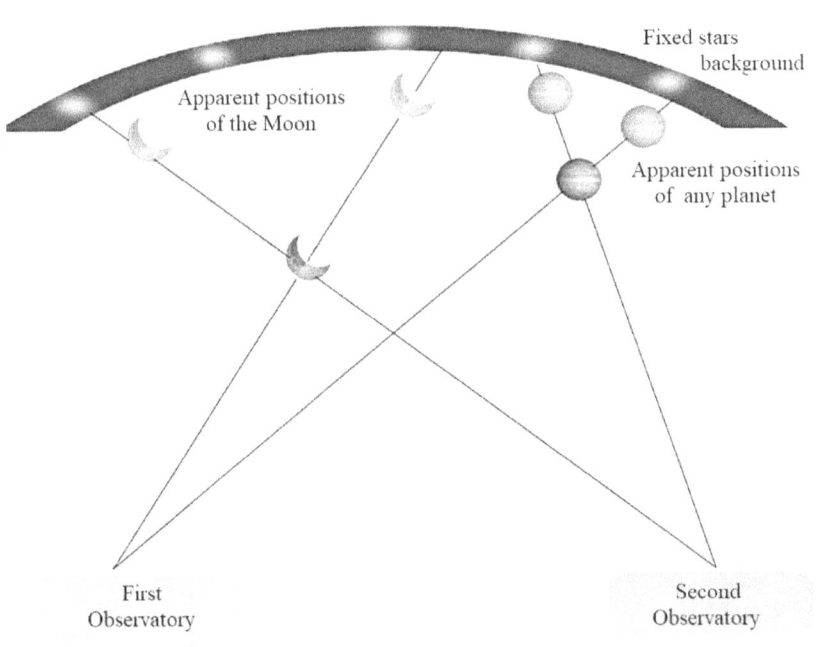

**Image 3.2
The Parallax:**

It is the apparent change in position of a star with respect to the fixed stars, when it is observed from two points separated by an appreciable distance; it is measured in units of angle.

Hipparchus carried out astronomical observations in Rhodes for a period of 30 years, applying a consistent method based on a simple instrument: an articulated rod that could be turned over the vertical and directed towards the star under study; while a circle marked in degrees served to measure the angle to the vertical. He is also credited with the invention of the *astrolabe*, a portable instrument for measuring the altitudes of celestial bodies, based on the same principle and which is the precursor of the *sextant*. In this way he mapped the positions of 850 stars and gave estimates of their brightness; equally he made for the five known planets: Mercury, Venus, Mars, Jupiter and Saturn. His data would form the basis of astronomical studies over the next 17 centuries, before they were eventually replaced during the Renaissance period by those of another great observational astronomer, the Danish noble named Tycho Brahe.

The *Heliocentric Theory of the Universe*, which had been initially proposed and developed by Aristarchus of Samos, was subsequently adopted, demonstrated and defended by the Mesopotamian astronomer Seleucus of Seleucia (190 BC), who according to some modern historians would have been able to formulate a justification based on the solar component of the tides for the heliocentric dynamics;[5] but by no other ancient astronomer. This general rejection was mainly due to Hipparchus who, facing the heliocentric hypothesis of the philosopher of Samos, better adopted and advanced both the Aristotelian geocentric theory and that of the eccentric and epicyclic deferents that had been previously proposed by Apollonius; and it was a later development of this theory what later came to be known as the Ptolemaic System. (See Image 2.7)

Although Hipparchus knew Aristarchus' heliocentric vision, he favored the geocentric conception of Aristotle due to the non-detection of parallax caused by the immense distance of the stars. *Parallax* is the apparent displacement of an object when viewed from different positions (See Figure 3.2). Like most of his predecessors, Hipparchus assumed a stationary spherical Earth at the center of the universe and developed the idea of a convenient combination of uniform circular motions, using the theory of eccentrics and epicycles, to represent the movements of the Moon and the Sun; further he developed geometric procedures to calculate their actual distances that best explained the observations. Using his own data, he wanted to quantify the Aristotelian model but soon he learned that a sole single sphere couldn't explain the complex movements of the celestial bodies, particularly those of the strange planets. So he appealed to the concept of secondary circular movement through a circle called *epicycle*, which moves around another larger or *deferent*. The planet moves in the circumference of the epicycle at a constant speed, while the center of the epicycle moves through the circumference of the deferent also at a constant but different speed; this model would be adopted some 300 years later by the last of the great Greek philosophers, Claudius Ptolemy, in his brilliant quantification of Aristotelian cosmology.

The most important astronomical works of Hipparchus referred to the orbits of the Moon and the Sun, the determination of their sizes and distances from Earth and the study of eclipses. He had a clear understanding of the three-dimensional arrangement of such stars and the Earth during these natural phenomena and was particularly interested in finding their periods, by themselves and because they could help him obtain the exact positions and movements of the Sun and the Moon. He was fortunate to be able to compare his own data on eclipses with those of the Babylonians and was practically the first Greek astronomer to have made use of such materials whit this purpose.

Towards the year 150 BC, following the methodology of the eclipses, improving Aristarchus's method and considering the data provided by Eratosthenes for the diameter of our planet, Hipparchus determined the sizes and distances for the Moon and the Sun for the first time in absolute values and with minimal errors, in such a way that no previous astronomer had approached so much to the correct value.

In the case of the Sun it was relatively easy to find an orbit that would meet the observations, since the different duration of the seasons was the only irregularity had to explain. According to the geocentric conception of the Greeks, in addition to the daily movement in the East-West direction the Sun traces each year an apparently circular path in

the opposite direction West-East in relation to the fixed stars. Hipparchus had good reason to believe that such a path, now known as the *Ecliptic,* was a big circle contained by a plane that passes through the center of the Earth and additionally it is inclined with respect to the terrestrial Ecuador. The two points in which the ecliptic and the equatorial plane intersect, known as *the vernal and autumnal equinoxes*, together with the two ecliptic points farther north and south from the equatorial plane, known as *Summer* and *Winter solstices*, they all divide the ecliptic into four parts corresponding to the four climatic seasons. However, the passage of the Sun through these sections is not symmetrical, therefore the astronomer tried to explain how the King star could travel yearly with a uniform speed along a homogeneous circular path and, nevertheless, produce seasons of heterogeneous duration.

Once Hipparchus recognized the movement of the King star along the ecliptic, he posed possible explanations to justify the seemingly irregular movement of the Sun: he developed the original idea of Apollonius and imagined a convenient combination of uniform circular motions using the eccentric deferent and epicycle model. His final hypothesis, by assuming that the Earth was not in the center of the Sun's orbit, and hence the name *Eccentric*, explained well the apparent variability of the solar movement: so that by drawing a line across the Earth's center and the real center of that solar orbit two *Apsides*, or points at which the distances from the Sun are, respectively, the smallest and the largest, are determined; they are also known as *Perigee* and *Apogee* and in them such Sun's movement will, consequently, appears faster and slower.

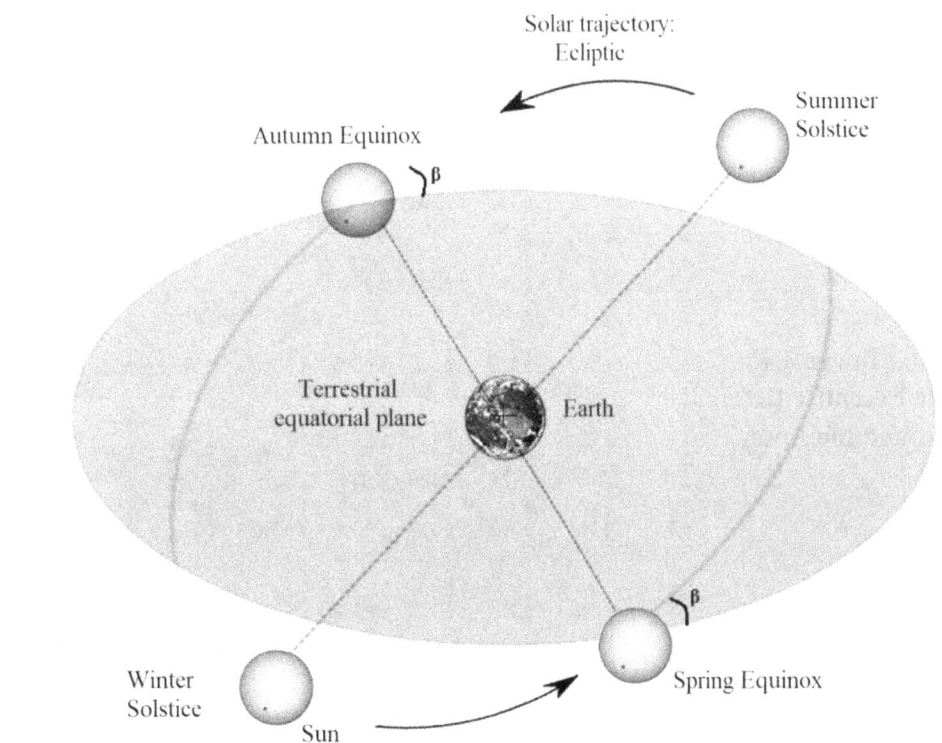

Image 3.3 Movement of the Sun: The Ecliptic, the Equinoxes and Solstices

According to the Geocentric Model of the Universe, the Sun makes an annual circular revolution around the Earth; such trajectory, known as *Ecliptic*, is inclined at an angle β with respect to the plane of the terrestrial equator. The two points of intersection determine the *Equinoxes*; while the highest and lowest points on the equator determine the *Solstices*.

After having established the hypothesis, Hipparchus proceeded to build the first solar tables to find the position of the Sun between the Fixed stars at any time and, hence, to find the position of perigee in the sky as well as the time of year when the Sol was there. Similarly, he proceeded to determine the *Eccentricity of the solar orbit* or distance from Earth to the true center of the Sun's orbit. He found that the time interval between the spring equinox and the summer solstice was 94,5 days, and from there to autumn equinox 92,5 days; consequently, from these two periods he determined both the eccentricity and the apsides with very good precision. On the basis of his observations, the necessary eccentricity of such orbit could be used to represent the annual course of the Sun along the ecliptic. His great contribution was to discover a method for using the observed dates of two equinoxes and a solstice to calculate the size and direction for the displacement of the center of the Sun's orbit.

Next, Hipparchus dedicated to studying the most complicated movement of the Moon with the purpose of elaborating both a lunar theory as another for the eclipses. In addition to varying its apparent velocity, the satellite diverges to North and South over the ecliptic and the periodicities of these phenomena are different. He adopted values for such periodicities that were known to the contemporary astronomers of Babylon, and confirmed their accuracy by comparing recorded observations of lunar eclipses separated by intervals of several centuries. Although the Moon does not stop for a while in its trajectory, as the planets apparently do, Hipparchus had discovered that its speed in the respective phases was variable, which indicated that its displacement was also to some extent dependent on its distance from the King star.

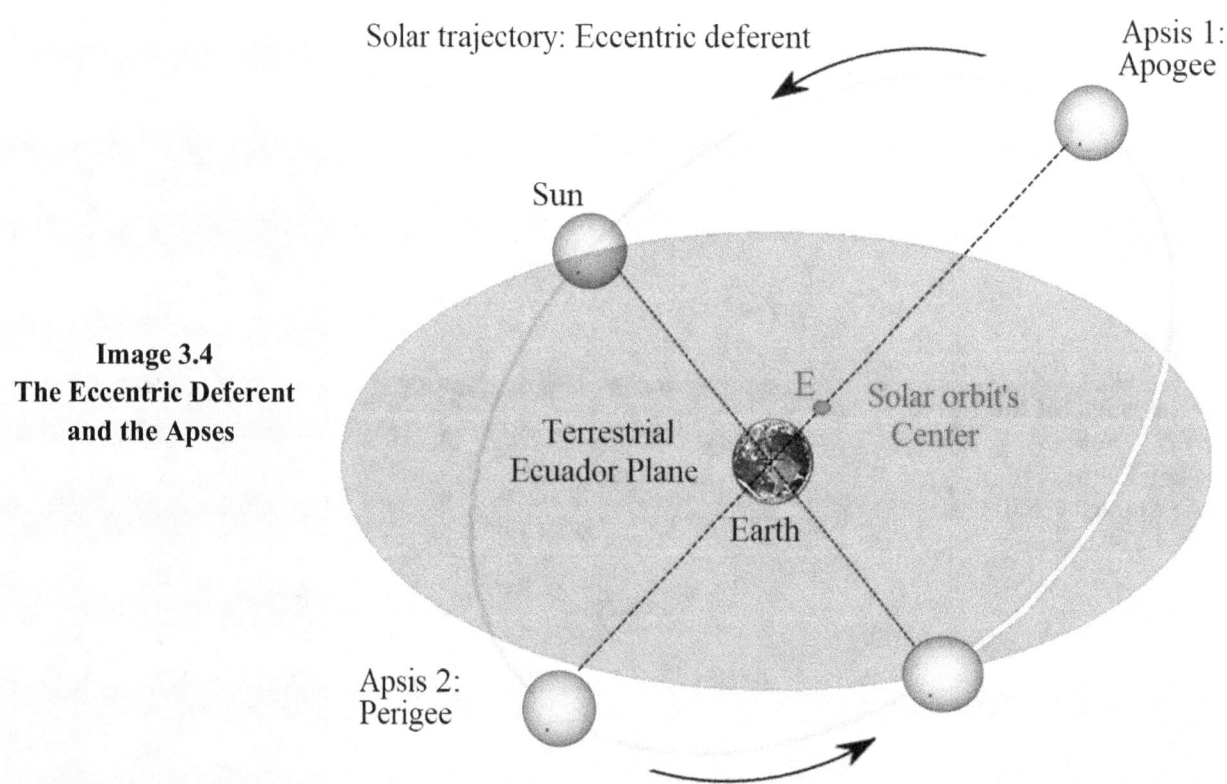

**Image 3.4
The Eccentric Deferent
and the Apses**

In the solar model of Hipparchus, the King star moves in a deferent circle that does not have its center on Earth, but in another point E, the *Eccentric*. This gives origin to two points denominated Apses: the furthest from our planet is called Apogee, while the closest is the Perigee.

The Nicaea astronomer made great attempts with regard to the movement of the Moon but he was confronted with even greater problems; so he had to resort to Babylonian sources to obtain extremely accurate values for its main movements, for he understood the importance of determining the period in which the Moon returns to the same position in relation to: 1) the fixed stars, or sidereal month; 2) the Sun, synodic month, 3) the perigee with respect to the Earth, or anomalous month, and finally 4) the nodes in the ecliptic or draconitic month.

Since the motion of the Moon is much more irregular than that of the Sun, its theory was much more difficult to develop; but it was still possible to represent the *First irregularity* of the lunar motion either by means of an eccentric or an epicycle. Hipparchus first assumed for the lunar orbit a deferent circle inclined at 5° with respect to the ecliptic and turning in a retrograde East-West direction around the axis of the latter, so that the lunar nodes make a complete revolution in 18,67 years. In this circle it moves in a right direction, that is West-East, the center of an epicycle in which the Moon moves in a retrograde direction along its circumference. The respective periods of revolution of the deferent and the epicycle are different; and the relationship between the radius of the epicycle and the deferent was found by the greater difference between the average and apparent positions of the Moon, which Hipparchus set at 5°1', whose sine is the sought reason, or 5,25/60 = 0,0875; which accounts for the so-called *First inequality of the lunar movement*, or the *Equation of the center*. Today it is known that it's actually caused by the elliptical shape of the lunar orbit. Hipparchus founded his theory on the Babylonian and Alexandrian observations of lunar eclipses, and therefore it represented the lunar movement in New and Full Moon well enough.

In his work *The sizes and distances*, which has been lost, Hipparchus measured the orbit of the Moon in relation to the size of the Earth. In a first method, using the visually identical sizes of the solar and lunar discs in the sky and the observations of the Earth's shadow during the lunar eclipses, Hipparchus found a relationship between lunar and solar distances that allowed him to determine that the average distance of the Moon to the Earth is approximately 29,5 times greater than the Earth's diameter.

In the second method he used an observation of a solar eclipse that was total near the Hellespont, but only partial in Alexandria. He supposed that the difference could be attributed entirely to the existence of the lunar parallax against the fixed stars, which is equivalent to supposing that the Sun, like the stars, is infinitely far away. Hipparchus calculated by this method that the average distance from the Moon to Earth is 33,67 times the terrestrial diameter. With an average of 31,585, he was not far from the true value of 384400 Km /12742 Km = 30,168 times.

He was not sufficiently satisfied with this, so he examined whether in other points of his orbit, such as in the quarters, the moments of the first and last quarter, the satellite would fitted such a model. He found that sometimes the observed place of the Moon fulfilled the theory, while other times not; but although it was clear that there must be some other inequality dependent on the relative positions of the Moon and the Sun, Hipparchus had to leave the investigation for future astronomers.

Hipparchus designed an epicyclic lunar model notable for the way in which he synchronized his theoretical movements with those observed for the star. He made the displacement of the epicycle around the Earth to represent the known average movement of the Moon in the ecliptic length, and that the rhythm of the satellite within the epicycle will be realized in harmony with the observed anomaly movement of the Moon itself. He found a geometric procedure that allowed him to derive the relative sizes of the circles and the movements around them, which was based on the observations of the periods of two trios of different lunar eclipses. His calculations were flawed, but the method itself was excellent and it had great originality; and allowed to explain very well the returns of the Moon to the opposition and the conjunction.

With respect to the movement of the five planets, he failed to formulate a satisfactory theory. According to Ptolemy, Hipparchus investigated only the movements of the Moon and the Sun, showing that it was possible to perfectly explain his revolutions by means of combinations of uniform circular

movements; while for the five planets he did not attempt to sketch a theory, limiting himself to systematically collecting observations and pointing out that they did not agree with the mathematical hypotheses of his time.

Among his other notable contributions to astronomy are his systematic and critical comparisons of ancient observations with his own, which had the purpose of discovering variations of small magnitude in the positions of the stars that may arise for long periods of time. Thus, through the systematic observation of eclipses and their comparison with the Babylonian records, Hipparchus recognized and calculated the movement of the Earth's rotation axis, or *Precession*, which is due to forces exerted by the Sun and the Moon on our planet, as it is more bulky at the equator; which allowed him to determine the duration of the months and years with greater accuracy, broadening the knowledge base that would eventually lead to our modern calendrical systems; a considerable achievement and only possible by the millimeter regularity of the eclipses.

El astrónomo encontró que los puntos de intersección de la eclíptica y el ecuador estaban cambiando de posición, retrogradando y desplazándose en sentido contrario al de rotación de la Tierra. Como la oblicuidad de la eclíptica y las latitudes de las estrellas no revelaron ninguna variación en el curso del tiempo, Hiparco concluyó que el eje de rotación del planeta había cambiado su dirección en el espacio; fenómeno que actualmente se conoce como la *Precesión de los equinoccios*; y que lo llevó a definir dos tipos de años diferentes: Uno es el *año tropical*, el período que toma el Sol para regresar a la misma posición con respecto a los puntos equinocciales, y el otro es el *año sidéreo*, el tiempo demandado por el astro Rey para volver a la misma posición con respecto a las estrellas fijas. Hiparco calculó la duración de estos años con destacable precisión y discutió los posibles errores de sus observaciones. El astrónomo de Nicea calculó que el eje de la Tierra se mueve describiendo un ángulo de unos 45'' cada año; el valor verdadero es aproximadamente 50,27'', así que su estimación fue de una gran precisión para la época. Dividiendo 360° entre 50,27'', nos permite estimar en unos 26000 años el tiempo que le lleva al eje de la Tierra efectuar una revolución completa para volver al punto de partida y recomenzar el ciclo.

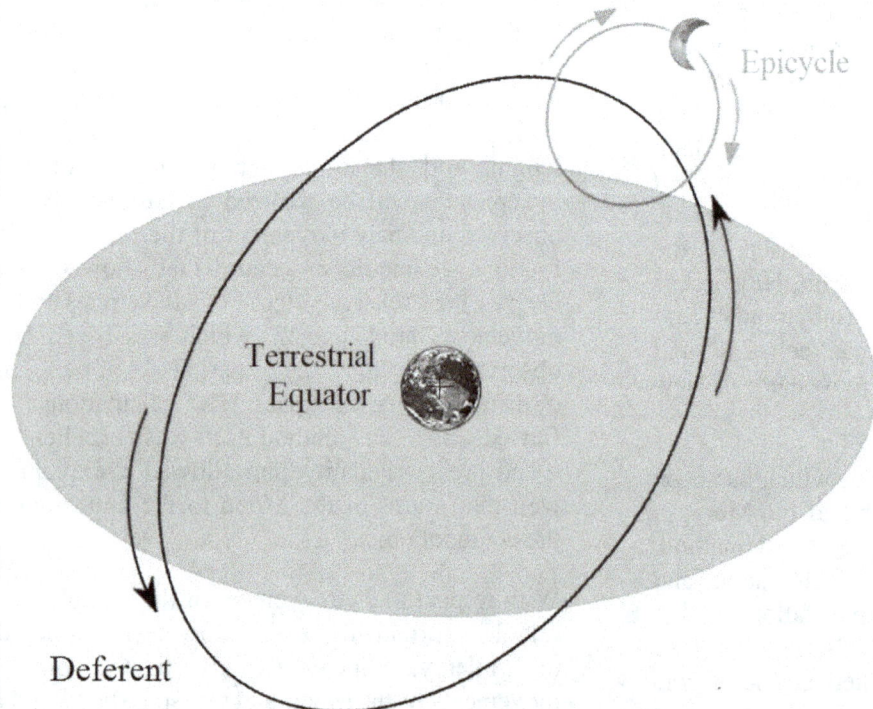

**Image 3.5
The Lunar model of Hipparchus**

To explain the movement of the Moon, Hipparchus elaborated a model according to which the star moved in an Epicycle in the East-West direction, while the center of it moves in the circular trajectory of a Deferent in the opposite direction; the Earth being the center of this deferent.

By way of conclusions, following the method of eclipses, previously devised by Aristarchus, Hipparchus determined the size and distance of the Moon and Sun. He concluded that the average distance of the Moon was 31,585 terrestrial diameters, and also appealing to the equally remarkable measurement of the diameter of the Earth by Eratosthenes, he determined for the first time in history that the Earth-Moon distance was in absolute values of 31,585 * 14,722 = 464994 kilometers, which constitutes an error of 21% in excess of the real value. He also indicated that the distance from the Sun was 1050 times the diameter of our planet, approximately 19 times less than the true value. As for the sizes of the stars, he found that the diameter of the Sun was 10,167 times that of the Earth, and that of the Moon was 0,294 times the same. So, we can see that Hipparchus had a very correct idea of the distance and the size of our nearest celestial neighbors.

His improved theories of the movements of the Moon and the Sun enabled him to predict more accurately than his predecessors the respective eclipses. Knowing the duration of the synodic and draconitic months, he was able to calculate the period after which the eclipses are repeated; the Saros period, for example, contained almost 223 synodic months or 242 draconitic months. Lunar eclipses could be predicted with precision within one or two hours; but the prediction of solar eclipses was more difficult since one had to consider the position of the observer on Earth, that is, to calculate the lunar parallax; therefore they were predicted with much less accuracy, especially with respect to the determination of small regions in the Earth where they would be visible as total or partial eclipses.

We must mention another development that we owe to Hipparchus, which is nothing less than the invention of trigonometry using *chords* where *sines* are now used. His most important contribution to mathematics was to work out, if not really invent, a trigonometry based on a table of chord lengths in a circle of unitary radius and tabulated as a function of the angle subtended to the circle center. This table would allow, for the first time, a systematic solution of general trigonometric problems, so recurrent in astronomy, and clearly Hipparchus used it extensively for his astronomical calculations.

Finally, as a geographer, Hiparco sought to extend the accuracy of astronomical methodology to the field of geography, for which he transferred his system of *Celestial coordinates* to the *Longitude* and *Latitude* counterparts of *Terrestrial cartography*, which we still use today. He emphasized that the exact location of any place on a terrestrial map was based on data obtained by astronomical observations; he proposed a parallel-meridian system, or latitude-longitude, of even intervals and from the point of view of scientific cartography. By latitude he described the celestial phenomena for each individual degree of the ninety degrees that run from Ecuador to the North Pole, giving for each one the longest day and the position of the visible stars. Regarding the geographical longitude, he was the first to determine it by means of the lunar eclipses, recognizing that such phenomenon, although it is visible in several places at the same real moment in which it is happening, it will not be visible at the same *local time* if the places differ in length.

For all its developments, achievements and contributions, Hipparchus of Nicaea is credited with a privileged place in the history of science, astronomy and knowledge in general.

Posidonius (135 - 51 BC) was a astronomer, Stoic philosopher, geographer and Greek historian born in Syria; at the time he was compared as a polymath similar to Aristotle and Eratosthenes; around the year 95 BC he settled in Rhodes. Many sections of his works on astronomy are known thanks to a text by the Greek astronomer Cleomedes, *The circular movement of the celestial bodies*; book considered as the original source of the well-known story of how Eratosthenes and Posidonius measured the Earth. As an astronomer, Posidonius made measurements of the relative sizes of the Sun and the Moon, and of their distances from Earth, but in this respect he didn't make significant improvements over those of his predecessors. While as a geographer he dedicated to measuring the globe but, unlike Eratosthenes who was based on solar measurements, Posidonius was based on observations of the altitude of Canopus, the second brightest star in the sky, and selecting as measurement points the cities of Rhodes and Alexandria; in such a way he obtained a result of 180000 stadia, or its equivalent of 33300 kilometers

for the circumference of the Earth, which expressed in terms of its diameter would give 10600 km, or 16,8 percent less than the actual real value.

The Roman Empire

The period of the Roman Republic came to end in the midst of several political-military conflicts and civil wars, after which *Gaius Julius Caesar Octavius Augustus* emerged as victor after having defeated the Roman soldier Mark Antony and his ally and lover, the last Egyptian queen Cleopatra VII Philopator, in the naval battle of Actium in 31 BC. Fact that would lead first to the suicide of the two lovers and then to that in the year 27 BC the Roman Senate granted Octavius, in an unprecedented way, the title of *Augustus* and ask him to assume once again the control of all the provinces. This makes it possible for Octavius to proclaim himself as the first Roman Emperor under the name of Caesar Augustus, thus giving starting the period of the great Roman Empire.

The Moon according to Plutarch

Everything that the most enlightened minds of antiquity could understand with respect to the constitution of the Moon is contained in an extremely charming dialogue written by Plutarch: *"On the face that appears in the orb of the Moon"*. This text refutes the opinion of the Stoic philosophers according to which the satellite is a polished mixture of air and a soft fire that gives it its light. Plutarch argues that by not borrowing all of its light from the Sun and not depending on it, the Moon should be visible in the New Moon, which doesn't happen; and with this he also proves that it is not formed by a substance like crystal that admits the passage of light, since solar eclipses would then be impossible. The way in which sunlight is reflected from the Moon and the absence of a bright image reflected from the Sun and the Earth serve him as proof that the substance of the Moon is not polished but it is rough as the terrestrial material. In the work Plutarch also exposes the correct explanation for the fact that the satellite remains dimly visible during a lunar eclipse. The general scheme of Plutarch's argument is that in order to reflect the light of the Sun, the Moon should be a solid body very similar to our planet, and additionally that those dark or shaded areas visible on the face illuminated from the satellite would correspond to shadows produced by lunar mountains. It is very important to note that this author, to refute that idea that the Moon cannot be like the Earth since it is not in the lowest place, boldly affirms that it is not proven that our planet is in the center of the universe, since the space is infinite and therefore has no center. With which Plutarch already puts into question the geocentric theory of the universe.

Giving meaning to the calculations made by astronomers about the distance to the Moon from Earth compared to the Sun, in this work Plutarch writes: *"So far has it been established from the Sun because of its weight, and almost borders with the Earth, that if we distribute the properties according to the localities, the portion and inheritance of the Earth invites the Moon to join it, and the Moon has a next claim for movable property and people on Earth in the right of kinship and closeness."* In addition he declares: *"Because it often fails to overcome the shadow of the Earth, rising little, because the illuminating body is so vast. But it almost seems to graze the Earth and is almost in it lap as it turns, being isolated from the Sun unless it raises high enough to clear that earthy shadowy region, dark as night, which is the Earth's inheritance. Therefore, I think we can say with confidence that the Moon is within the limits of the Earth when we see it blocked by its contours."*[6] This is a clear reflection of the level of understanding that the people of Greco-Roman society had about the proximity at which the Moon is located from Earth compared to that of the Sun.

Plutarch also tells us the way in which the educated people of his time understood eclipses: *"For my part, however, I still need to be convinced; I have only heard that when the three bodies, the Earth, the Sun and the Moon, enter a straight line eclipses occur, the Earth withdraws the Sun from the Moon or the Moon the Sun from the Earth; that is, the Sun is eclipsed when the Moon, the Moon when the Earth is in the middle of the three, the first case occurs in the New Moon, the second in its fullness."*

In the two hundred and sixty years elapsed between Hipparchus and Ptolemy, and excepting the works and contributions of Posidonius, astronomy seems to have stalled; in both theoretical and experimental astronomy no truly significant advances were made

until the Egyptian scholar Claudius Ptolemy developed this science considerably, and presented for posterity the first complete treatise covering the entire spectrum of astronomy.

Claudius Ptolemy

The last advances in the Greek worldview are associated with the name of Claudius Ptolemy (100 - 170 AD) astronomer, mathematician, geographer, astrologer and Greco-Egyptian writer; who lived primarily in Alexandria, in the then Roman province of Egypt where he was most probably born, although he had Roman citizenship and worked within the framework of the Roman Empire towards the year 150 BC. He was the author of several scientific treatises, three of which were of considerable importance for the development of Byzantine science, Medieval Islamic and of the European Renaissance of later centuries; and his work had a greater influence on cosmology than any other figure in history, presenting himself as a colossus in the astronomical and geographical knowledge of Greco-Roman society. His writings represent the culminating achievement of the knowledge and science of his time; with his works *Syntaxis Mathematica*, a treatise on astronomy also called Almagest, his text the *Geography* and most especially with his *Geocentric Model of the Universe*, now known as the Ptolemaic system, it can be said that Ptolemy tends to dominate both astronomy and geography for more than fourteen centuries. Even so, not much is known about his life and personal data.

This Egyptian astronomer occupies a prominent place in the history of knowledge mainly due to the mathematical and geometric methodologies applied to astronomical problems. His contributions to trigonometry were especially important; for example, his data of chord lengths in a circle constitute the oldest and most accurate table that survives from a trigonometric function. Additionally, he made extensive use of fundamental theorems in spherical trigonometry, probably developed half a century earlier by Menelaus of Alexandria (70 - 140 AD), for the solution of many basic astronomical problems.

Ptolemy inherited from his Greek ancestors a set of models and a series of mathematical and geometrical tools to predict the position of the stars in the sky. Aristarchus had calculated the distances and sizes of the Moon and the Sun. Apollonius of Perga had introduced into astronomy the geometrical concepts of the deferent and the epicycle, very useful to describe the celestial movements. Finally, Hipparchus had a good knowledge of Mesopotamian astronomy, and argued that Greek models should coincide with the Babylonians with greater precision; and he had created mathematical models for the movements of the Sun and the Moon too, but he could not create accurate models for the five remaining planets.

Claudius Ptolemy developed a good theoretical model of the *Aristotelian Geocentric Universe*, but not enough good in his eyes, since it didn't agree well with the data provided by Hipparchus' observations; reason why he introduced two slight deviations to the Aristotelian theories. He added two more imaginary points to each orbit called the *Eccentric* and the *Equant* points; in such a way that the Deferent has as its true center the Eccentric point instead of the Earth itself; while the Equant point is in the opposite direction and at the same distance as the Earth with respect to the Eccentric.

For the Greek astronomers the celestial bodies had to move in the most perfect possible way, and for them the geometric shape so excellence was the circumference. Thus, in the Ptolemaic geocentric model of the universe the Moon, the Sun, each planet and all the stars orbit circularly around a stationary Earth, a circular trajectory represented by the geometric concept of the deferent. To retain that perfect movement and continue explaining the seemingly erratic trajectories observed by the stars, Ptolemy moved the center of the orbit of each body to one side of the Earth, placing it in another imaginary point: the *Eccentric*; and also used a second orbital movement, known as *Epicycle*, to account for the retrograde movement. The *Equant* is the point from which each body sweeps equal angles along the deferent in equal times. The center of the deferent, the eccentric, is halfway between the equant and the Earth.

In his system a star rotates uniformly in a circle or *epicycle* whose center additionally moves at uniformly speed in the path of another circle called the *deferent,* and which revolves around

an *eccentric* point that does not coincide with the Earth; which together with its eccentric nature makes it seem that the planet moves with a variable angular speed when viewed both from Earth and from the equant point; exhibiting direct conflict with the Aristotelian doctrine that established that the celestial movements had to be perfectly uniforms. By appropriately adjusting speeds of the two circles and the proportions their radius, the location of the eccentric point and the velocity of the respective star, he managed to accurately represent the main irregularities in the movements of the planets, especially the stationary points and the retrogradations. (See Image 3.6)

Such adjustments, which were arbitrary and introduced only to match theoretical predictions with experimental observations and historical record s, would be understood many centuries later when the astronomer Johannes Kepler developed his laws of planetary motion. In any case, with such arrangements, Ptolemy constructed a planetary model that could give the positions of celestial bodies, in the past and in the future, with approximately 1° of precision with respect to observations.

Ptolemy wrote several scientific treatises, three of which were of great importance for Byzantine, Medieval Islamic, Renaissance European and Modern science. The first is the astronomical treatise originally titled in Greek Μα θημ ατικὴ Σύντ αξις, which translates into English as a *Mathematical treatise* or also *Syntaxis Mathematical;* in the Arabic language it is known as *Al-majistī*, from which the English version derives the *Almagest* or the *Great treatise* too. Its original name is due to the fact that the author thought that his subject, the movements of the celestial bodies, could be explained well in purely mathematical terms. The work constitutes a compendium of all the cosmological knowledge of his time; and in it is proposed what is now known as *Ptolemaic system*: a unified geocentric system in which each star is attached to its own celestial sphere having the Earth as its center. His cosmological model gave an order of distance for the celestial bodies with the Moon as the closest to our planet and then, in order, Mercury, Venus, the Sun, Mars, Jupiter and Saturn; beyond the reach of Saturn's epicycle he placed the limit of the Universe, the Sphere of fixed stars. The second text is *Geography* and it consists of an exhaustive presentation of the geographical knowledge of the Greco-Roman world. The third work presents his astrological thought through a treatise in which he pretends to adapt horoscopic astrology to the natural Aristotelian philosophy of his time, and commonly known as *Tetrabiblos*, which means *Four books*.

Syntaxis Mathematica or Almagest

The culminating work of Greek cosmology is the *Almagest* by Claudius Ptolemy, which is based on the works of his predecessors, especially Hipparchus; work so elaborated and successful that it made the previous cosmological treaties superfluous, and prevented them from being copied and read anymore. Its main astronomical work was largely a compendium of contemporary astronomical knowledge in which he described the now called *Ptolemaic system*, a geocentric universe with the Earth fixed at the center and the Moon, the Sun, the five known planets and the set of stars revolving around it. The previous Greek astronomical theory, simpler and based on epicycles and deferent, was modified by Ptolemy for the purpose of better reproducing the apparent motions of the planets and including the retrograde loops; doing it so well, that it remained in force until the revival of the heliocentric theory by Nicolaus Copernicus in the sixteenth century.

The work is one of the few ancient treatises on astronomy that survives and it is one of the most influential scientific texts of all time. Its rich geocentric model was accepted for about 1.400 years from its origin in the Hellenistic Alexandria, passing through the medieval Byzantine and Islamic worlds, by Western Europe during the Middle Ages and by the Early Renaissance until Copernicus; becoming the fundamental source of information about ancient Greek astronomy and mathematics. Babylonian astronomers had developed arithmetical foundations to calculate astronomical phenomena such as eclipses; Greek astronomers such as Aristarchus and Hipparchus had created geometric models to explain and determine the movements of celestial bodies. Finally, Ptolemy acknowledged having derived his geometric models of the universe from hypotheses, astronomical observations and data provided by those predecessors.

The Almagest contains an inventory of the observations made and a description of the mathematical procedures that the astronomer used to deduce the parameters of his model. It also presents a catalog of around 1000 stars, probably very based on the previous catalog of Hipparchus, but with additions and modifications. Additionally, it includes tables that allow the reader to determine the position of a star for any desired date and according to the theory, with the advantage that it presents data for trigonometric calculations. Finally, it also contains improvements to the lunar and solar theories of the Nicaea astronomer.

The work presents both philosophical and empirical arguments for the basic cosmological framework within which Ptolemy worked, and which can be summarized as follows: The celestial kingdom is spherical and moves like a sphere; the Earth is an absolutely immobile sphere and located in the center of the cosmos; in relation to the distance of the fixed stars, our planet does not have an appreciable size and must be treated mathematically as a point. The celestial sphere rotates at a perfectly uniform velocity around the Earth, transporting the stars with it and thus provoking their respective nocturnal configurations. In the same direction to the rotation of the celestial sphere the Sun traces slowly in the course of a year a great circle called the *Ecliptic*. In a similar way the Moon and the planets move, this is why the latter were also known as *wandering stars,* as opposed to *fixed stars* that basically do not change their relative position in the sky. The fundamental hypothesis of the Almagest is that the apparently irregular movements of these stars can be explained by combinations of uniform circular movements.

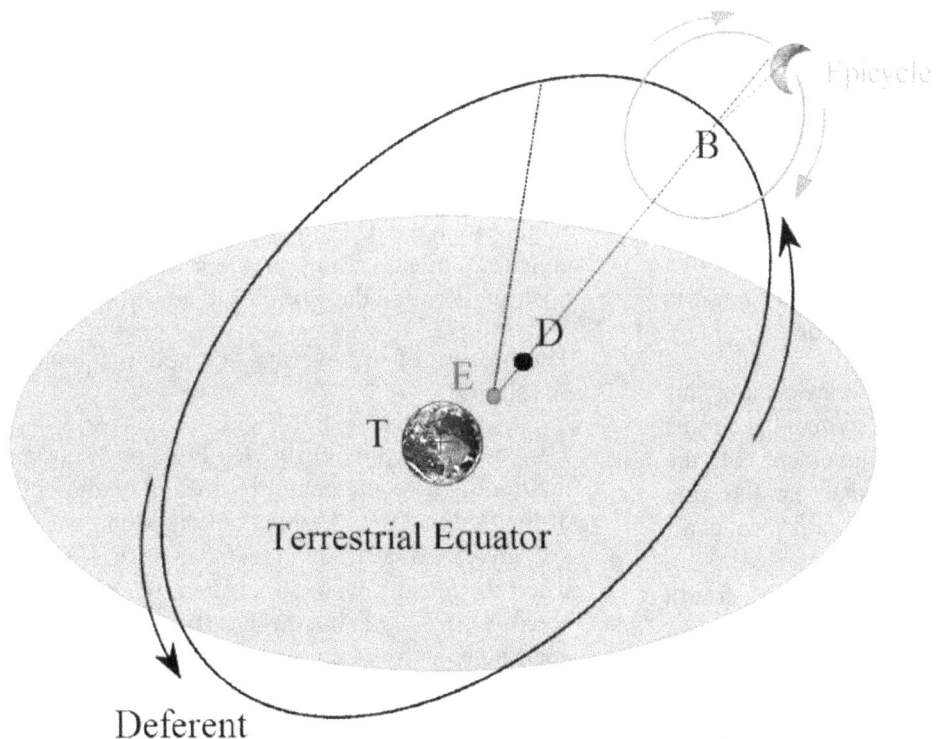

Image 3.6

The Ptolemy's Lunar Model.

Unlike Hipparchus, the lunar model of Ptolemy is eccentric with respect to our planet: the center of the Deferens is the eccentric point E, while D is a point symmetric to the Earth that he named Equant.

The text is divided into 13 sections known as books. The first one contains a description of Aristotelian cosmology: it provides the arguments for a spherical and geocentric cosmos, with the spherical shape of the heavens and a spherical Earth that lies motionless as its center, with the fixed stars and the various stars revolving around our planet. Next it presents the necessary data for the fundamental trigonometric calculations, as well as an introduction to spherical trigonometry; all of the above allows him to explain and predict the movements of the Sun, the Moon, the planets and stars in later books.

Book II uses spherical trigonometry to explain the celestial cartography and the astronomical phenomena characteristic of several localities, such as the length of the longest day at the solstices. It covers problems related to the daily movement attributed to the heavens, namely, the rising and setting of the stars, the duration of daylight, the points at which the Sun is at vertical, the shadows of the gnomon at the equinoxes and solstices, the determination of latitude and other measurements that change with the viewer's position.

Book III deals with the movement of the Sun and the duration of the year; and presents the calculations of the position of the King star in the zodiac for different periods. Based on the works of Hipparchus related to the discovery of the precession of the equinoxes and in his own observations on the position of the vernal and autumnal equinoxes, he discusses the irregularities of the course of the Sun, and explains them by means of the circular eccentric movement hypothesis. This chapter concludes with a clear statement on the factors on which the *Equation of the center* for solar movement depends. .

While in Books IV and V the author develops his own *Lunar Theory* and deals with the most complicated problems of the movement of the Moon: the determination of the lunar parallax and the movement of the lunar apogee; and also deals with the sizes and distances of the Sun and Moon in relation to the Earth. The lunar theory of the fourth book, based on eclipses, explains well the movement of the satellite in conjunction with the Sun, that is, in the New Moon, and in opposition, the full Moon; but it is inadequate for intermediate positions in the lunar orbit. His observations showed him that the Moon changes in apparent size, which he explained by a kind of unstable mechanism that operates in the epicycle of the satellite and that eventually leads him to his important discovery of the *Second inequality* of lunar motion, known as *Lunar evection*. Additionally, this fourth book deals with the construction of instruments for carry out such investigations.

Ptolemy made very substantial improvements on the lunar theories of his predecessors. For the satellite Hipparchus had simply used an epicycle that moved on a deferent around the Earth. But Ptolemy discovered that the outstanding errors of this theory, already vaguely noticed by Hipparchus, peaked at the time of the quadratures and disappeared completely in the syzygies. And an additional difficulty was that the error did not always return in each quadrature, sometimes disappearing completely and sometimes ascending as much as 2° 40', its greatest value. Eventually it turned out that when the Moon was in quadrature and at the same time in the perigee or apogee of the epicycle, so that the equation of the center was zero, the place of the Moon coincided perfectly with Hipparchus' theory, while the error was greater when the equation of the center reached its maximum at the time of quadrature.

By *Syzygy* is meant any of the two opposing points in the orbit of a celestial body, specifically the Moon, in which it is in conjunction with or in opposition to the Sun; or the almost linear configuration of three celestial bodies, such as the Sun, Moon and Earth during a solar or lunar eclipse. While quadrature refers to an intermediate position between the syzygies, that is at 90°.

The process of discovering this new inequality was as follows.

The ancients previously to Ptolemy determined the *Equation of the center* by means of the eclipses of the Sun and the Moon: when the Moon was in the syzygies, and also in agreement with Hipparchus when the apsides were so situated, the equation was equal to 6° 20'; but when the satellite was in quadrature it was only exactly five degrees. On the other hand, Ptolemy observing the Moon when it was in quadrature discovered that if the apsides were also in quadrature the equation was 6° 20'; while if they were in syzygy it amounted to 7° 40'. Therefore the difference between the average state and any of

the extremes, which is equal to 1° 20', he called the *Second lunar inequality;* and this is the equation that was subsequently designated as *Lunar Evection*. This inequality becomes null when the satellite is in the syzygies, and then the equation of the center coincides with that initially determined by Hipparchus. But if the apsides are in syzygy or quadrature, and if the Moon is in the last position, the Evection takes a maximum value and then the Equation of the center differs by approximately 1° 20' from the values assigned to it by that ancient astronomer.

Hipparchus had discovered the *First Inequality* or *Lunar Equation of the center,* which corrected the average movement in the syzygies, and had noticed that another correction in the quadratures was necessary; but he didn't manage to find it. In this sense Ptolemy completed Hipparchus' work finding that the eccentricity of the lunar orbit was subjected to an annual variation dependent of the movement of the line of apsides, whose variation in its position creates an inequality of lunar motion in quadratures, which is called *Evection*. To explain this new inequality Ptolemy advanced the hypothesis of the epicycle, a circle described by the Moon around an imaginary point that slides along the deferent circle, also called eccentric because it does not have the Earth as its center. To explain other discrepancies between the theory and the observation of the lunar movement, Ptolemy introduced a small oscillation of the epicycle that he named *Lunar Nutation*. His final theory agrees so well with observation that the error in his tables, in which it was possible to calculate the position of the Moon for any epoch, was rarely greater than a degree.

Ptolemy discovered that the effect of the second inequality was always to increase the absolute value of the first one, particularly in quadratures. The obvious deduction was the distance from the epicycle to Earth should vary so the Moon could appear at different angles at different times; in other words, the center of the epicycle should move through an eccentric circle but, for the angular velocity being variable, not with respect to the geometric center of the deferent but with respect to the Earth. In short, our planet is not the center of the deferent.

In Figure 3.6, the distance from B to Earth T will be, therefore, greater in the syzygy and lower in the quadrature. This is the *Second inequality* which is caused by the fact that the epicycle is not in the position it would have been if it would move in a concentric circle, and is equal to the angle formed between the lines from Earth to the two places where the Moon would be according to the two different hypotheses. This angle will be zero in the syzygies because the centers of the epicycle and eccentric deferent (B and E) are aligned with the Earth; while in quadrature the centers are separated 90°. However, if at this moment the Moon is in the perigee or apogee of the epicycle then it will be in the ED line and the angle representing the second inequality will remain null; while it reaches its maximum value of 2° 39' if the line joining the Moon with B is at right angles to the EB line, that is, when the anomaly of the Moon is 90° or 270°. The maximum value of the sum of the two inequalities is 7° 40'.

Ptolemy took a big step on discovering the second lunar inequality now known as *Evection*, setting its magnitude in 1°19'30'', very close to the true value, and adapting Hipparchus' theory to his own. But his continuous observations showed that the theory wasn't yet sufficiently developed, since some outstanding error still persisted. Baffled, the Egyptian astronomer addressed again the problem but failed to discover the third inequality, the *Variation;* and, on the contrary, he only made the theory even more complicated than it already was.

Continuing with his text, the fifth book focuses on a detailed exposition about the apparent sizes, movements and relative distances between the Moon and the Sun. It also deals with the theory and measurement of the lunar parallax, and the theory, calculation and tabulation of the true positions of the Moon; he obtains the distance of our satellite by the method of parallax, as it is still done today.

In Book VI the theories developed up to this point are applied to the calculations and tabulations of the syzygies, and of the oppositions and conjunctions between the Moon and the Sun. Finally, it presents the methods for calculating and tabulation for both lunar and solar eclipses, including their respective durations.

Books VII and VIII deal fundamentally with fixed stars and present a catalog of 1022 stars, largely based on that of Hipparchus, in which data of ecliptic coordinates and stellar magnitudes are included, as well as a respective classification in 48 constellations. Additionally such books are about a description of the Milky Way, as well as about the construction of a stellar globe that takes into account the precession movement.

The last five books present data and detailed geometric models for the movement of the five known planets with the purpose of calculating their positions at any given moment. There he deals the stations or stops of the planets and their retrograde movement which occurs when the planets seem to stop and then briefly reverse their movement on the background of the fixed stars. Additionally, he deals with the latitudinal movement, that is, the deviation of the planets from the ecliptic. These last subjects constitute the most remarkable original contribution of Ptolemy to astronomy.

The man who was able to advance so much in the lunar theory was not willing to leave the theory of planetary movements in the unsatisfactory state in which he found it. He always referred to the *five wandering stars* to distinguish these bodies that show their movements much more notoriously irregular compared to those of the Sun and Moon.

In the planetary theories of the Hipparchus' days, a planet travels around its epicycle while the center of it travels around the Earth in a deferent circle, with the particularity that both movements are uniform, of constant speed. But in Ptolemy's theory the movement of the center of the epicycle is not uniform but it accelerates and slows down, which constitutes a radical departure from Aristotelian physics. Although this seems to be an unnecessary complication, it was just what an explanation of the planetary movement required. With Ptolemy the Greek geometrical planetary theory finally achieves a real numerical precision; his hypotheses predict planetary behaviors approaching widely to reality and dominate astronomical practice for 1400 years. Ptolemy assigned the following order to the planetary spheres beginning with the innermost ones: Moon, Mercury, Venus, Sun, Mars, Jupiter, Saturn and finally the Sphere of Fixed stars.

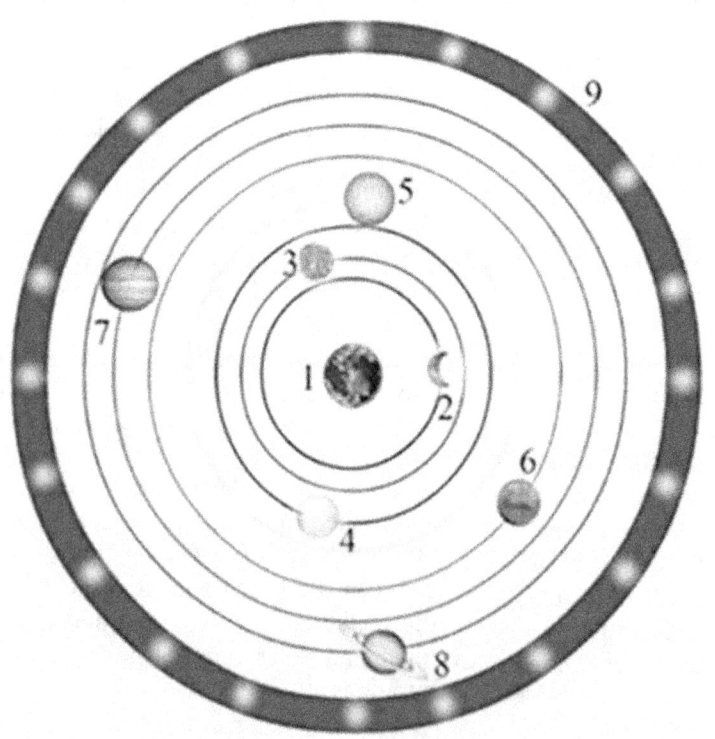

Image 3.7

Order of the stars in the Ptolemaic Geocentric Universe

1 The Earth
2 The Moon
3 Mercury Planet
4 Venus
5 The Sun
6 Planet Mars
7 Jupiter
8 Saturn
9 Fixed stars
The epicyclic circles are not indicated.

Assuming there was no lost or empty space in the cosmos, something completely consistent with Aristotelian physics, the Egyptian astronomer assumed that the mechanism for Mercury would be immediately after the mechanism of the Moon, and then the mechanism for Venus, and so on for Mars, Jupiter, Saturn, and finally the sphere of the fixed stars. Taking as a basis the known distance of the Moon, his calculations reported the distances of the fixed stars of approximately 10000 terrestrial diameters. For the people of that era this was a truly enormous cosmos.

The ancient astronomers when estimating the sizes of our planet, the Moon and the Sun, as well as the distances of our satellite and the King star, used rationally valid theories and methodologies but they were hampered by the lack of instruments of good precision; however, many of their results were surprisingly well. Eratosthenes estimated the diameter of Earth at 14722 km, fifteen percent in excess of the modern value. Hipparchus concluded that the distance of the Moon was almost 29,5 times the diameter of our planet and combining it with Aristarchus's values, indicated that the distance from the Sun was 600 times the Earth's diameter , or approximately 19 times smaller than the true value.

Like Hipparchus, Ptolemy estimated the average distance of the Moon at 29,5 times the diameter of the Earth; while the straight figure is about 30,168 times. Once known the distance to the Moon he was dedicated to calculate the distances of the main celestial bodies. He concluded that the Sun was 20 times more distant than the Moon, while the real value is approximately 400; as for the planets he calculated that Saturn, the most distant, was almost 300 times farther than our satellite, while the real value is close to 4000.

The apparent position of the Moon at the moments of syzygy and quadratures could be determined well according to Ptolemy's theory and with a level of precision that was very acceptable for its time; since they only possessed elementary instruments that could not fix the position of any celestial body with an error lower than perhaps 10'.

None of the ancient astronomers came close to the size and distance of the Sun, which everyone underestimated. Their computations for the sizes and distances of the satellite and the Sun, in terms of the diameter of our planet, were those indicated in Table 3.1, which was elaborated with data taken from *Aristarchus of Samos, the ancient Copernicus.*[7] While Table 3.2 presents data for the size and distance of the Moon in absolute values and according to the terrestrial diameter calculated by Eratosthenes as 14722 kilometers.

The correct figure for the distance of the Sun is of 11726 times the Earth's diameter. Except for Ptolemy's, it will be seen that these estimates improved continuously; the Posidonius' value is about half the correct number. In general, the image of the Solar System of the Egyptian astronomer was not so far from the truth.

The great merit of Ptolemy consists in transmitting, by means of the *Almagest,* the astronomical doctrines of his predecessors, especially preserving and advancing those of Hipparchus; and also to have introduced new and remarkable contributions to the knowledge of his time. With Ptolemy the history of the Greek astronomy finishes, that was very fruitful in the development of theories and of mathematical methods, especially geometric, and which presented sufficient explanation and concordance between the theories and the phenomena observed.

Historians have long debated how much credit to give to Ptolemy and how much to assign to his predecessors. Although he was quite sincere in attributing much of its content to his antecessors, particularly Hipparchus, he did not always mention the origin of his ideas. He attributed to the Nicaea astronomer essential elements of his solar theory as well as parts of his lunar theory, but denied that Hipparchus had developed and explained planetary models. There are strong arguments to doubt that Ptolemy had independently observed the more than 1000 stars listed in his own catalog. In any case, what is not debatable is the clarity and mastery of the mathematical, geometric and astronomical analysis that he exposed in the Almagest.

It is now a recognized fact that the Ptolemaic epicyclic theory was simply a means of calculating the apparent places of the stars without pretending to represent the authentic universal system; and it certainly fulfilled its objective satisfactorily and in a very elegant way from the mathematical and geometric point of view. As Robert Wilson states:

"The Almagest was an epic that was used for 14 centuries to predict the positions of the Sun, Moon, planets and stars with an accuracy of about 1°, the accuracy of Hipparchus' observations."[8]

The historical development of reliable methodologies for the determination of the positions of the stars, especially our satellite, was a transcendental step in the evolution of knowledge, because if one day men aspire to reach the Moon, this is the first question that must be solved: ¿Where and how far away is the lunar globe?

TABLE 3.1
Diameter and Distance of the Moon and the Sun, according to some ancient astronomers, and expressed in terms of the Earth's Diameter D_T

Astronomer	Moon Diameter	Moon Distance	Sun Diameter	Sun Distance
Aristarchus	0,360	9,50	6,75	180,0
Hipparchus	0,333	33,67	12,33	1245,0
Posidonius	0,157	26,20	39,25	6545,0
Ptolemy	0,294	29,50	5,50	605,0
Modern Value	0,270	30,25	108,90	11.726,0

TABLE 3.2
Diameter and Distance of the Moon according to some ancient astronomers, calculated according to the Earth's diameter estimated by Eratosthenes in 14722 km. The respective errors are also included.

Astronomer	Diameter Km	Error %	Distance Km	Error %
Aristarchus	5299,92	52,56	139859,00	-63,6
Hipparchus	4902,43	41,12	495689,74	29,0
Posidonius	2311,35	-33,47	385716,40	0,3
Ptolemy	4328,27	24,59	434299,00	13,0
Modern Value	3474,00		384400,00	

The *Geography* of Ptolemy

Finally, Ptolemy was the author of an important treatise on cartography, called more precisely *Geography*, which compiles the geographical knowledge of the Roman Empire during the second century. The work was used many centuries later by the Italian explorer Christopher Columbus as the guide for his Western route in his pretensions to reach Eastern Asia.

Many maps had been made based on scientific principles since the time of Eratosthenes in the third century BC, but Ptolemy improved them considerably by designing various methods to create maps and introduce cartographic projections. He knew several ways to draw a grid of lines on a flat map to represent the circles of latitude and longitude of the globe, which we now know as *Geographic projection;* and he shared and divulged such methodologies in his works. He provided all the information and techniques required to draw maps of the entire portion of the world known to his contemporaries, reporting the respective geographical longitudes and latitudes in degrees for approximately 8000 locations on his world map. His maps projected in grids provide an impressive and detailed view of the spherical surface of the Earth, and also, to a certain extent, preserve the proportionality of the distances. The more sophisticated projections of these maps, which use circular arcs to represent both the parallels and the meridians, anticipated the bases of later projections to conserve the area.

His maps are seeing distorted when compared to modern maps because Ptolemy's data were inaccurate. One reason is that he calculated the size of the Earth as too small: while Eratosthenes found 700 stadia as equivalent to a degree of circumference on the globe, Ptolemy used 500 stadia in his Geography. It is very likely that these were the same stadia, since Ptolemy changed from the previous to the last scale between the Syntaxis and the Geography, and severely readjusted the degrees of length accordingly.

Bibliographic Citations

---

[1] Internet Archive. https://archive.org/details/Cleomedes-DeMotuCirculari-TheHeavens

[2] Internet Archive. Plutarch on the face which appears on the orb of the Moon.
https://archive.org/details/plutarchonfacewh00plut

[3] Plutarch. The Parallel Lives. Book two - Pau lo Emilio

[4] Veyne, Paul. *The Greco-Roman Empire.* Madrid: Ediciones Akal, SA 2009. Pág. 7

[5] Russo, Lucio. *The Forgotten Revolution.* New York: Springer- Verlag Berlin Heidelberg. 2004.

[6] Internet Archive. Plutarch on the face which appears on the orb of the Moon.
https://archive.org/details/plutarchonfacewh00plut

[7] Internet Archive. Heath, Thomas Little. *Aristarchus of Samos, the ancient Copernicus.* P. 350
https://archive.org/details/aristarchusofsam00heatuoft

[8] Wilson, Robert. *Astronomy through the Ages. The story of the human attempt to understand the Universe*. London: Taylor & Francis. 1997

Chapter 4

A Medieval Look at Our Satellite

The Crescent Moon as it was seen from Earth on Saturday, January 27, 2018, at 06:00 UCT, 10 days and 3 hours and 30 minutes since New Moon; the 75,1% of its illuminated part is now viewed, which continues growing, and then it can be drawn with two curves both convex and looking like a hump or bulge, why it is usually called Gibbous Moon.

Courtesy of NASA's Scientific Visualization Studio: https://svs.gsfc.nasa.gov/4604

"The Moon, which left late and almost at midnight, it made the stars look scarcer: similar to a lighted cauldron, it ran against the sky along that road that inflames the Sun when the inhabitant of Rome sees him fall between Corsica and Sardinia."

The Divine Comedy[1]
Dante Alighieri (1265- 1321), Italian poet

"One says that the moon turned back at the passion of Christ and interposed herself, so that the light of the Sun reached not down; and others that the light hid itself of its own accord, so that this eclipse answered for the Spaniards and for the Indians as well as for the Jews."

The Divine Comedy
Dante Alighieri (1265- 1321), Italian poet

The Middle Age

The fourth and fifth centuries AD mark a sharp turn in the history of knowledge. In the year 306 AD Constantine I *The Great* (272 - 337 AD) becomes the new Roman emperor, and he was the first to convert into Christianity and to legitimize this religious cult through the Edict of Milan of the year 313, also known as the *Tolerance of Christianity*; adopting it almost as a substitute for official Roman paganism. Towards the year 330 he re-founded the ancient Greek city of Byzantium giving it the name of *New Rome* and the category of imperial residence; it was also known as Constantinople and from 1930 it is the current Istanbul in Turkey. In the year 395, and for administrative purposes, the colossal Roman Empire is divided into two: the Western one with its seat in the millenary Rome, and the Eastern Empire centralized in the renewed Byzantium. Due to a strong political-administrative crisis of multiple causes the Western Roman Empire finally collapsed around 476 AD, year in which Rome was submitted to the authority of the Eastern Empire; with which Constantinople becomes the new political, cultural and scientific development pole of the world.

After the great Greek period and during the Roman boom, all intellectual manifestations decreased in Europe and the Middle East. All the Mediterranean countries and a good part of Europe were under the military political dominion of Rome for about six centuries; and when the empire definitively collapsed in 476 AD, the intellectual developments practically stagnated in all areas of erudition. The period that followed in Europe is now known as the *Middle Age* or Medieval period: for some five hundred years intellectual and cultural efforts, including science and astronomy, fell into minimum levels in all regions, and for another five hundred additional years in other areas. But fortunately the eastern region remained well organized under the *Byzantine Empire*, which lasted another thousand years after the fall of Rome; and at this juncture such intellectual activities were revived and sustained in the Middle East, where Arabic mathematicians and astronomers led the world from the 9th to the 13th century.

Additionally, during this period and together with Judaism, two new great monotheistic religions were developed that would have important implications, not only for astronomy and knowledge in general, but also for all human activities. Christianity, which was based on the belief that Jesus Christ was the son of God, the same God as that of the Jews, was derived from Judaism very early in the first century AD and quickly spread throughout Europe. The second new monotheistic religion was Islam which had the same ancient Sumerian roots as Judaism: the belief in one almighty God. His great prophet was Muhammad who wrote the Koran on the basis of his communications with God. These three great monotheistic religions, in chronological order Judaism, Christianity and Islam, have, therefore, the same common ground: the belief in a unique and all-powerful God, the creator whose will and teachings must be obeyed as wrote on the Holy Scriptures.

Within the set of previous events, the Ptolemy's writings and the Greek cosmological tradition were lost, they went astray from the western world for more than a thousand years. In the intermediate millennium between the ancient and modern worlds, from the 5th century to the 15th century, period known as the Middle Age, and due to the extraordinary heyday of Christianity, a cosmology widely influenced by Christian ideology developed in Western Europe. But this religion didn't stimulate the intellectual developments with religious requirements during the first centuries of its evolution. The belief was established that the whole truth was found in the *Holy Scriptures*, as revealed in the Gospels; and consequently, being unnecessary, philosophical and scientific studies weren't encouraged. This led to a fundamental conflict between the Christian religion and Western science that would continue much for centuries.

In the cosmological aspect, within the society of primitive Christians the shape of our planet caused deep divisions and sharp debates. The *Theological School of Alexandria* is characterized by its proximity to the Greek philosophical legacy, in a good marriage between reason and faith; to it belonged Origen of Alexandria (185 – 254 AD) and Cyril of Jerusalem (315 – 386) as exponents of the terrestrial sphericity, and Eusebius of Caesarea (263 – 339) who supported the idea that the Earth is

79

located in the center of the cosmos and it is surrounded by the ocean. On the contrary, the *Theological School of Antioch* takes distance from Greek philosophy and rather defends the literal interpretation of the Holy Scriptures; what in the field of cosmology meant to reject one of the most firmly established convictions: the sphericity of the Earth and of the stars in general. Saint John Chrysostom, or John of Antioch, (347 – 407) was one of the Christian thinkers who most emphatically argued against the spherical shape of the world; additionally, Theophilus of Antioch (183 AD) argued that while the Heavens were domed, the Earth was flat.

In the previous controversy participated actively Cosmas Indicopleustes (490 AS), a sailor and Greek merchant of Alexandria who became a monk of Nestorianism, a variant of Christianity. Desiring to spread what he considered to be true Christian teaching, he composed a Geography and an Astronomy, both now lost; and a work with the very curious name of Christian Topography, a mixture of Hellenistic and biblical geographies, of which three manuscripts have been preserved. He sharply attacked pagans or infidels whom claimed that the world was spherical and whom mocked his cosmology. Cosmas considered that the true form of the world had been revealed by God in the Sacred Scriptures, and that it was not spherical or cylindrical, but that of a chest. He maintained that the Earth was flat and that the cosmos had the shape of a huge rectangular vaulted box divided in two by the sky, which serves as a screen that separates them: The lower part represents the visible world in which the men live, and this inhabited or earthly world he presented as a rectangle surrounded by an ocean with a rectangular frame; while the upper part represents the invisible world, the Kingdom of God.

Cosmas acted at the time when the Christian beliefs were tenaciously superimposed on the Greco-Roman or pagan ones and they questioned the most firmly established beliefs regarding the spatial conception in the ancient world, and therefore represented an important turning point in the history of cosmology. The character and theological content of Christian Topography are reflected in the subtitle: *"For those who want to be Christians and in front of those who from the outside believe in the spherical sky and glorify it."*[2] Thus, the work is indicative of the debate and tensions between the schools of Alexandria and Antioch, between Aristotelian and anti-Aristotelian, and those who wanted to host the pagan thought and those who definitely rejected it and best welcomed to the sacred teachings. Completely dominated by theological supernaturalism, a large part of the Medieval Christians philosophers made no serious attempt to show the world as it really was.

But fortunately the ancient Greek Hellenistic tradition was preserved in the Byzantine and Arab societies of the East, and later it came to light again in Western Europe in the fifteenth century. From the 9th century the conscious reactivation of the Hellenistic traditions and the importance of the Ptolemaic cosmology were taken more actively by the Byzantine and Arab scholars; with which a variant in medieval cosmology was developed in which the influence of Greek philosophy eventually re-dominated over theological considerations. Through the Byzantine Empire the works of Ptolemy had been reintegrated into the mainstream of cosmological thought in Europe, which led to another form of medieval cosmography in which the Earth was conceived as a sphere. The rediscovery of Ptolemy can be considered the fundamental contribution of Byzantine scholarship to the long-term development of cosmology.

Islamism and the Arab astronomers

The other new religious current, Islam, developed from the sixth century AD and was essentially established by a single individual, the prophet Muhammad, who wrote the Koran as the word of his God, *Allah*, as it was divinely communicated to him. Like the Christian Bible, the Koran expresses the will of the Almighty and, therefore, is inviolable and cannot be questioned. But unlike Christianity, the Muslim faith greatly stimulated scientific efforts due mainly to the strong intellectual demands of their religious practices, which posed very difficult problems for the mathematical astronomy of the period.

In Arabia and during the sixth century AD, the new Islamic religious movement would become a powerful one and led to the ascendancy of the Arab people with their vision of worldwide conquest by

the Islam. With incredible speed a great Arab Empire was formed by Muhammad, the self-proclaimed prophet of the unique God. A holy war was preached and millions of Arabs converted to the cause. In a few decades after the death of the Prophet Muhammad in 632 AD, Muslims had established a community that stretched from China in the East to Spain in the West, and gave the world a new culture. After the initial stage of the Arab conquest, when the Koran was considered to contain a complete code of conduct and a comprehensive philosophy, a belief that was responsible for the destruction of the remnant of the Library of Alexandria when the city was plundered by the Arabs in the year 642 AD, it continued spreading throughout the Arab world and established centers of culture in Baghdad, Cairo and Cordoba in Spain, etc.

Image 4.1
Medieval astrological celestial map:

In the two outermost circular sectors, the 28 *Lunar mansions* and the lunar phases are schematized. In the central circular sector the 12 signs of the solar zodiac are represented. Image taken from the Ottoman manuscript *Zubdat - al Tawarikh* of 1583, preserved in the Museum of Turkish and Islamic Art in Istanbul.

Medieval Arab scholars stood out for their great methodological precision. But, like other Eastern nations, they almost failed in the development of speculative philosophy and many of them devoted their intellectual efforts rather to astrology, or divinatory arts, than to astronomy.

The pre-Islamic Arabs were influenced by Hindu astrology through their contacts with the Persians. This influence is evident in the division of the ecliptic into 28 zones called *Manazil-Al-Qamar*, or the *Lunar Mansions* or *Houses of the Moon*; which are comparable to the 28 Nakshatras of the Hindu astrology that in turn derive from the Mesopotamian tradition's *Way of the Moon*. In the traditional Arab astrological system it was considered that the Moon moved through distinct 28 Manazil during the normal solar year. A detailed list of Manazil-Al-Qamar is presented by the famous Arab astronomer and mathematician Al-Biruni (973 - 1048 AD) in his work *Kitab al-tying al-baqiya 'an al-Qurun al-khaliya*.

In the societies that adopted it, the lunar calendar is based on the phases of the Moon; but it is also possible to draw the background of the fixed stars along the sidereal revolution of the satellite of approximately twenty-seven days, as seen from Earth. This can be assimilated as a *Lunar zodiac* in which the satellite seems to be located or stay between different stars every night. Since the Moon's path is five degrees away from the ecliptic, the stars in which it is found are almost all belonging to the constellations of the traditional Solar zodiac. The first Islamic scholars described a formal system of twenty-eight *Lunar mansions*, or *Manazil-Al-Qamar*. Each of these zodiacal sections has duration approximately equal to 13 days, 365 days divided into 28 Mansions; and it was usually named after the brightest star inside it. Additionally, astronomical texts often define each Mansion as an equal amount of arc along the course of the Moon: each covers 12° 51', that is, 360 degrees divided by 28 Mansions. This results in a coordinate system that could be applied in maritime navigation and also to indicate the hours of the night.

A remarkable revival of the study of astronomy was in the Middle East during the seventh century due to the rise on power of the Abbasid caliphs, who, influenced by the precepts of the Koran and the traditions of the Sunna, gave considerable value to knowledge. After its territorial conquests and once established very close to the ruins of ancient Babylon in the year 762 AD, the second Abbasid caliph Al- Mansur (712 - 775) founded Bagdad as the capital of Islam; which would quickly become a cultural center of great importance. Greek texts were brought and translated, among them the Almagest; with which the Arabs felt the influence of the ancient Western civilization. As well some works from the Far East were imported, especially from India. In Baghdad and Damascus astronomical observatories were founded with instruments similar to those of the Greeks, but improved and larger. Continuous observations were made of the main stars, as well as of the eclipses of the Moon and the Sun.

Very notable among the Arabs for the encouragement he gave to the progress of knowledge, particularly astronomy, was the Caliph Al Mamun, who lived during the eighth century AD. He founded in the city of Baghdad a *House of Wisdom*, which would become a center of learning similar to the great schools of ancient Greece, and which included a library made up of books related to all the disciplines appreciated in his day: natural sciences, literature, mathematics and logic. There all the most important scientific and philosophical works of the ancient world were permanently translated into Arabic, especially those from Greece and Egypt, and they were granted the status of truth. In its translation into Arabic, Ptolemy's *Syntaxis Mathematica*, now known as the Almagest, was to form the basis of Arab astronomy.

The Almagest and the Lunar astronomy of the medieval Arab world

The name of Almagest appeared as a Latin corruption of the Arabic expression for *al- Majistī*; and the work became the fundamental guide for the Byzantine, Islamic and European astronomers until the beginning of the 17th century. Due to its great reputation it was in great demand and was translated into both Arabic and Latin: The first translations from Greek into Arabic were made during the ninth century, and most probably the mathematician, astrologer and astronomer Sahl ibn Bishr al-Israili (786 - 845) performed the first; later it would be translated from Arabic into Latin during the last half of the 12th century.

The religious scholar Henry Aristippus (1105 - 1162), made the first Latin translation directly from a Greek copy, but was not as influential as a later translation from Arabic into Latin made by the Italian Gerardo de Cremona around 1175. Cremona (1114 - 1187) was a famous translator of the twelfth century and one of the most prolific of his time, with some seventy works translated from Arabic into Latin; he translated the Arabic text of the Almagest, *Kitab al- Medjisti*, into Latin when he was in the Toledo School of Translators in 1175; this version definitely introduced the Almagest in the medieval European scientific tradition.

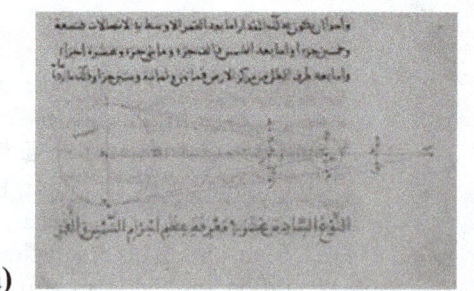

Image 4.2

Kitāb al- Majistī: the medieval Arab Almagest

Details taken from an Arabic copy of the Almagest, Kitāb al- Majistī, kept at the Qatar Digital Library's. Image a) corresponds to the general geometry of an eclipse of the Sun. Image b) corresponds to an argument about the circular movement of the stars: the epicycle circle, and the deferent circle including its center, the eccentric point and the equant point.

In the 13th century a Spanish translation was produced under the patronage of King Alfonso X of Castilla, also known as The *Wise*. Later, Johannes Müller von Königsberg (1436 – 1476), better known as Regiomontanus, a mathematician and astronomer of the German Renaissance who worked in Nuremberg, Vienna and Buda, he made an abridged Latin translation of a Greek version of the Almagest that appeared in Western Europe at the time. Additionally, his contributions in cosmology were fundamental for the development of Copernican heliocentrism in the decades after his death.

During the medieval period in the Middle East, North Africa and throughout Europe it was the maximum text on astronomy and its author became practically a mythical figure. The work was preserved, like most existing texts on classical Greek knowledge, in Arabic manuscripts. In his works, Ptolemy left instructions for later astronomers on how to use quantitative astronomical observations with recorded dates to revise cosmological models. The numerical tables in the Almagest, which allowed us to calculate the planetary positions and other celestial phenomena for arbitrary dates, had a profound influence on medieval astronomy; largely through a revised and separate version of the tables that Ptolemy published as *Practical Tables*.

From the 9th to the 15th century Muslim scholars excelled in all areas of scientific knowledge and their contributions in Mathematics and Astronomy were quite important. They brought with them their own popular astronomy which was then mixed with local teachings and later they incorporated the astronomical and mathematical traditions of the Indians, the Persians and the Greeks; which dominated and adapted to their needs. The early Islamic astronomy was thus a mixture of pre-Islamic, Indian, Persian and Hellenistic Arab wisdom; and by the tenth century it had acquired very distinctive characteristics of its own; becoming so dominant that many astronomical words in use today are of Arabic origin instead of Greek, including names of stars such as Aldebaran, Vega, Betelgeuse, Rigel and Deneb; and technical terms such as the zenith. In addition, the great cosmological work of Ptolemy, the *Syntaxis Mathematica*, is now better known by its English name derived from Arabic, the *Almagest: The greatest*.

Astronomy was the most important of the Islamic sciences, as we can judge from the volume of the associated textual tradition, and it flourished in such society at two different levels: popular astronomy devoid of theory and based solely on what it could be seen in the sky; and scientific astronomy which involved theoretical considerations, systematic observations, calculations and mathematical predictions. The knowledge of the passing of the stars through the twelve signs of the zodiac, the phases of the Moon, the associated meteorological and agricultural phenomena, and the calculation of time using shadows on day and the lunar stations at night, formed the basis of Islamic scientific astronomy.

Given the Ptolemaic models, the tables of the middle movement and the equations for the Sun, the Moon and the planets, available for the Muslim astronomers in the Almagest and other works, from the ninth to the sixteenth centuries they tried to improve the numerical parameters in which these tables were based. They compiled ephemerides or almanacs with the solar, lunar and planetary positions for every day of the year; as well as information about the New Moons and astrological predictions resulting from the relative position between the Moon and the planets.

One of the greatest challenges posed by Islam to Arab astronomers was related to their calendar. In the times of Muhammad the other two great monotheistic religions, Judaism and Christianity, established the dates of their great religious celebrations, such as Easter and Resurrection, according to the early Jewish calendar that was based on the Moon's phases. This led to the number of months in a year was approximately 12,33, and this, not being a whole number, demanded cumbersome adjustments to keep the religious and civil seasons well synchronized with the celestial cycles. This offended Muhammad, who decreed in the Quran that, in the sight of God who created everything, there was a total of twelve exact months in a year. When establishing the Muslim calendar, this was interpreted in the sense of 12 months determined by the respective lunar cycles. Therefore, the lunar year of the Muslim calendar is approximately 11 days shorter than a year determined by the solar cycle, which means that the seasons migrate through the calendar to make a complete cycle every 30 years.

According to this, the Muslim or Islamic calendar is a lunar one where the years are composed of 12 lunar months. It began when Muhammad, the prophet of Islam, had to flee from Mecca to Yathrib, later renamed Medina by him, due to the persecution of his adversaries, a trip known as Hegira and which occurred in 622 of the Christian era. The Muslim calendar is based on lunar cycles of 30 years and equivalent to 360 lunations according to the Sumerian tradition. The 30 years of the cycle are divided into 19 years of 354 days and 11 years of 355 days. The 354-day years are called simple years and are divided into six months of 30 days and another six months of 29 days. The 355-day years are called intercalary and are divided into seven months of 30 days and another five of 29 days. For a total average of about eleven days shorter than the solar year and the seasonal round. So, its main use is to establish dates for religious celebrations instead of seasonal civil activities on an annual solar basis.

The calendar being strictly lunar, the beginnings and endings of the respective months, in particular of the Ramadan sacred month, and of several festivals throughout the twelve months of the year, are regulated by the first appearance of the Crescent Moon, *Al Hilal*. Medieval astronomers knew that the determination of the possibility of sighting such phenomenon on a given day was a complicated mathematical problem, involving knowledge of the positions of the Moon and the Sun among themselves and additionally with the local horizon. In most cases, the conditions required to ensure the visibility of this dim lunar light can be determined by observational records and practices. But the theoretical formulation of a definitive set of such conditions has challenged even modern astronomers; and the most fervent of them can be denied the thrill of seeing the first rays of the crescent Moon at the predicted time if local weather conditions, clouds and fog, restrict his sight.

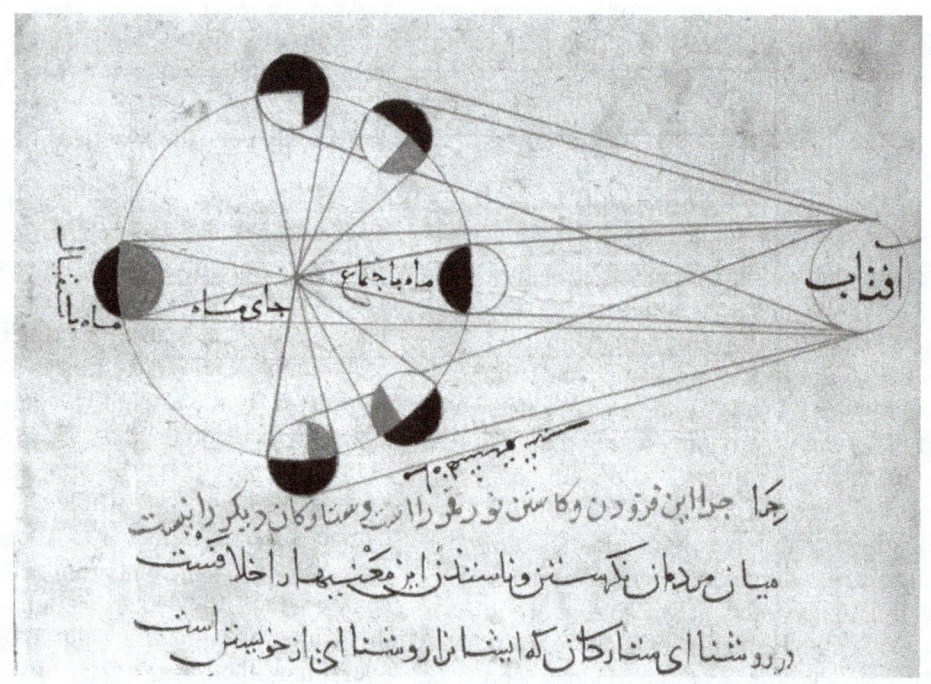

Image 4.3 Lunar phases according to medieval astronomy.
Medieval graphic representation of the Lunar Phases, according to the work *Kitab al- Tafhim* of the Arab astronomer Al- Biruni (973 – 1048). It was scanned from: *Islamic Science: An Illustrated Study*, Seyyed Hossein Nasr (1976), World of Islam Festival Publishing Company

Some of the main Muslim astronomers proposed conditions that involved three different parameters, such as the apparent angular separation between the Sun and the Moon, the difference in their times of concealment or laying on the local horizon, and the apparent angular velocity of the moon. The *Ephemerides* or annual almanacs, that is, tables showing the positions of the Sun, the Moon and the five planets with the naked eye for each day of a given year, used to provide information about the possibility of sighting at the beginning of each month. In order to improve the data, at the beginning of the 9th century the Abbasid caliph *Al-Mamun* sponsored the astronomical observations, first in Baghdad and then in Damascus, gathering the best available astronomers to follow the Moon and the Sun.

Using such observations, the Persian mathematician, astronomer and geographer *Muhammad ibn Musa al-Khwarizmi* (780 - 850), also known as *Algorithmi*, compiled a table showing the minimum distances between these stars to ensure its visibility throughout the year in the geographic latitude of Baghdad. From his treatise on astronomy, *Zīj al-Sindhind*, which is also based on Indian astronomical works, there is a Latin translation of the original version. The main topics covered in this work are the calculation of the true positions of the Sun, the Moon and the planets; spherical astronomy, tables of sines and tangents, calculations of parallaxes; visibility of the Moon and eclipses; the calendars and astrological tables.

Abū Abd Allāh al-Battānī

Abū Abd Allāh al-Battānī (858 - 929) was a Middle Age prince, mathematician, astrologer and astronomer, also known as *Albategnius*. He was born in 858 in *Harran*, traveled to *Raqqa* to receive higher education, and later at the end of the ninth century he moved to *Samarra*, where he lived and worked the rest of his life. He made many and very important contributions to astronomy: he determined with great precision the duration of the solar year, establishing it in 365 days, 5 hours, 46 minutes and 24 seconds, with only a difference of 2 minutes and 26 seconds with respect to the current measurement; he calculated the inclination of the ecliptic at 23°35' and described its relationship with the climatic seasons. He also calculated a value for the precession of the equinoxes at 54,5'' per year, or 1° in 66 years, and found that the solar apogee, the maximum distance between Earth and the Sun, is variable. Using trigonometry he corrected the orbital calculations realized by Claudius Ptolemy, and developed astronomical tables for the movements of the Moon, the Sun and the planets, which improved the accuracy of those of that one. Finally, he made excellent observations of lunar and solar eclipses, which led him to discover the existence of annular solar eclipses.

Much of his astronomical knowledge he depicted in his text *Kitāb az-Zīj* or *Book of the Astronomical Tables*, of which many translations have been made to both Latin and English, the last known is that of Bologna in 1645. Another important work of al-Battānī has come to us in a twelfth-century Arabic codex named *Opus Astronomicum*; which was translated into Latin by the orientalist Carlo A. Nallino. Al-Battani explains in the introduction the reason that led him to write this work:

"For a long time I have been dedicated to astronomy, and I spent a lot of time studying it. I have observed many differences in books dealing with celestial movements, and I have even seen that some authors have been wrong in laying the foundations. Therefore, after much reflection, I have thought about correcting and establishing better all these things using the methods of Ptolemy in his Almagest, following his steps and following his precepts......... I have corrected the position and the movements of the celestial bodies in the ecliptic that I have found from observations, from eclipses calculations and from other operations; and I have added other necessary things."[3]

In this work the Arab astronomer deals with the lunar parallax and the distance from the satellite to the Earth, based on the eclipses of the Moon and the Sun; he also deals with the positions of the five known planets on the ecliptic at different times of the year. It also presents the results of his observations and calculations in the form of astronomical tables, which contain data for calculating the calendar, for the movement of the Moon, the Sun, and the coordinates of the constellations. Finally, a catalog of the known fixed stars concludes the book.

Al Battani used his observations of eclipses to show that Ptolemy's statements about the diameters of the Sun and the Moon, determined with data on two lunar eclipses, were unsustainable.

The contributions of Al-Battani in the plane of Mathematics consisted in to propound techniques for the solution of trigonometric problems using the methods of orthographic projection. He was the first to replace the Greek *cords* with modern *Trigonometric functions*, and introduced the Cotangent concept.

Image 4.4
A medieval look for the Moon
Astronomers observing the Moon and the stars; Seventeenth century Ottoman miniature painting; It is preserved in the Library of the University of Istanbul.

During the tenth century the great Arab Empire began to crumble due to a process of political and religious disintegration what led to many provinces going up revealed and separated; which conduct to the emergence of multiple independent regional dynasties in Spain, Morocco and Egypt. The Arab scientific tradition was then transferred to the area of the Western Mediterranean where academies and libraries were established in Cordoba and Toledo in Spain; it was mainly through and from these centers that the Arab tradition subsequently spread throughout Western Europe. The Greek works that had been first translated into Arabic, are now translated from Arabic into Latin. Fortunately, the Arab influence in Western Europe was consolidated before the great resurgence of Christian religious fervor, materialized in the Crusades (1095 – 1290), which proceeded to sweep and eradicate Islam and *infidels* throughout Europe.

Islamic cosmology was guided mainly by the preserved works of Aristotle and Ptolemy, which were translated into Arabic and studied in depth. The Muslim astronomers would be even more faithful to Aristotle, whose cosmological statements they accepted totally and without question. Ptolemy's criticism began in the 11th century when a prominent Cairo philosopher, Ibn al-Haytham, wrote a book called *Doubts Concerning Ptolemy*. In the following century there was an even greater criticism on the part of Ibn Rushd of Andalusia, which at that time was part of Islam.

Ibn Yunus

If we consider now the Moon, we find that Arab astronomers did not considerably advance on the work of the Greeks. Several of them noticed that the inclination of the lunar orbit was not exactly 5° as was thought. The Egyptian Arab astronomer *Ibn Yunus* (950 – 1009) wrote that he has found 5° 3' or 5° 8', while other observers say they have found from 4° 58' to 4° 45' for such lunar inclination. Yunus further elaborated a record of Arab observations spanning nearly two centuries and where he included three observations of eclipses, two solar and one lunar, made by himself near Cairo in 977, 978 and 979. But the lack of perseverance and of precise instruments made him lose a remarkable discovery, that of the variation in such lunar tilt. Anyway, the observations and the valuable data provided by Yunus on eclipses and conjunctions of the Moon and the Sun were used in the calculations on the *Secular acceleration of the Moon* that made, later and separately, the astronomers Richard Dunthorne and Simon Newcomb during the XVIII and XIX centuries respectively.

Ibn al- Haytham

Ibn al - Haytham (965 – 1040) was a Muslim mathematician, astronomer and physicist born in the Basra city in today Iraq; in his time he was usually known as the *Second Ptolemy* and also as the *Physicist*; now he is usually known in West as *Alhazen*. He is considered the founder of optics due to his works and experiments on the reflection and refraction of light using lenses and mirrors, and for writing the first comprehensive treatise on these topics, with which he paved the way for modern science of physical optics. In his texts on astronomy he gave his readers the impression that he had found the true physical configuration of the universe, whereas Ptolemy could not have achieved it.

Alhazen is considered a great promoter of the *Scientific method* since he argued that all theoretical hypotheses should be stolen with empirical evidence; what distanced him from the primitive Greek belief that natural phenomena could be discovered only through reason. For him, the experimental exercise could not be waived to determine if the proposed theoretical and mathematical developments made sense and described reality well. This is considerably closer to what we now understand as the Scientific method, and what is supposed to have its origin in the seventeenth century. This scholar philosopher wrote almost a hundred works of which fifty-five remain.

In his work *Al-Shukūk ' alā Batlamyūs*, translated as *Doubts Concerning Ptolemy*, published towards 1025, he argues that the judging of existing theories occupies a special place in the development of knowledge; and in it Alhazen severely criticized the *Almagest* and *Planetary Hypotheses* of Ptolemy, pointing out several contradictions that he found in such cosmology. He considered that some of the geometrical elements that the Egyptian introduced in astronomy, especially the *Equant*, they did not satisfy the physical realities of uniform circular motion, and declared as absurd to relate the physical movements of real bodies with imaginary points, lines and geometrical circles. According to Alhazen: Ptolemy supposed an arrangement that cannot exist, and if this arrangement reproduced in his imagination the movements belonging to the stars, that would not free him from the mistake he made in his theories.

According to the above and believing that there was a true configuration of the planets that Ptolemy had failed to unveil, Alhazen had the intention for resolving the contradictions and of completing and repairing Ptolemy's system, although not to replace it completely. He developed a new planetary model in which he describes the movements of the stars in terms of spherical geometry, infinitesimal geometry and trigonometry. Like Ptolemy, he embraced to the system of a geocentric universe and assumed the uniform circular motion of celestial bodies, which required the inclusion of epicycles to explain the observed movements, but he managed to remove the point equant employed by the Egyptian astronomer. In general, in this model he did not try to establish a causal explanation of the celestial movements, but instead he devoted to derive a complete geometric description that could explain the observed movements without the contradictions inherent in Ptolemy's model.

The work *Theory of the planetary movement for each one of the seven stars* of Alhazen was written towards 1038 and only one deteriorated manuscript containing the introduction and the first section of the work has survived. Moreover, in his book *On the Configuration of the World* he presented a detailed description of the physical structure of our planet: Earth as a whole is spherical and its center is the center of the world; it is fixed in it and always at rest; it does not move in any direction, nor does it move with any of the varieties of movement.

Abū al-Wafā Būzhjānī

In his works Ptolemy had written about the first and second irregularities of lunar movements; but a Muslim astronomer who lived in Cairo and observed in Baghdad in 975 claimed to have discovered a third inequality now known as the *Lunar Variation*. His name was Abū al-Wafā Būzhjānī (940 - 998), a mathematician and astronomer born in the Persian

city of Buzghan, now Iran; but who worked primarily in Baghdad. Among his works there is *Kitāb al- Majisṭī*, or *The Almagest Review*, a simplified version of the Ptolemy's work which was widely read by medieval Arab astronomers in the centuries after his death. He also studied widely the movements of the Moon, and for his work in this field it was decided to name in his honor Abū al-Wafā a lunar impact crater located near the equator on the hidden face of our satellite. Finally, he made important innovations in spherical trigonometry and he is credited with the compilation of the tables of sinuses and tangents at intervals of 15', as well as the introduction of secant and cosecant functions.[4]

The al-Wafā's Almagest has never been published in its entirety, but there are three translations in which some chapters only differ in certain trivial points. He starts by describing the first inequality, the Equation of the center, continues with the second, the Evection, and then he states when they reach their maximum values. He then says that we have found a third inequality which occurs when the center of the epicycle is between the apogee and the perigee of the eccentric, and reaches its maximum when the Moon is near a *tathlith* of the Sun; while it is insignificant in syzygies and in quadratures. He states that such variation is caused by a deviation of the line of apsides respect epicycle and describes quite correctly s geometric constructions adopted by Ptolemy, whose name he doesn't mention, causing the line of the apses to be directed not to the Earth, but to another point in the line of apsides of the eccentric. Even so, it is difficult not to grasp that Abū al-Wafā was simply copying the Egyptian astronomer. If he would have made a new discovery, the later Arab astronomers should have pointed it out; but none of them offers more than interpretations of the lunar theory of Ptolemy and expressions very similar to those used by al-Wafā. Therefore, the Arab astronomer knew nothing about the movement of the Moon that he had not borrowed from the Egyptian one.

In conclusion, everything seems to indicate that Wafā perceived a third inequality of the Moon, but as in his time the Arabs distrusted their own abilities, filling themselves with a veneration too great for those of the ancient Greek scholars, he did not study it in depth and didn't establish its mathematical formulation nor its magnitude; in such a way that the discovery went unnoticed. On the contrary, in Europe this third inequality of the Moon's movement, according to which the star moves faster when it is in its new or full phases and slower in the first and third quarters, was rediscovered and well explained by Tycho Brahe six centuries later, around 1580.

The intellectual legacy of the Arabs smoothed out the way for a great flourishing of knowledge in Western Europe. The first European universities beside the monastic orders known as Franciscans and Dominicans were founded at about the same time; and these academic and religious centers had a great influence on the development of Western science.

Medieval Europe

The first universities that deserve such a name are those of Bologna in Italy founded around 1090, that of Oxford in England of 1096 and that of Paris in France of 1150; which provided the social elites of the moment with the knowledge necessary to serve the Church and the State; and at the beginning of the 12th century there was a generalized rise in its social and intellectual importance. The primitive universities acquired some of its patterns of organization of the old monastic orders and cathedral schools; but now they had a particular feature: their international recognition for the privileges and protection provided by local rulers and the Pope. The ultimate purpose of the universities was, in principle, to provide the Church with a well-educated and educated clergy, but they no longer had the same religious concerns as the old cathedral schools. The general outline of medieval formal education in the West was based on the seven liberal arts, which constituted the main element or university curriculum. The vast majority of students were required to study the *Quadrivium* or the *Four Sciences:* Arithmetic, Geometry, Astronomy and Music. The faculty of arts was in the lower level and preceded by the faculties of law and medicine. Apart and with the highest rank was the theology faculty that only gave access to a small fraction of the students. The seven liberal arts served as an introduction to the study of theology which was their goal and to which they were subordinate. The work of Aristotle *On The Heavens* was studied for its cosmological content and was the subject of many

commentaries; its validity was maintained for a long time because it was linked to the study of natural philosophy and metaphysics.

Johannes de Sacrobosco

With the passage of time new texts were needed for teaching, so Johannes Sacrobosco, who was probably a teacher in Oxford and certainly taught in Paris, circa 1230 wrote what was to become one of the astronomy books most widely studied in the medieval period: *Tractatus de Sphaera*, *De Sphaera mundi* or *The Sphere of the World*. Work inspired to a great extent in the Ptolemy's *Almagest* but resorting to ideas of the Arab astronomy and very ornamented with citations of the classic poets; it addressed only elementary spherical astronomy, geography, and tangentially the planetary theory. Initially circulated in the Codex form, a kind of manuscript notebook, folded and stitched; but after the invention of the printing to 1450, its first printed edition appeared in 1472 in Ferrara, and more than 90 editions were printed in the following two centuries.

John of Holywood, or Johannes of Sacrobosco in his Latinized form, was born around 1195 in Halifax, England, and died in 1256 in Paris; he was an English scholar educated at the University of Oxford. His text *The Sphere of the World* was constituted in the most influential work of the pre Copernican astronomy published in medieval Europe. It is a book composed of four chapters of which the fourth is an introduction to the theory of motion of the Moon, the Sun and the planets; equally it also deals with the causes of lunar and solar eclipses. The book was first used at the University of Paris, and widely throughout Europe from the middle of the thirteenth century as the fundamental text of astronomy, up to the seventeenth century.

Of this work we are now interested in two graphic representations. The first is the lunar eclipse diagram corresponding to Figure 4.5.[5] According to modern astronomical theory, it can be seen that Sacrobosco's contemporaries had a very good level of illustration and understanding of such eclipses, since the representation made in this diagram is highly coincident with modern schemes for such phenomena. First, in the cosmological aspect, it is clear that astronomers of the time worked within the framework of the geocentric universe, understood well the concept of the relative sizes of the Moon, Earth and the Sun, and they were perfectly clear that the three stars should be sufficiently aligned, or in the best case they must be just in a straight line, for the lunar eclipse to be total. We can also see how they understood the fact that it is the shadow of the Earth, or penumbra, which does not allow sunlight to reach the satellite and that it is directly responsible for the eclipse.

The other interesting diagram is the one shown in Image 4.7 which corresponds to a representation of the lunar phases. This figure also explains itself and allows us to conclude that medieval astronomers also had a perfect understanding of those appearances of the satellite that we now call Lunar phases.[6]

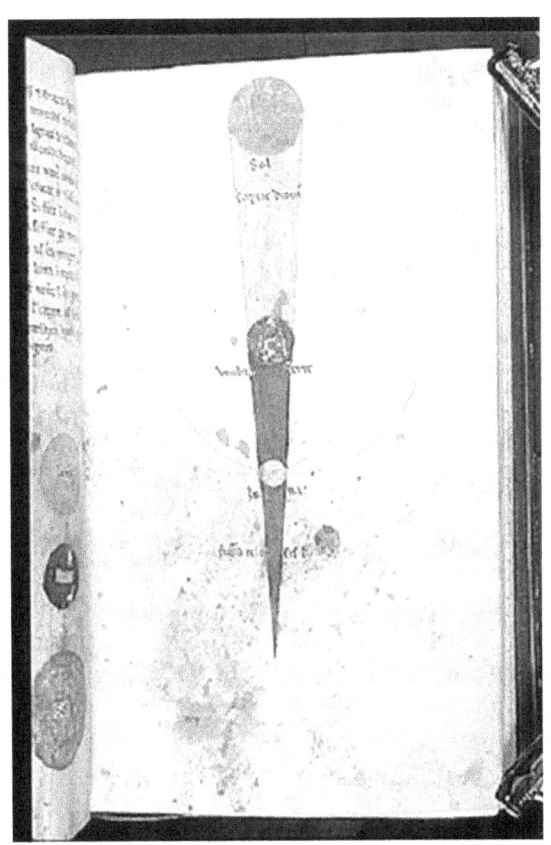

Image 4.5
Medieval diagram for a Lunar eclipse
Diagram of a Lunar eclipse present in the work *Tractatus de sphaera* by Johannes de Sacrobosco, published in the third quarter of the thirteenth century. Preserved in the *British Museum* since August 13, 1840.

Medieval Cosmology and Literature

Medieval cosmology also greatly stimulated the general literature of the time. One of the summits of universal literature and one of the fundamental works of the transition from medieval thought, of theocentric characteristics, to Renaissance thought, essentially anthropocentric, *The Divine comedy,* was written towards 1304 by the famous Italian poet *Dante Alighieri* (1265 – 1321). Within it the author synthesizes all the vast cosmological knowledge collected over the centuries, from ancient civilizations to the European medieval world; as well as he expresses his moral convictions, religious beliefs and his philosophical doctrines. In this work Alighieri describes us with amazing realism his wonderful journey in the course of which he meets the souls of both magnificent and terrible characters of all time.

The entire text is presented in three sections which describe the poet's journey through the three different levels of the spiritual world, and are full of symbols that refer to medieval thought and knowledge: theology and religion, astronomy and astrology, philosophy, mathematics, etc.; materializing them in places, characters and actions. The Hell, which could be written towards 1308, symbolizes man in front of his sins, which basically fall into three categories: incontinence, violence and malice, and their deplorable consequences. The Purgatory, written by 1314, represents the slow healing of such human sins until the final liberation of the soul. And finally, the Paradise, written between 1314 and 1321, date of Dante's death, symbolizes both knowledge and science, and where the three theological virtues: faith, hope and charity are rewarded. The whole work is a poem to humanity that only in faith and Divine grace will find its authentic happiness.

As a quick synthesis of the poem, we would say that once Dante has lived in this hell that is the world, and has traveled the seven terraces of purgatory, he finally arrives at the celestial Paradise at the top. This Paradise is described by the poet in a way very similar to the cosmological system of Ptolemy: a central star and nine more planets revolving around it. Then Dante is allowed to climb through the nine celestial spheres: first that of the Moon, where the souls are found who consecrated themselves to the service of God leading a monastic life, but who later were forced by different circumstances to break their vows; and to which a faint blue air comes. Then there are the spheres of Mercury and Venus, to which the shadow of the earth comes; then those of the Sun, Mars, Jupiter, and Saturn. The eighth sphere is that of the fixed stars, and finally the ninth corresponds to the *Primum Mobile*, where the souls are that achieved a greater understanding of God and, therefore, expressed a maximum love towards Him. The nine spheres are moved by three triads of angelic intelligences: the Seraphim guide *Primum Mobile*, the Cherubim the fixed stars, the Thrones the sphere of Saturn, and so on until the inner sphere of the Moon, which is in charge of the Angels.

Latest contributions of Arab astronomy

Now approaching the end of the medieval period, a renewed interest in mechanical representation of the universe and more accurate measurement of time manifests itself; which materializes in the design and construction of gigantic *astronomical clocks*. Richard of Wallingford (1292 – 1336) was an English mathematician and monk who made important contributions to both astronomical science and the technique of time measurement or horology. He spent six years studying at the University of Oxford before becoming abbot in San Albans about 1327. He is best known for his astronomical clock or *Horologium astronomicum*; which was the most complex clock mechanism in the British Isles at the time: it indicated the hours and minutes of the day, the movements of the Sun and the Moon, as well as the ebb and flow of the tides. Likewise, Wallingford designed and built other devices for astronomical calculations such as the *Rectangulus*, the *Torquetum* and the *Equatorium*.

A little later, *Giovanni Dondi dell'Orologio* (1330 - 1388) an astronomer, mechanical engineer and Italian doctor from Padua; who inherited his interest in astronomy and watchmaking from his father; he became a famous pioneer in the art of design and construction of watches. He developed an ambitious program to describe and model the solar system with sufficient mathematical precision and technological sophistication; towards 1384 he designed and built a complex mixture of astronomical and planetary

clock: the *Astrarium*. The apparatus comprised a seven-sided structure, and a quadrant for the Sun, the Moon and for each of the planets; each planetary system was essentially a set of gears imitating a Ptolemaic mechanism. In this respect it can be affirmed that it was more faithful to the Almagest than to the Planetary Hypothesis.

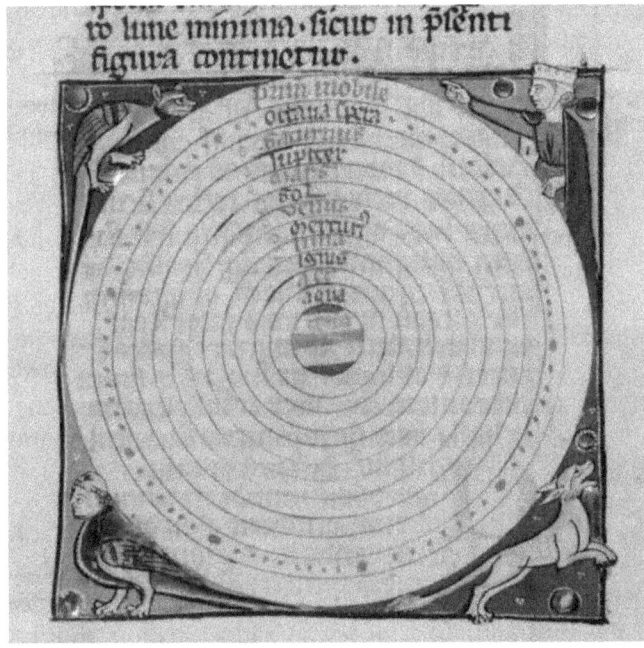

Image 4.6

Medieval outline of the geocentric Universe

Detail of a graph that shows the Solar System as concentric circles: with the Earth in the center and then the water, the air and the fire; then the planets, including the Moon and the Sun; the sphere of fixed stars follows, and finally the *Primum Mobile*.

Ibn al-Shatir (1304 – 1375) was an Arab astronomer who lived and worked in Damascus, where he built a magnificent sundial for the towers of his mosque in 1371. He also wrote an astronomical treatise called *Kitab nihayat al- sul fi tashih al- usul*, or *The Final Quest for Rectification of Principles*; in which he drastically reforms the Ptolemaic models for the Sun, the Moon and the planets, eliminating both the eccentric deferent and the equant point, and better introducing extra epicycles. For the movement of the Sun the additional epicycle obtained no additional advantage to Ptolemy's method. While for the Moon this new configuration appreciably corrected the main defect of the Ptolemaic lunar theory, since it greatly reduced the variation of the lunar distance. Finally, for the movement of the planets the relative sizes of the primary and secondary epicycles were chosen so that the models were mathematically equivalent to those of the Egyptian astronomer. Although he was based mainly on a geocentric system, the mathematical foundations of his cosmology were identical to those that Copernicus established in *On the Revolutionibus* a century and a half later; but historians of science do not agree on whether Copernicus read al- Shatir's work previously or not.

Muayyad al-Dīn al-Urḍī (1.200 – 1.266) was an engineer, architect and astronomer of Damascus. He was the author of three astronomical treatises: *Book of Astronomy,* which deals with the movement of the planets; *Treaty on the construction of the perfect sphere,* and *Treaty on the determination of the distance between the center of the Sun and the apogee.* He was a member of the group of Islamic astronomers who, during the 13th and 14th centuries, actively participated in the criticism of the astronomical model exhibited by Ptolemy in his Almagest, and whose works and activities must have been known in fifteenth-century Europe and, ultimately, would greatly influence the thinking of Copernicus.

During the great period of Islamic astronomy efforts to scientific research were basically absent in Christian Europe, largely because of the lack of stimulations for such efforts of the Christian Church; which stated that truth and spiritual guidance could come only from the Holy Scriptures. Islam, unlike Christianity, greatly stimulated scientific activities, primarily the mathematics and the astronomical, during the first centuries after its establishment. Therefore, it is quite paradoxically the really great scientific revolution which was coming, and that should start and establish the *Modern Western Science* with all its vast implications for the development of mankind, occurred definitely within Christianity.

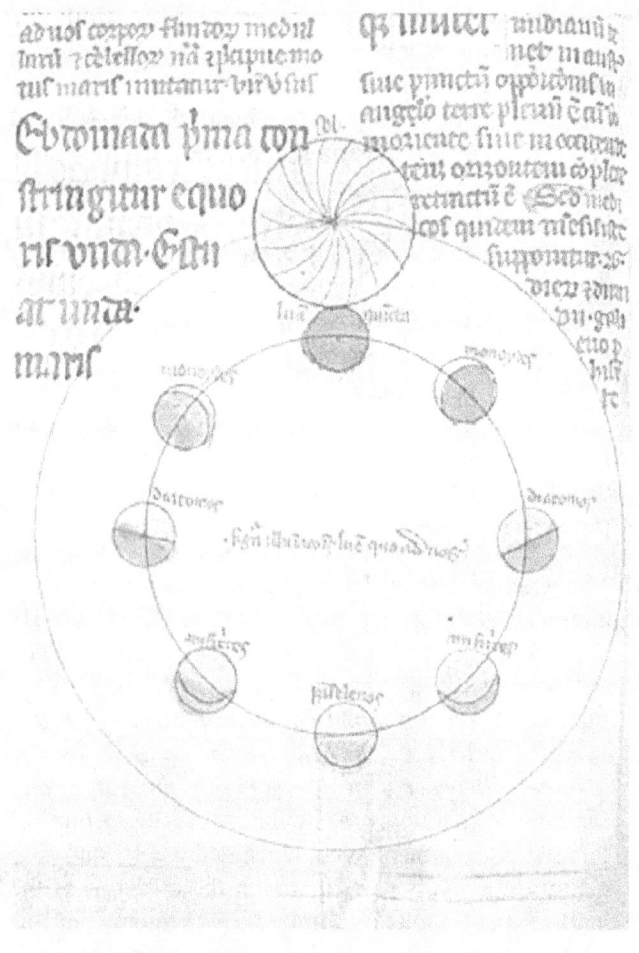

Image 4.7

The lunar phases according to medieval astronomy.

Medieval graphic representation of the Lunar Phases, according to the work *Tractatus de sphaera* by Johannes de Sacrobosco, appeared in the third quarter of the thirteenth century. Preserved in the *British Museum* since August 13, 1840

The Lunar Eclipse of Constantinople

This fourth chapter was opened with the ancient city of Constantinople, and for forced historical reasons we will also close it with that.

We must remember in the year 330 the Roman Emperor *Constantine I The Great* re-founds the ancient Greek city of Byzantium, he gives it the name of *New Rome* and the category of imperial residence. The city was also known as Constantinople, and after the fall of Rome and the final collapse of the Western Roman Empire in 476 AD, it became the continuation of the Roman tradition in the East for another thousand years.

The history of the Turkish Empire or Ottoman Empire goes back to the year 1299 when the Ottoman dynasty headed by its leader Sultan Osman I Gazi, latinized as Osmán I, declared its independence from the Seljuk Turkish dynasty and began its territorial expansionism taking advantage of the weakness of the Seljuk and Byzantine empires to establish and strengthen its own emirate, which historically gives beginning to the Ottoman Empire: With Osman I began the territorial expansion of the Turks forming an empire that would last about seven hundred years.

As time passed, in 1451 the seventh sultan of the Osman dynasty came to power: Mehmed II, also known as el-Fātih or the Conqueror, who would exercise until 1481 and who had as his top priority the military defeat of the Byzantine Empire and the take the glorious Constantinople.

For this purpose, on Thursday April 5, 1453, Mehmed II, at his twenty-one, started a harassment or siege of the city with Ottoman elite troops, the Janissaries, and backed by ordinary infantry attacks and large cannon fire and other light artillery. After forty-seven long days of siege, the Byzantines saw their Moon to put out: on May 22 a partial eclipse of the Moon occurred that lasted three hours and today it is known that such eclipse was effectively visible in the region.[7] Some writers and historians maintain that the Byzantines remembered with terror an ancient prophecy according to which the city would never fall while the Moon, the symbol of ancient Byzantium, was illuminating it. Anyway, and considering the persistent superstition of that time, it is obvious that the phenomenon could have deeply moved the spirits of both the Byzantines and the Turks: each interpreting such a dramatic moment either as a bad omen or an extraordinary warning to come.

The truth of the case is that the siege lasted another seven days until the Turkish troops broke through the strong Byzantine walls, entered and seized the city on Tuesday May 29, 1453. After which Sultan Mehmed transferred the capital of his empire from Edirne to Constantinople and established there his court: an empire died, the Byzantine, and a new one was born, the Turkish or Ottoman Empire.

The fall of the city marked the end of the Byzantine Empire. In addition, it was a turning point in military history, as the walls and fortifications of Constantinople, hitherto impassable, had been a model of defense followed by the cities of the Asian, European and Mediterranean regions. But the Ottomans finally managed to tear them down with the use of gunpowder that powered their mighty cannons. The conquest of the city of Constantinople and the end of the Byzantine Empire was a key event in the late Middle Ages and for a large part of historians marks the end of such Medieval period.

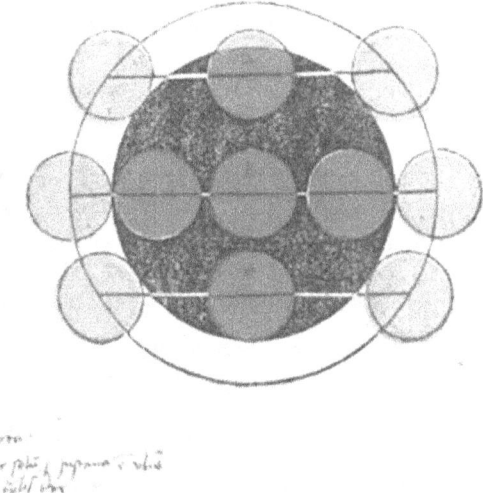 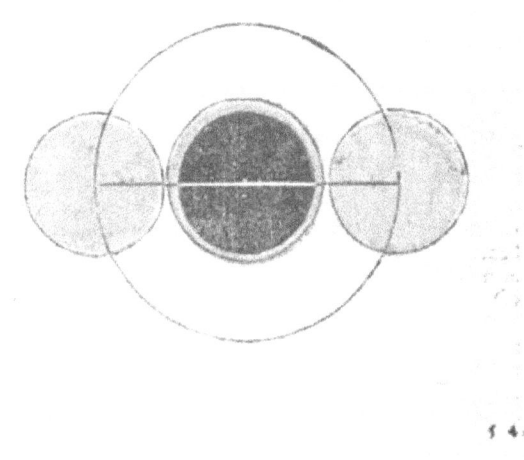

Image 4.8

The eclipses in medieval astronomy.

Schematic representation on the eclipses of Moon and Sun contained in a 1485 edition of *Tractatus de sphaera* by Johannes de Sacrobosco; preserved at the Metropolitan Museum of New York.

Bibliographic Citations

[1] Project Gutenberg. Dante Alighieri. The Divine Comedy. Translation by The Rev. H. F. Cary, M.A http://www.gutenberg.org/files/8800/8800-h/8800-h.htm

[2] Molina M., Antonio I. La geografía bizantina: Cosmas Indicopleustes. Revista Antigüedad y Cristianismo; N° 27 Año 2010, http://www.um.es/cepoat/antig%C3%BCedadycristianismo/?page_id=1231

[3] Abetti, Giorgio. *The history of astronomy*. London: Sidgwick and Jackson. 1954

[4] Muslim Heritage. Abu al- Wafa al- Buzjanî . http://www.muslimheritage.com/article/abu-al-wafa-al-buzjan%C3%AE

[5] The British Library. Catalgue of Illuminated Manuscripts. Ergeton 843 f.25 Diagram. http://www.bl.uk/catalogues/illuminatedmanuscripts/ILLUMIN.ASP?Size=mid&IllID=10144

[6] The British Library. Catalgue of Illuminated Manuscripts. Egerton 844 f. 37v Diagram. http://www.bl.uk/catalogues/illuminatedmanuscripts/ILLUMIN.ASP?Size=mid&IllID=10159

[7] NASA. NASA Eclipse Web Site. https://eclipse.gsfc.nasa.gov/LEhistory/LEhistory.html

Chapter 5
The Moon in Renaissance Astronomy

With the passage of time, the portion of the satellite's illuminated hemisphere that we can appreciate has been growing, finally comes the moment when we can contemplate the whole of such enlightened hemisphere. This Moon phase is known as *Full Moon*, and we can actually see our satellite completely glistening. The Moon as seen from Earth on Wednesday January 31, 2018, at 10: 00 UCT; 14 days 7 hours and 45 minutes after New Moon; now we can contemplate 100.0% of its illuminated hemisphere.

Courtesy of NASA's Scientific Visualization Studio: https://svs.gsfc.nasa.gov/4604

"Now, the reason for all this was their ceremony, because they feared the world would fall asleep when one of its eyes began to blink, and therefore they did what they could with loud sounds to get it out of its drowsiness and keep it awake by means of bright torches to grant it that light that it began to lose. Some of them thought by that means to keep the Moon in its orb, because otherwise she would have fallen on Earth and the world would have lost one of its lights, for credulous people believed that the Enchanters and the Witches could lower the Moon."

The Discovery of a World in the Moon, 1638. [1]
John Wilkins (1614 - 1672) English religious and naturalist.

The European Renaissance

The historical backgrounds of the period normally known as the *European Renaissance* have their roots in the late fourteenth and early fifteenth centuries, and they are related to the generalized decadence of the medieval world, which derived from a deep socio-economic crisis of the feudal system characteristic of that period, the weakening of the Catholic Church because of heretical movements and internal divisions or schisms; as well as also derived from stagnation of arts and knowledge, biased by theological approaches.

The main European academic centers of the mid-fifteenth century reacted positively to this decline, and sought to regenerate themselves through the return to the intellectual values of Greco-Roman classical culture. Additionally, the rediscovery of ancient texts was accelerated because much Byzantine scholars who had to seek refuge in Western Europe, especially in Italy, due to the fall of Constantinople, the seat of the Eastern Roman Empire, into hands of the Ottoman Turks in May of 1453. With which this century witnessed the beginning of the Renaissance socio-cultural movement and the emergence of a more anthropocentric world view, detached from medieval religion and theology; vision in which man and its cognitive advances established a new way of valuing the world: *Humanism*.

Renaissance is the name given to a vast cultural movement of transition between the Middle Ages and the beginnings of the Modern Age, which developed in Western Europe in the course of the fifteenth and sixteenth centuries. Its main manifestations were basically in the fields of arts and sciences, both natural and human. This process was the result of the diffusion of the set of ideas that today is known as *Humanism* and that determined a new conception of both man and the universe. Geographically, the Italian city of Florence was the birthplace and development site of this movement, which later spread throughout Europe. The term *Renaissance* is applied to refer to the traditional values of Greco-Roman classical culture, and means a bring back, a return, to these values; as opposed to a more rigid, theological and dogmatic type of mentality established within the general framework of the medieval world. In this great period, a new way of understanding the Universe as a whole as well as the human being in its complexity was developed; and new approaches were established in the fields of arts, philosophy, science, economics and politics; with the primordial characteristic that medieval theocentrism was replaced by modern anthropocentrism.

In ancient Greece there was a transition from the primitive mythological thought to the classical philosophical reasoning; during the Middle Ages the theological points of view and conceptions were developed and strengthened; finally, during European Renaissance there was a transition from the theological worldview to the rational and scientific way of thinking. The new philosophical approach of these times was the *Humanism*, which influenced all areas of knowledge, seeking to fully describe the position of the human being in the world. Humanist thought was, above all, a moral and literary movement that promulgated the value and importance of human beings in the universe as a whole, in stark contrast to medieval philosophy, which was grounded in theology and always placed God at the center of the world. *Humanism* is far more interested in man and nature rather than in the divine and spiritual matters; and is based on a marked opposition to medieval culture and on the return to classical Greek antiquity, which has Plato and Aristotle as its maximum exponents.

Thus, Renaissance culture was a special return to the influence of classical Greco-Roman philosophy, with its rationalism and its study of nature through empirical research. The three fields of knowledge that had the most attention and development were natural philosophy, humanism and political philosophy.

In a world already directed towards modernity, Renaissance philosophy was characterized by its marked distance from theology, since this distracted it from the rational sphere and circumscribed it exclusively to the spiritual and doctrinal spheres; although it didn't completely

renounce to religion. The new way of undertaking universal themes was rationalism: reason as a tool for the study of man, society and nature. Thus, the natural philosophy of the Renaissance ended the medieval supernaturalist conception of the world in terms of divine purposes and ordinances, and instead developed a thought grounded in terms of physical causes, mechanisms, and forces.

During this period of transition science presented a great boom due to the new anthropocentric vision of humanism, and it was favored by the advent of a new generation of learned philosophers as consecrated and incisive as the ancient Greeks, by the invention of the printing press and by the great voyages of geographical discovery that took place in this era. Factors all these that favored and led to the development of what we know today as the *Scientific Revolution* which developed between the 16th and 17th centuries. The fundamental feature of science, the *Scientific method*, is based on empiricism and begins by a hypothesis of rational, theoretical and mathematical origin to arrive by a process, or method, to the verification or subsequent demonstration of this preliminary hypothesis. The English philosopher and writer Francis Bacon (1561 - 1626), promoter of philosophical and scientific empiricism, in his work *Novum organum* presents science as a rational, inductive and experimental process, capable of giving human being knowledge and control over nature.

The scientific method strongly establishes that all scientific propositions, hypotheses and theories must be always subjected to a first probe consisting of explaining all the available information on the matter, and additionally, to a test of prediction of some derived effect or new phenomenon that can be investigated and measured. That being the case, the scientific method is completely opposed to blind belief or dogma.[2]

The scientific discipline that most developed during the Renaissance was astronomy: the astronomers Nicolaus Copernicus, Tycho Brahe, Johannes Kepler, Galileo Galilei and the physicist Isaac Newton were first order characters in this scientific revolution; and Bacon provided the philosophical foundations to justify and support the scientific method that would characterize it.

Coincidentally, during this period of mentality change began to emerge a new social class, the bourgeoisie, which established the foundations of capitalism with a pre-industrial and mercantile economy; which finally led to an accelerated growth of trade between the Mediterranean and European nations; and additionally to the exploration of new commercial routes to the east and west, which would eventually lead to the discovery of America by Europeans.

The first Lunar eclipse documented on American lands[3]

At a very early moment in the Renaissance period we found a total lunar eclipse historically very important, as it is practically the first of which there is historical documentation that it was seen in the newly discovered American lands: the lunar eclipse of Christopher Columbus.

After the discovery of America on October 12, 1492, Christopher Columbus made three other exploratory trips of American lands in the following ten years. After the preparations in Seville, the fourth and last voyage of discovery left Cádiz on May 9, 1502, and reached the port of Santo Domingo on the Caribbean island of Hispaniola, now the Dominican Republic, on June 29.

During the trip he discovered the Caribbean coasts of the current countries of Honduras, Nicaragua, Costa Rica and Panama; as well as another series of small islands. Due to such long maritime voyages in the midst of a climate with inclement storms and hurricanes, and after suffering serious damage to their ships, Columbus and his men finally shipwrecked on the island coast of Jamaica on June 25, 1503. The expeditionaries found themselves with an island not yet colonized; then they set up camps with the hulls of the ships and tried a positive coexistence with the natives of the island, who initially received them well and offered them lodging and food. There they would have to stay more than a year until the arrival of a rescue boat.

Their livelihood depended to a large extent on the cassava and the rodent meat provided by the natives, with whom they had a complicated relationship that soon turned into serious disagreements. After 6 months of stay, and after many disagreements between the sailors and the natives, these refused to continue providing their services. Given the refusal of the Indians to continue helping them, the Admiral decided to use their knowledge in astronomy and the superstition of the Indians for his own benefit.

When consulting the book of the German mathematician and astronomer Johann Müller Regiomontanus that brought with him, Columbus found that on February 29, 1504, a total lunar eclipse would occur that would be visible in this region. Thus, three days before the eclipse, the Admiral requested to meet with the Cacique, the indigenous leader, to intimidate him and pressure him by saying that the Christian God was very angry with them and that they would suffer the consequences for refusing to help the Europeans; and for showing his anger, in three days the flames of his wrath would make the Moon disappear from heaven.

Later, at the appointed time, the lunar eclipse and the reddish moon took place as predicted by Columbus, and the indigenous people were supremely impressed and frightened when they saw the wrath of the Europeans' God materialized. Ferdinand, the Admiral's son, would write later that: the Indians with great howls and laments came running from all directions towards the castaways loaded with provisions, and begging the Admiral to would intercede for them and by all means with his God for returning the Moon to its original radiance and so that its wrath will not fall on them.

Then the admiral locked himself in his cabin for about fifty minutes, supposedly to talk to his God. But in reality he devoted himself to recording the phases of the eclipse using his hourglass, and just before the phenomenon came to an end, he announced to the Indians that the Mighty One agreed not to punish them and give them back his Moon. Thus, the grateful natives continued feeding and helping them until their departure for the Caribbean on June 29, 1504. Knowledge had won the battle to superstition.

Image 5.1

Lunar eclipse of Christopher Columbus, 1504

Image included in the work *Astronomy Populaire*, written in 1879 by the French astronomer Nicolas Camille Flammarion (1842 -1925).

Actually the natives were not so scared by the lunar eclipse itself, it is certain that they had seen enough. What should have surprised them, and too much, was the fact that it happened just when the god of Columbus, apparently, arranged it. They must have thought that it was a product of divine power, or of magic of the white man. But it was only science and some knowledge.

Today it is known that indeed a total lunar eclipse was visible on the night of February 29 in the American territories: the phenomenon was visible after sunset from most of North America and all of South America. As also it was visible on the morning of March 1 through Europe, Africa and Western Asia.

Finally, on June 29 1504, two years after their departure, the shipwrecked expeditionaries are rescued from Jamaica on a ship sent by Diego Méndez and taken to Santo Domingo in the Hispaniola Island, where they arrive on August 13; they leave there on September 11, 1504, and arrive in Sanlúcar de Barrameda, Spain, on November 7.

Nicolaus Copernicus and the Heliocentric Theory

Everything explained so far in this text has been made within the general framework of a geocentric universe established by Aristotle and perpetuated by Ptolemy. The philosophical weight provided by Aristotle and the fact that this model was adopted, imposed and defended historically by the Catholic Church, mainly because it agreed well with the Holy Scriptures in the sense that both man and Earth occupied the center of the divine creation, contributed to the fact that for a long period of time the Aristotelian- Ptolemaic model of a *Geocentric Universe* would continue in effect. More specifically until the sixteenth century: even around 1502, when Christopher Columbus made his last discovery trip to the American continent, the vast majority of people thought, believed, that the Earth was the center of the universe. The effective discrediting and the total abandonment of the Aristotelian- Ptolemaic conception of the geocentric universe, after being in force for almost two millennia, began to take shape during the sixteenth century, in plenty Renaissance, with the entry on scene of the astronomer Copernicus and his rethinking of *Heliocentric theory*.

But this drastic change in cosmology basically did not affect the way of perceiving and understanding both the Moon and its eclipses, since the intrinsic nature and the primordial cause of these phenomena remains the same: the three stars align and one of them, the Moon or the Earth, blocks the passage of the solar rays preventing them from reaching the other, indistinctly from that the Sun is located in the center of the Universe or not. From the point of view of eclipses, the positions and relative movements of the three bodies are what matters.

Of Polish origin, the scientist Nicolaus Copernicus (1473 – 1543) made his first studies at the University of Krakow and in 1.501 he was appointed canon in the Cathedral of Frombork in Poland. Later, he traveled to Italy where he studied mathematics, philosophy, medicine, astronomy and canon law at the Universities of Bologna and Padua. In 1523 he returned definitively to his country and held some public offices, such as the administration of the Diocese of Warmia; there he also practiced medicine and made his invaluable contribution in the field of Astronomy. He died on May 24, 1543, in Frombork, Poland.

His fundamental contribution to modern astronomy was to have retaken, improved and put back into circulation in Europe the ancient Greek theories of the Heliocentric Universe. His famous book *De Revolutionibus Orbium Coelestium*, or *On the revolutions of the celestial spheres*, in which he established his system of circular planetary movements around the Sun, was published the same year of his death in 1543, and with it he became creditor of the title of father or founder of modern astronomy. In the absence of evidence, many historians think that Copernicus was reluctant to publish his book earlier essentially because of two fears: being judged as heretic by the Church and being

harshly criticized by the scientists on turn.[4] When Copernicus was fully convinced of his heliocentric theories as to publish and disseminate them, he had to confront three authorities of great weight in his time: the Christian Church, the Aristotelian orthodoxy of the universities and the astronomers themselves of the time, entities that were still working within a widespread Ptolemaic geocentric tradition.

The *Commentariolus*

Previously around 1514, Copernicus wrote a small treatise called *Nicolai Copernici of hypothesibus motuum coelestium a constitutis commentariolus*, or *Brief exposition of Nicolaus Copernicus' hypotheses about celestial movements*; text fortunately better known with the reduced name of *Commentariolus* and that was never printed during his life, although the manuscript became known among his friends and contemporaries. There he enunciated the three fundamental presuppositions of his *Heliocentric Cosmology*: All celestial movements occur around the Sun; the Earth is one more of the planets that revolve around the King star, and the universe is limited by the sphere of the Fixed stars. Likewise, he enunciated the basic postulates of his *heliocentric cosmological system*: the stars revolve around the Sun and this one is very close to the center of the universe, although not all celestial bodies revolve around a single point; the Earth moves in a sphere around the Sun, which causes the apparent annual movement of this star, and additionally our planet has more than one movement. He also exposed that the stars are immobile because their apparent daily movement is actually due to the daily rotation of the Earth; that the distance between our planet and the Sun is an insignificant fraction of the distance of these bodies to the fixed stars, and therefore no parallax is observed in such fixed stars. That the orbital motion of the Earth around the Sun causes the seemingly retrograde movement of the planets; and finally, that the center of Earth is the center of the lunar sphere, or the orbit of the Moon around our planet.[5]

In this set of postulates we find, for the first time in history, a recorded statement by a philosopher that the Moon orbits around a planet that is not the center of the universe but it revolves around the Sun like the other stars, what constitutes the first formal definition of what a *natural satellite* is.

Considering that a heliocentric theory of the universe was never definitively established in classical antiquity, it is understandable that no ancient philosopher, except perhaps Aristarchus, had managed to conceive this characteristic of the Moon; very much in spite the fact that they knew, as Plutarch expressed, their considerable closeness to Earth in comparison with the other stars. This is why Copernicus is the first scholar duly documented to give our lunar globe its true position within the Universe: a satellite of an Earth that orbits around the Sun.

It was in the *Commentariolus* that Copernicus established the basic hypotheses for his Heliocentric Universal Model, and there he made it clear that he decidedly embraced the principle of uniform circular motion and ruled out the use of eccentric deferents and the equant point. Also it is obvious that he understood that his model agreed well with the observed celestial phenomena and at the same time it had a greater intellectual appeal than that of Ptolemy. In general terms, his models for the movements of the stars follow the pattern established a century and a half earlier by the Arab astronomer Ibn al- Shatir, who avoided the use of both the eccentric deferent and the equant point; but they are geometrically transformed to bring the center of the model of each star to a common point, which is not exactly the Sun but the center of Earth's orbit. The primary objectives of the *Commentariolus* were to make a general description of a heliocentric Universe with a static Sun that is settled very approximately at the center; as well as to make it very clear that the daily movements that we observe of the Moon, the Sun and the planets are effects, optical illusions caused by the translational movement of the Earth around the Sun, that of rotation on its own axis and also the precession of this same axis; and finally to stablish that the movement of the stars isn't

103

irregular, but that they move with uniformity around the Sun following perfect circles.

In this work the Polish astronomer limited to enunciate and exposing his *Heliocentric System* without having the purpose, or worrying, of carrying out the respective geometric or mathematical demonstrations. The parameters on which his models were based are very similar to those of the Almagest because they were re-derived from the Ptolemaic observations; and they gave reasonable results for the positions of the Sun, the Moon and the planets, although they did not have the necessary precision for the acceptable calculation of the respective conjunctions of these stars nor for the eclipses. Therefore, the canon astronomer devoted himself to work for collecting the set of observations he would need to make his cosmological models more precise and more acceptable than those they were trying to replace. He made such observations between the years 1512 to 1529, and they included positions of the Moon, the Sun, the equinoxes, lunar and solar eclipses; the determination of the conjunctions and oppositions of the planets, and of altitudes or distances to the zenith for several stars. Considering all this material, it is easy to see that we are facing one of the few astronomers who decided to build a cosmological model based on observations from others as well as his own and well-grounded in theoretical principles.

On the revolutions of the celestial bodies

The results and final conclusions of his astronomical researches he presented in his greatest work: *De Revolutionibus Orbium Coelestium*, or *On the revolutions of the celestial bodies*; which was finally printed in 1543, but the only copy he saw was presented to him on his deathbed and he never opened it. The text is divided into six books: The first one presents a general view of its cosmological system and includes two chapters on geometry and trigonometry. Although it is not very advanced, the second text deals with the theory of spherical astronomy. The third deals with the movements of the Earth, that of translation around the King star, of rotation on itself and with that of the precession of its axis. In the fourth book Copernicus presents his lunar astronomy and covers the respective movements of the satellite. In the fifth he deals with the position of the planets according to their celestial length, while in the last he deals with their latitudes.

In his work *De Revolutionibus* Nicolaus Copernicus takes up and exposes the heliocentric ideas initially proposed by Aristarchus of Samos, and establishes a system in which the Sun is motionless very close to the center of the universe, and around it the planets circle in circular orbits. In this work the Polish astronomer, like his Greek colleagues, is influenced by philosophical and aesthetic reasons, and axiomatically assumes that all celestial movements should be circular and uniform. Whatever finally forced Copernicus to also include epicycles in his system in order to adjust his theory with the observations he made. Although in the *Commentariolus* Copernicus avoids using the eccentric deferent to represent the movement of the stars around the Sun, in *On the revolutions* he is forced to use them, both to be able to demonstrate his models as for making them coincide with the observations; still he resorts to Epicycles to elaborate his system of the world, although the centers of the deferents are not exactly in the Sun, but rather in any point close to it and that corresponds to the center of the Earth's orbit: The Earth is a planet that orbits circularly around the Sun, but the center of its orbit does not coincide with the center of the Sun, but it is the center of the other planetary orbits. The latter slightly tarnished the simplicity of his cosmology. Even so, what is really important in his work was having dethroned of the Earth from its preeminent geometric and cosmological position and having placing the King star in that place.

The only big difference from the *Copernican heliocentric model* compared to the *geocentric Ptolemaic model* was to place the Sun at the center of the Universe instead of the Earth; in all the other concepts Copernicus was faithful to Aristotle, with the celestial bodies, including the Earth, being spherical and moving in concentric circles and at constant speed. But, like Ptolemy,

The Polish astronomer discovered that his theories did not fit well with the observations, and he had to make small corrections and chose to be the same ones developed by the Alexandrian astronomer almost a millennium and a half before, but accommodated to his heliocentric system. To adjust satisfactorily to the observations, Copernicus had to use a smaller number of little epicycles and additionally to place the Sun at a point slightly displaced from the geometric centers of the planetary orbits. His improved model with epicycles showed that the assumption of the annual movement of the Earth around the Sun would explain in a very simple way the most notorious irregularities in the movements of the planets: well immediately explained the biggest problem facing a geocentric system, that is, the retrograde movement of the planets.

Additionally, the Copernican system made the specific prediction of the stellar parallax, or the apparent change in the position of a star when the Earth moves from one end of its orbit to the other; whereas the Ptolemaic Aristotelian geocentric model, with a static Earth, could not predict any, and therefore it could not detect anything. Now we recognize that this fact is due to the immense distance to the closest stars, which makes its parallax so thin and difficult to measure that it was not appreciated until three centuries later, when the telescopic techniques developed sufficiently.

The planetary movement viewed by Copernicus

The movement of the planets is analyzed in the fifth and sixth books of *De Revolutionibus*; the fifth book includes the best of his work and deals with the celestial positions of the higher planets. Copernicus placed an immobile Sun in the center of the sphere of the fixed stars, which is also stationary; and the planets revolve around the King star moving in circles at constant speeds on the same plane, today known as the ecliptic's plane. In the development of his planetary theories, the Polish astronomer had advantages over the Egyptian: for each planet the model is related not to the true position of the Sun, but to the center of Earth's orbit, so he had only to consider the first inequality for explaining the revolution of the planet with respect to the stars.

After some preliminary considerations, Copernicus proceeded in the tenth chapter to fix the order of the planetary orbits, because even up to his time the respective order of the planets had never been established conclusively. Until now there was perfect unanimity regarding the Moon, which is the closest body to Earth and makes its revolution in the shortest time of 29,53 days; whereas Saturn, having the longest period of 29,45 years, is the most distant; following Jupiter with almost twelve years and Mars with about two years. But the case was different with

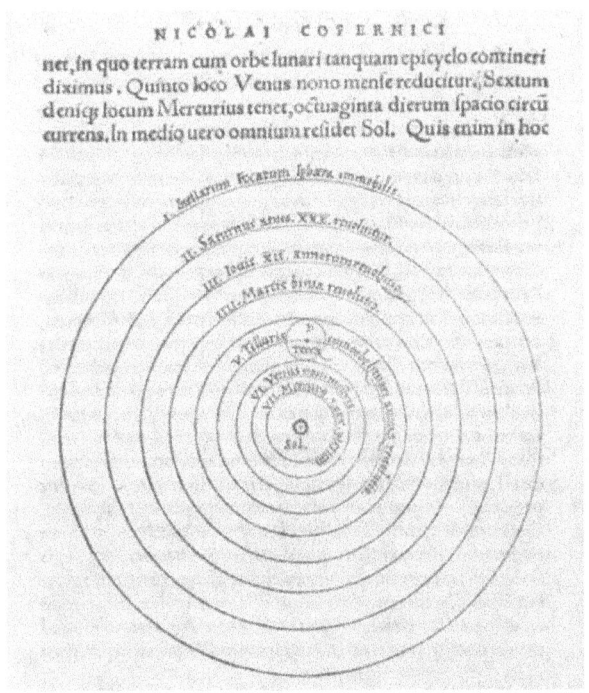

Image 5.2
The Copernican heliocentric system

Detail of the first 1543 edition of the work *De revolutionibus orbium coelestium*, by Nicolaus Copernicus. It is a geometric representation of the complete Solar system as conceived by the Polish astronomer. It shows the Sun in the center and the planets, including Earth, orbiting it; but with the Moon orbiting around the Earth.

Mercury and Venus which always were very problematic. The Pythagoreans had placed these two bodies above the Sun, Ptolemy and most of the later astronomers placed them below; while some Arab astronomers put Venus on top and Mercury below. For these two stars it was easy to move from the Ptolemaic to the Copernican theory, since the Sun itself could become the center of the respective epicycles.

In the Copernican planetary model, Mercury and Venus each one revolving around the Sun with its own distance and speed, will make an apparent movement similar to that explained by the epicyclic and deferent mechanism of Ptolemy. The canon astronomer evaluates the synodic and sidereal periods of these two planets, and also the relative sizes of their orbits compared to those of the Earth, and finds values very similar to the modern ones.

Arguing over the positions of the planets Venus and Mercury, Copernicus declares: *"Therefore, the Sun is the center of its orbits, and the orbit of Mercury is enclosed within that of Venus which is more than the double. If we take advantage of this to refer to Saturn, Jupiter and Mars to the same center, taking into account the great extension of their orbits that enclose these two planets, as well as the Earth, we will not fail to find the true order of their movements, since it is true that these are the closest to the Earth when they are in opposition to the Sun, being the Earth between them and the Sun, but they are further away from us when the Sun is between them and the Earth, which proves that their center belongs rather the Sun and it is the same as that of the paths in which Venus and Mercury move."* [6]

Given the hypothesis that they revolve around the Sun, the explanation of the movements for the higher planets Mars, Jupiter and Saturn becomes difficult: not always the center of the deferent is in the position of the Sun, but it can be anywhere between that and the earthly orbit's center. In any case it can be shown that the movement of a superior planet is very similar to that of a lower planet, provided that the radius of its epicycle is greater than that of its deferent. From the observation of the oppositions of the higher planets, the canon astronomer obtained very precise values for his synodic and sidereal periods.

The general order of the celestial bodies within the Solar System is summarized by Copernicus in the following way: *"Consequently, with the first principle intact, since no one will propose a more adequate principle than the size of the spheres is measured by the duration of the period of revolution, the order of the spheres beginning with the most distant one is the following. The first and the highest of all is the sphere of Fixed stars which contains itself and everything, and therefore is immovable. It is undoubtedly the place of the universe to which the movement and position of all other celestial bodies are compared. Some people think that it also changes in some way. A different explanation of why this seems to be so will be adduced in my discussion about the Earth's motion. It is followed by the first of the planets, Saturn, which completes its circuit in 30 years. After Saturn, Jupiter achieves its revolution in 12 years. Then Mars rotates in 2 years. The annual revolution takes the series' fourth place, which contains the Earth, as I said, together with the lunar sphere as an epicycle. In the fifth place Venus revolves in 9 months. Finally, the sixth place is occupied by Mercury which rotates in a period of 80 days. At rest, however, in the middle of everything is the Sun. For in this beautiful temple, ¿Who would place this lamp in a better position than the one from which it can illuminate everything at the same time? For, the Sun is not invoked inappropriately by some people as the lantern of the universe, its mind by others and its ruler by still others. Hermes Trismegistus labels it as a visible god, and Sophocles' Electra, the all-seeing. So, in fact, as sitting on a royal throne, the Sun governs the family of planets that revolve around it. In addition, the Earth is not deprived of the Moon's assistance. On the contrary, as Aristotle says in a work about animals, the Moon has the closest kinship with Earth. Meanwhile, the Earth has sexual relations with the Sun and is fecundated for its yearly parturition."* [7]

The discussion about the stationary positions of the planets is quite elaborate. Copernicus concluded that such positions have to exist, and showed how they can be calculated accurately; which evidences how much more effective is the Copernican system compared to the Ptolemaic: considering the movement of the Earth and that of the outer planets with their respective speeds around the Sun, it is easily seen that during a certain period of time, and immediately after of the respective stations, there must be an investment in the direction of planetary movements. This is the *Apparent retrograde movement*: the planet will appear to temporarily reverse the direction of its movement through the sky. This movement consists of an optical illusion, a mirage caused by the difference in the speeds of translation between the Earth and the respective planets. According to this phenomenon, the planets seem to slow down, stop, change its direction and move so for a short period of time, to stop again and change to the original direction, and then continue their orbital courses.

Copernicus finally presented a clear and specific heliocentric system in which the order of the planets around the Sun was: Mercury, Venus, Earth with the Moon, Mars, Jupiter and Saturn; and at the bottom of everything, the set of Fixed stars. And with this order, he also explained the relative sizes of the Arcs of the retrograde route for the outer planets, which decrease from Mars to Saturn that is the smallest; he also argued why the exterior planets are much brighter in opposition. In Table5.1 the average distances of the planets from the Sun relative to the diameter of the Earth are listed, which were derived from the work of Copernicus.

With respect to the distance from the Earth to the Sun, Copernicus had to adopt the value of the solar parallax given by Hipparchus but only making it a small correction due to the values of the apparent diameters of the Sun and the Moon adopted by him. The average solar parallax used by Copernicus was 3' 1", and the average distance of 571 terrestrial diameters.

The Earth according to Copernicus

In his work, Copernicus makes the statement of whether the Earth is at the center of the world or if it is another planet revolving around the Sun. That it is not the center of all celestial movements is shown by the seemingly irregular movements of the other planets, and for its different distances to Earth. In this regard, Copernicus decidedly argues that the same physical conditions are developed both in the celestial bodies and on Earth: *"For my part, I believe that gravity is not more than a certain natural need, that the Divine providence of the Creator of all things has implanted in the parts, so that they come together as a unit and a whole when combined in the form of a globe. This impulse is present, we can suppose, also in the Sun, the Moon and other bright planets, so that through their operation they remain in that spherical form that they exhibit. However, they revolve around their circuits in different ways. If, then, the Earth also moves in other ways, for example around a center, its additional movements must be equally reflected in many bodies outside of it. Among these movements we find the annual revolution; for if this is transformed from a solar movement to a terrestrial movement, with the Sun recognized as at rest, the rises and sunsets that bring the zodiacal signs and the fixed stars to sight every morning and every evening will appear in the same way."*[8] In this last sentence the astronomer is arguing that the observed movement of the stars, their ascents in the morning in the East and their sunset in the afternoons in the West, would be well explained both by the geocentric Ptolemaic model, as for his heliocentric model. All these ideas are enunciated to establish an analogy between the nature and the movements of the Earth and the planets, and thus be able to show that it is reasonable to suppose that the Earth is endowed with orbital movement, like the other celestial bodies, and then to be able to remove it from de world's center.

The Polish astronomer argues the position of the Earth and its satellite within the whole Solar System in the following terms: *"Therefore, the Sun is the center of its orbits, and the orbit of Mercury is enclosed within the Venus, which is more than double. If we take advantage of this to refer to Saturn, Jupiter and Mars to the same center, taking into account the great extension of their orbits that enclose these two planets, as well as the Earth, we will not fail to find the true order of their movements, since it is true that these are the closest to the Earth when they are in opposition to the Sun, being the Earth between them and the Sun, but they are further away from us when the Sun is between them and the Earth, which proves that their center belongs rather the Sun and it is the same as that of the paths in which Venus and Mercury move. Then it is necessary that the space left between the orbits of Venus and Mars be occupied by the Earth and its companion, the Moon, and everything that is under the Moon. For we cannot, in any way, separate the Moon from the Earth, to which is undoubtedly closer, particularly because there is a lot of room for it in that space."*[9]

With respect to the Earth's orbit around the Sun, Copernicus did not have much to add to the simple eccentric circle that Ptolemy had used to represent the movement of the Sun: *"Therefore, we are not ashamed to maintain that everything below of the Moon, with the center of the Earth, describes among the other planets a large orbit around the Sun that is the center of the world; and that what appears to be a movement of the Sun is indeed a movement of the Earth; but that the size of the world is so great, that the distance of the Earth from the Sun, though appreciable in comparison with the orbits of the other planets, is nothing compared to the sphere of the fixed stars."*[10] However, the Polish astronomer finally introduced a complication in his model; by making the center of the Earth's orbit displace in relation to the true position of the King star.

Copernicus saw that the daily revolution of celestial bodies could be explained very well by resorting to both the rotation of the Earth itself from West to East, as well as the rotation of the entire universe from East to West. But having to choose between the simplicity of the first option, which Ptolemy had rejected as contrary to common sense, and the obviousness of the latter, which the Egyptian astronomer had deliberately adopted despite the number of celestial movements involved, Copernicus argued for the first and went to work to show that it was the real one, to demolish the objections to it and, ultimately, to make it prevail. Then, once he argued and admitted that the Earth moved in any way, it became more understandable that it had other types of movement: by a tenacious and diligent application, Copernicus developed his cosmological system by which all planets, including Earth, revolve around the Sun; and additionally our planet have a second earthly movement: that of rotation around its axis.

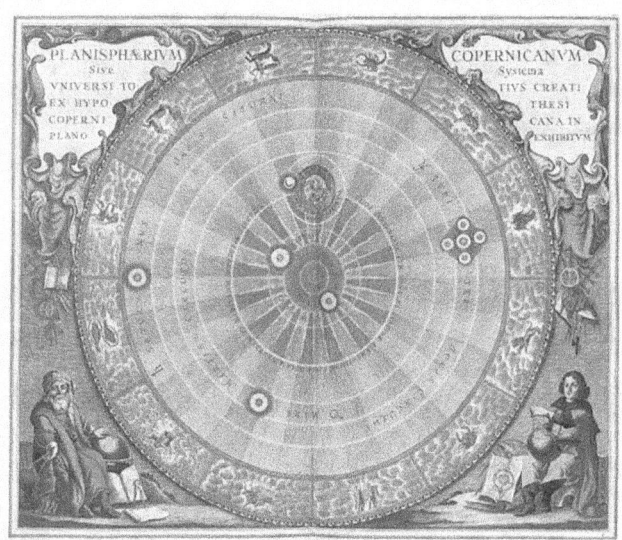

Image 5.3 Copernican Planisphere

Taken from the star atlas *Harmonia Macrocosmica* by the German cartographer Andreas Cellarius (1596 - 1665), published in 1660.
It shows the Solar System according to the Copernican heliocentric theory with the planets known until then.

By making the Earth give an annual revolution to the Sun, which would account for the discrepancies observed in the movements of the other stars, and, additionally, with a terrestrial rotation axis inclined with respect to the plane of its orbit that gave a simple but effective explanation of the annual change in climate seasons, Copernicus had laid the foundations of a much simpler system than the Ptolemaic. But unfortunately he was forced to spoil the simplicity of his model, because in his time a pure heliocentric system was not enough to explain the variable speeds of the planets, the first inequalities. There was no salvation for that and finally he was forced to make use of eccentrics as well as epicycles.

But for Copernicus it wasn't enough to endow the Earth with a double movement, the annual trajectory around the Sun and the rotation in 24 hours on itself; he still had to explain the fact that, despite the annual movement, the earthly axis always points to the same place on the celestial sphere.

Therefore he had to assume a third movement of the Earth, that of *Declination*; and according to which the terrestrial axis annually describes the surface of a cone, moving in the opposite direction to that of the planet's rotation, that is, from East to West; with which such axis would always continue pointing to the same direction in space and time. In this aspect Copernicus was deceived by the distance from the point to which the earthly axis is directed, which is practically infinite so the terrestrial axis does not need any additional movement to continue signaling it as it would if the point were close.

Copernican Lunar Astronomy

The theory of the Moon is studied in the fourth book and has the fundamental purpose of reducing the disagreement between theory and observation; for which Copernicus elaborates a new system that became much simpler, but also much more effective than the previous developments. Here the Polish astronomer reviews the theories for the movements of the Moon proposed by the ancient astronomers and argues the insufficiency of their hypotheses: *"This combination of circles was assumed by our predecessors to be in agreement with the lunar phenomena. But if we analyze the situation with more care, we will find that this hypothesis is not convenient or enough adequate, as we can demonstrate by reason and by the senses. Because while our predecessors declare that the movement of the center of the epicycle is uniform around the center of the Earth, they must also admit that it is not uniform in its own eccentric (which it describes)."*[11]

In consequence, Copernicus proceeds to enunciate his own theory about the revolutions of the Moon and its particular movements; and proceeds to the demonstration of both the *First irregularity* of the Moon's movement, which occurs in its phases of New Moon and Full Moon; as of the *Second irregularity* of the lunar trajectory. Equally, he proceeds to establish the methods of calculation for the course of the Moon, which were made based on data on lunar

Table 5.1

Distances of the planets to the Sun according to the Earth's diameter D_T, according to the work of Copernicus.

Planet	Copernicus	Modern value
Mercury	214,8 D_T	4545 D_T
Venus	410,7 D_T	8493 D_T
Earth	571,0 D_T	11742 D_T
Mars	867,8 D_T	17891 D_T
Jupiter	2980,0 D_T	61091 D_T
Saturn	5238,0 D_T	112006 D_T

eclipses. In his system the Moon revolves around our planet and in the same ecliptic plane.

The *Copernican lunar model* had already been briefly exposed in his Commentariolus, and it is very similar to that of Ibn al-Shátir. In its development, Copernicus appealed to parameters that fit to the data and to the known ephemeris, but it is evident that it was based on a brilliant work of Ptolemy: his determination of the second lunar inequality. The purpose of the Copernican Lunar Model was to correct the most notable defect of the Egyptian astronomer's model: the great variation in the distance of the Moon; for which he showed that it is basically a mirage, an illusion, as proven by rigorous observations. Copernicus succeeds in representing the main irregularities of the lunar movement by means of a special disposition of two epicycles: in his model, the Moon also moves according to a pair of epicycles that have different size proportions and speeds of movement.

The movement of the Moon was explained by Copernicus using constructions much simpler than those of the Egyptian astronomer. He explained the First inequality, or *Equation of the center*, by means of an epicycle. But for the Second inequality he rejects the eccentricity of the deferent and uses rather a second epicycle: in his model the deferent is concentric with the Earth, and in its circumference the center of the first epicycle moves from West to East with the average sidereal movement of the satellite. The center of the second epicycle moves in the circular path of the first with the average anomalous lunar movement, but in the opposite direction. Finally, the Moon moves in the second epicycle also from West to East, twice in each lunation, which would account for the observed lunar phases; as shown in Figure 5.4. The Polish astronomer retained the ancient value for the sum of the two lunar inequalities: 7° 40', and therefore he determined that the ratio between the epicyclic sizes was 4,63.

Regarding the position of the Moon in the Solar System and its relationship with the Earth, Copernicus writes: *"And then they thought that the Moon was crossing its path in the shortest period of time because, being very close to Earth, it rotated in the smallest circle; but Saturn, which completes the longest circuit in the longest period of time, is the most distant. Under Saturn, Jupiter is. After Jupiter, Mars is. But since they all have a common center, it is necessary that the space between the convex orbit of Venus and the concave orbit of Mars should be seen as an orbit or sphere homocentric with them with respect to both surfaces, and that it should receive the Earth and its satellite the Moon, and whatever is encircled by the lunar globe. Because in no manner we can separate the Moon from the Earth, since she is unquestionably very close to Earth, especially since we find in this space a place for the Moon that is sufficiently proper and sufficiently large. Therefore, we are not ashamed to maintain that this totality encompassed by the Moon and the center of the Earth, also crosses that great orbital circle among the other wandering stars in an annual revolution around the Sun; and that the center of the world is around the Sun. The fourth place in order is occupied by the annual revolution in which we said that the Earth, together with the orbital circle of the Moon as an epicycle, is contained. In the fifth place, Venus, this circulates in nine months. The sixth and last place is occupied by Mercury, which completes his revolution in a period of eighty days. Furthermore, the Earth is not in any way deceived by the services of the Moon, but, as Aristotle says in De Animalibus, the Earth has the closest connection (cognation) with the Moon. The Earth, moreover, is fertilized by the Sun and conceives spring every year."*[12]

Copernicus' exposition of the parallaxes, distances, and apparent diameters of the Sun and the Moon leads to much better conclusions than those of Ptolemy. When it deals with the distance of the Moon from Earth, he explains the determination of the lunar parallaxes, and the construction of instruments for said measurements. The largest distance of the Moon that he found was between 227 and 876 Earth diameters, both occurred in quadrature. The apparent diameter of the Moon therefore varies between 28'45'' and 37'34'', a great improvement over Ptolemy's theory, according

to which the apparent diameter should be almost one degree in the perigee. He addresses the issue of the apparent diameter of the Moon and its relationship with the parallaxes: *"From this, it will now become evident how great is the distance from the Moon to the Earth; and without this distance a safe relationship cannot be given for the parallaxes, since they are mutually related."*; and states that this phenomenon appears in the case of the Moon as the closest body to Earth. Copernicus then proceeds to describe the experimental confirmation of the lunar parallax by experiments carried out in Bologna in March of 1497, and in which different measurements are taken when the Moon hides or blocks a star: *"Because we observed how long the Moon hid the bright star of the Hyades, we saw that the star came into contact with the dark part of the lunar body and that it lay hidden among the horns of the Moon at the end of the fifth hour of the night, when the star was closer to the southern horn by three quarters as if it were like the diameter of the Moon."*[13]

Finally, in this fourth book the Polish astronomer explains his theory on eclipses, addressing the issues of oppositions and average conjunctions of the Sun and the Moon; and the calculation of the magnitudes and durations of the lunar and solar eclipses. When he explains how to compute and predict eclipses, it is also evident that his method is much better than that of all his predecessors.

Of course there were serious difficulties for the acceptance and definitive establishment of the Copernican theory, the biggest of these was the non-detection of the stellar parallax; so the canon astronomer correctly deduced that the fixed stars should be considerably much more remote than the Sun. It was not until the late nineteenth century that the measurement techniques became accurate enough to observe and measure the stellar parallax, although it was only for some of the closest stars.

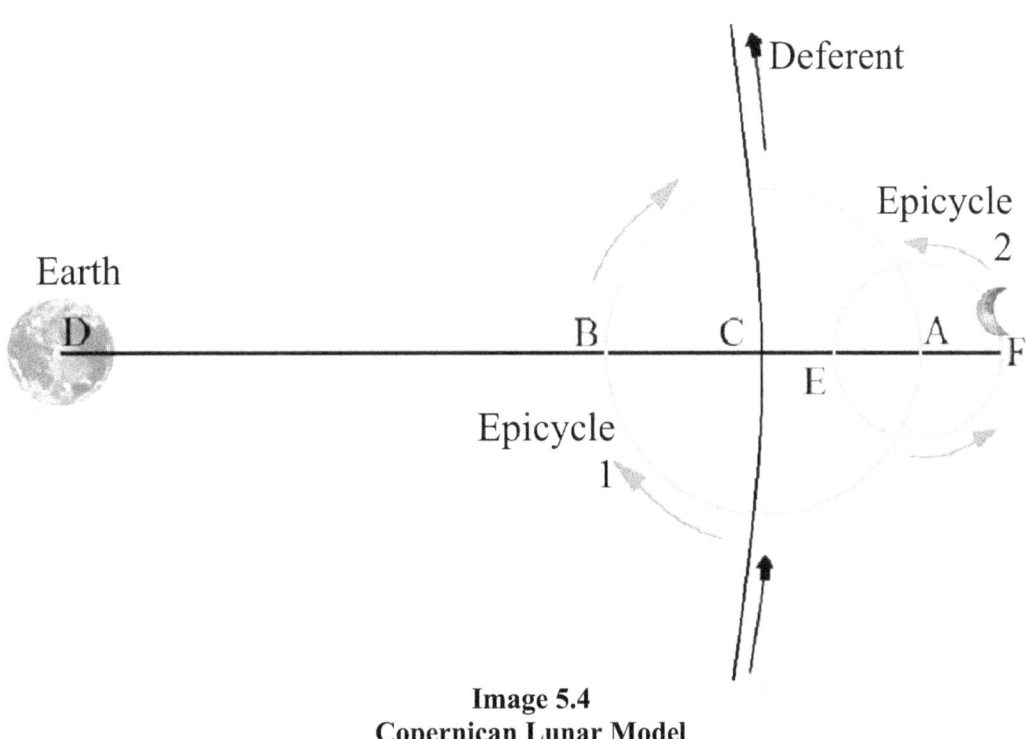

Image 5.4
Copernican Lunar Model

To explain the movement of the Moon, Copernicus resorted to a model consisting of a deferent concentric with the Earth, and two epicycles.

The Copernican theory that apparently opposed the theological dogma and in general was too difficult to understand, did not progress rapidly after the first publication of *De Revolutionibus*. The heliocentric system gradually extended abroad, especially in England; but before it could be accepted without restriction, the fundamental principles of celestial dynamics had to be established as Kepler, Galileo and Newton did. It was not until these brilliant scientists established the natural laws for the general movements of the bodies that the new heliocentric theory acquired its great solidity and its universal validity.

Tycho Brahe and the Observational astronomy

Copernicus was a brilliant theorist, but in reality his observations were not very useful for he was not in a position to give any truly conclusive proof in favor of his hypotheses, so for a long time astronomers questioned them. In Tycho Brahe we find the opposite: his observations, with the means he could use before the invention of the telescope, are superior in quantity and precision to those made previously. After Copernicus the next important astronomer was Tycho Brahe (1546 – 1601), who adopted an intermediate position, a Geo Heliocentric theory: he argued that the Sun and the Moon revolved around the Earth, but that the other planets revolve around the Sun. Unlike Copernicus, Brahe was an essentially empirical astronomer, passionately dedicated to the observation of the sky, but he was not exactly a brilliant theorist. Born in Denmark and of good fortune, he designed and ordered to build his own pre-telescopic instruments of observation and astronomical measurement. The importance of Brahe was not as a theoretician but as an observer, first under the patronage of King Frederick II of Denmark and then under the tutelage of the Austrian emperor and Archduke Rudolf II. He made a catalog of stars and observed the positions of the planets over many years. Towards the end of his life, Kepler, who then was a young man, became his assistant and for him the Tycho's observational records were invaluable.

Tycho Brahe was the eldest son of a noble, famous and wealthy family that had close relations with the Danish Crown. When he was still a child Tycho was adopted by a rich and without children uncle, a vice admiral that sent him in 1559 to study philosophy at the University of Copenhagen, he was then 13 years old. Tycho's interest in astronomy was an early development; staying in Copenhagen, at the age of 14, he witnessed a partial solar eclipse that caused him a fascination for this science that never left him. Later, in 1562, he left Denmark and went to the University of Leipzig with the intention of studying law, although most of the time he devoted to his passion for astronomical observations. It is known that there he acquired, studied and carefully annotated Latin editions of the Ptolemy' texts, and that he bought an astrolabe with which he made his first observations of a scientific nature; in August of 1563, while studying in Leipzig, a conjunction of the planets Jupiter and Saturn occurred which the young was able to observe.

These two events, the solar eclipse and the planetary conjunction, fascinated him and he was very impressed by the fact that the astronomers had predicted them using the Ptolemaic cosmology and updated tables. But Tycho Brahe realized that all predictions about the date of these phenomena were wrong in days, or even months. For which Brahe perceived the urgent need to develop new astronomical instruments with which to make new and more precise planetary observations, which in turn allowed him to realize more exact tables. In this way he was able to elaborate a precise stellar catalog of more than 1000 stars, whose positions he measured with accuracy much higher than that achieved until then. His best measurements reached accuracies of half a minute of arc, which allowed him to demonstrate that comets were not meteorological phenomena, of the terrestrial scope, but objects beyond our planet's influence.

After an extensive education, during which his official studies were much neglected due to his passionate astronomical practices, Tycho finally returned to his native Denmark at the age of 26

years. He returned as a rich man in his own right, because during his studies his uncle prematurely died leaving him a great fortune.

Next, Tycho settled with his maternal uncle Steen and temporarily diverted his attention to chemistry. Until November 11, 1572, when he was 26 years old, he observed a strange event in the constellation of Cassiopeia: a new and very bright star had appeared that was visible for about eighteen months. The astonishment of the astronomer was enormous given that at his time people believed in the immutability of the sky and in the impossibility of the appearance of new stars; but its radiance was inexplicable and it became as bright as Jupiter, being visible even during the day; but slowly it faded until it was no longer visible toward March of 1574. His observations on the star, now known as the supernova SN 1572 or *Tycho's Nova*, he summarized them in a book printed at 1573 and titled *De nova stella*, in which the word *nova* for the first time appears in the astronomical vocabulary; this fact instantly turned him into a respected astronomer. Today these stars are better known as *Supernovas*, and they actually refer an "astronomical" starburst, which constitutes the *final sigh* or death of a star, and which becomes perceptible, even to the naked eye, at points of the celestial sphere where none star had been detected before.

Before that and in view of his growing prestige, to retain him the King Frederick II of Denmark first offered him to settle in a royal castle and, in view of his refusal, he later agreed to give him the little island of Hven, to grant him an fixed income and to build a house for him. The final agreement was signed at 1576, and the young astronomer also built there what would become the observatory of *Uraniborg*, named in honor of Urania, the Greek muse of astronomy. It was also known as the *Castle of Heaven* and was located in the center of a large square garden surrounded by high walls like a fort. Another building, built later by Tycho as the number of his collaborators and students increased, was called *Stellaeburg* or *Castle of the Stars*. Brahe spent much of his life on the island of Hven making astronomical measurements and systematizing a large amount of data on the movement of stars and other celestial bodies.

In this way Brahe became the greatest observational astronomer since Hipparchus, and in the last of the great observing astronomers of the time prior to the invention of the telescope. The main work he developed in the two decades he spent working in Uraniborg, was the routine but all-important to measure and record the positions of the Moon, the Sun and the planets with respect to the fixed stars. Over a period of more than twenty years the positions of stars and planets were measured with unprecedented precision: with a minute of arc, about 50 times better than that achieved by the astronomer of Nicaea. His data were considered the highest quality in Europe in his moment and were widely accredited; so that when in November 1577 a comet was detected, it was his calculations that were considered as the definitive proof that its orbit ran between the space of the outer planets, and not between the Earth and the Moon, or in the sublunary region as Aristotle argued.

In the course of twenty-one years a series of rich and complete observations were made and cataloged, more accurate than all previous ones at that time. These observations and their results, together with those concerning the new star that gave him the opportunity to publish the text *Of Nova Stella*, induced Tycho to write a complete treatise on astronomy. The work consisted of several volumes in which he addressed the respective theories of the Sun, the Moon and the planets. The first introductory volume is entitled *Astronomiae Instaurasae Progymnasmata*, or *Introduction to the new astronomy*; where he exposes an intermediate model of the universe between those of Ptolemy and Copernicus, considering the Earth as fixed in space while the Sun revolves around it, but the King star was still the center of the orbits of the other planets; text that was started at 1588 and that would not be completed in Tycho's life, but by Kepler at 1602. The second volume, *De Mundi Aetherei RecentioribusPhaenomenis Bast Secundus*, was finished before and sent to his friends, colleagues and correspondents. Another text, *Astronomiae Instauratae Mechanica*,

includes detailed descriptions of the instruments that were designed and built by him, along with a report of his main discoveries, and with a brief autobiography.

The main characteristics of Tycho's work are the great precision of his observations, never reached by his predecessors, and his great continuity. It can be said that his observation errors were never greater than 1' or 2'; such a level of precision was due to the dimensions and the great stability of his instruments. Additionally, he repeated his observations under very different conditions with the purpose of eliminating accidental errors. Regarding the continuity of his observations, he evaluated the positions of the Sun and the Moon every day for more than twenty years, as well as other celestial bodies.

In relation to cosmological theories, Tycho didn't accept the Copernican system, probably for religious reasons and because Copernicus' arguments were still imperfect; and rather he proposed a new hypothesis that equally explained the observed celestial phenomena, leaving the Earth at the center of the world as in the Ptolemaic system. The *System of the universe* that Tycho proposed was a kind of transition between the *Geocentric theory* of Ptolemy and the *Heliocentric theory* of Nicolaus Copernicus. In this system the Sun and the Moon revolved around the immobile Earth in the center of the world, while all the other planets revolved around the King star.

Brahe adopted the simplification of making the planets revolve around the Sun, however, he assumed that the entire universe would revolve around the Earth. He considered, with very good reason, that if the Earth really moved around the Sun, the stars would show it by its apparent displacement. Brahe was convinced that the Earth remained static in relation to the rest of the Universe because, if it was not so, it should be possible to appreciate the apparent movements of the stars, or the parallax. However, the reason why he didn't detect it is that the phenomenon cannot be valued with direct visual observations: the stars are much further away than was thought reasonable at that time. This *Tychonic System* contains practically the same complications as the Ptolemaic, but, even so, it represents a considerable progress in the explanation of the celestial phenomena observed. It was superior to the Ptolemaic, but it did not possess the wonderful simplicity of the Copernican one.

Tycho Brahe's theory is partially correct; in a heliocentric system the Earth is considered to be rotating around the Sun because it is taken as a point of reference. But if a geocentric system is considered, the Earth is the point of reference and the Sun and the Moon revolve around it. On the other hand, Brahe put the other planets around the Sun and he thought that the orbits of the stars were all circular, when in reality they are elliptical. The actual shape of the orbits was proposed by Kepler in his first law, who was based on Tycho's observations.

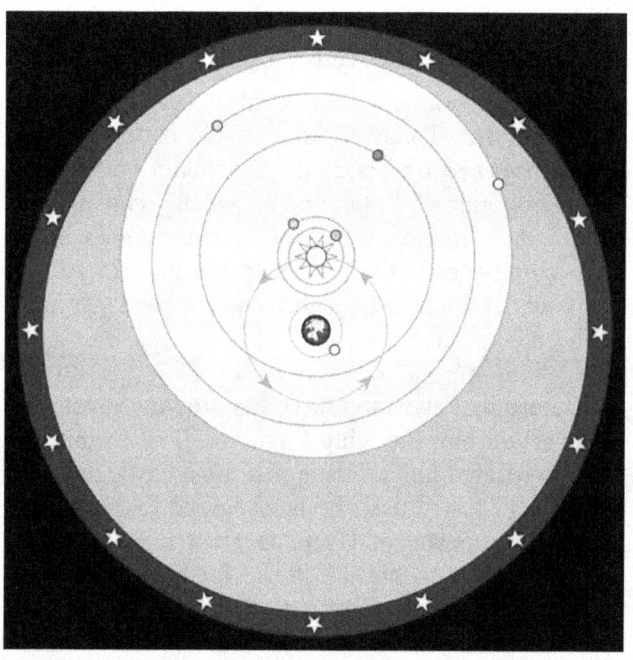

**Image 5.5
Tychonic model of the Universe**

Tycho Brahe adopted an intermediate model, a Geo-Heliocentric one, to describe the Universe.
In his system the Moon, the Sun and the fixed stars revolved around the Earth; but the other planets did it around the King star.

Brahe made important discoveries in relation to the Lunar theory: by observing the Moon in all its phases, instead of concentrating only on the quarters as his predecessors did, he discovered the longitudinal deviation of the lunar movement normally known as the *Third Inequality*: the *Variation*. He also found an *Annual equation*, which is a small inequality that depends on the position of the Earth in its orbit around the Sun. In addition, he discovered that the inclination of the lunar orbit to the ecliptic, as well as the movement of the nodes, were not fixed but varied regularly. So this great astronomer left the lunar theory considerably more advanced than he found it.

The position of Tycho Brahe in Denmark began to weaken in 1588 when King Frederick II died and the succession fell on his son Christian IV, with whom the astronomer didn't have a good understanding. As he was a rich man, at the beginning of 1599 Brahe decided to leave Hven Island, taking with him all the transportable instruments, the press to print his texts and even his assistants. After short period in Copenhagen and Rostock, Brahe finally arrived in Prague in June of 1599 on virtue an offer of the Holy Roman-Germanic's Emperor Rudolph II of Habsburg, who granted him the title of *Mathematician and Imperial Astrologer*, assigned him a considerable rent and gave him the castle of Benatek, at 35 kilometers from Prague, as a residence and site to install his astronomical observatory. All this with the purpose of Brahe elaborating for him horoscopes and astrological predictions.

Tycho was then fifty-three years old and would no longer make important discoveries; but just at this moment he met the character that finally could best take advantage of his huge data file, the young astronomer Johannes Kepler: to whom he would entrust the complete set of his observations and measurements on the movements of the Moon and the planets performed for decades. But soon difficulties arose between the two men and which became more serious by the fact that Tycho was fast approaching his death, that finally occurred in November, 1601. Thanks to this wealth of astronomical data Kepler would be able, a few years later, to find those which are now called Kepler's laws and that govern the planetary movement.

There is a character that is hard not to mention when it comes to the astronomy of this Renaissance period. Filippo Bruno, better known as Giordano Bruno (1548 – 1600), who was an Italian philosopher, theologian, mathematician and astronomer. This ambitious philosopher extended the newly promulgated Heliocentric Theory to apply it to the totality of the Cosmos: he held that, like God, the Universe was infinite; and that the so-called Fixed Stars were other worlds, other solar systems in which each star was likewise a sun surrounded by their respective planets; and as if that were not enough, he raised the possibility of existence of life in these worlds. Because of his scandalous cosmological and theological pronouncements he entered into strong conflicts with the Catholic Church of the time; finally he was judged as a heretic by the Roman Inquisition, found guilty and burned alive at the stake in Rome on February 17, 1600.

Bibliographic Citations

[1] Gutenberg Project. The Discovery of a World in the Moone by John Wilkins. http://www.gutenberg.org/ebooks/19103?msg=welcome_stranger

[2] Dolmage, Cecil G. *Astronomy of to-day*. London: Seeley and co. limited 38 Great RUssell Street. 1,910

[3] Fernández de Navarrete, Martín. *Colección de los viajes y descubrimientos que hicieron por mar los españoles* Tomo I. Madrid: Imprenta real, 1.825. Internet Archive: https://archive.org/search.php?query=Colecci%C3%B3n%20de%20los%20viages%20y%20descubrimientos

[4] Philip's Astronomy Encyclopedia. Londres: Philip's, 2002. www.philips-maps.co.uk.
John D. North, Historia Fontana de la Astronomía y la Cosmología. México: Fondo de Cultura Económica, 2005.

[5] Nicolaus Copernicus Thorunensis Proyect. http://copernicus.torun.pl/en/archives/astronomical/1/?view=transkrypcja&

[6] Dreyer, John LE *A History of Astronomy from Thales to Kepler* . Cambridge: Dover Publications, Inc. 1953.

[7] Polish Academy of Sciences. Nicholas Copernicus Complete Works II. Warszawa- Kraków : Polish scientific publishers 1,978. http://kpbc.umk.pl/dlibra/docmetadata?id=48792&action=ChangeLanguageAction&language=en

[8] Polish Academy of Sciences. Nicholas Copernicus Complete Works II.

[9] Polish Academy of Sciences. Nicholas Copernicus Complete Works II.

[10] Dreyer, John LE *A History of Astronomy from Thales to Kepler* . Cambridge: Dover Publications, Inc. 1953.

[11] Polish Academy of Sciences. Nicholas Copernicus Complete Works II.

[12] Internet Archive. Great Book of the Western World Vol 16. Nicolaus Copernicus. On the Revolutions of the Heavenly Spheres. https://archive.org/details/greatbooksofwest16hutc

[13] Internet Archive. Great Book of the Western World Vol 16. Nicolaus Copernicus. On the Revolutions of the Heavenly Spheres. https://archive.org/details/greatbooksofwest16hutc

Chapter 6

A Telescopic Look at Our Satellite

After having reached its *Full Moon* phase, the process reverses and the illuminated portion of the Moon that we can contemplate begins to diminish, to wane, and hence the name of the Waning Moon. In the graph we see the satellite as it looked on Tuesday, February 6, 2018 at 03:00 UCT, 20 days 0 hours 30 minutes after the New Moon and appreciating 65,3% of its illuminated hemisphere; this phase is called the *Waning gibbous moon* .

Courtesy of NASA's Scientific Visualization Studio: https://svs.gsfc.nasa.gov/4604

"I checked my impatience, and listened with all my ears to the wonders he related. He went on to inform me that the inhabitants of the moon resembled those of the earth, in form, stature, features, and manners, and were evidently of the same species, as they did not differ more than did the Hottentot from the Parisian. That they had similar passions, propensities and pursuits, but differed greatly in manners and habits. They had more activity, but less strength: they were feebler in mind as well as body. But the most curious part of his information was, that a large number of them were born without any intellectual vigor, and wandered about as so many automatons, under the care of the government, until they were illuminated with the mental ray from some earthly brains, by means of the mysterious influence which the moon is known to exercise on our planet. But in this case the inhabitant of the earth loses what the inhabitant of the moon gains; the ordinary portion of understanding allotted to one mortal being thus divided between two; and, as might be expected, seeing that the two minds were originally the same, there is a most exact conformity between the man of the earth and his counterpart in the moon, in all their principles of action and modes of thinking."

A Voyage to the Moon (1827)[1]
Joseph Atterley (George Tucker (1775 - 1861))

"I have now enumerated," said Barbicane, "the experiments which I call purely paper ones, and wholly insufficient to establish serious relations with the Queen of Night. Nevertheless, I am bound to add that some practical geniuses have attempted to establish actual communication with her. Thus, a few days ago, a German geometrician proposed to send a scientific expedition to the steppes of Siberia. There, on those vast plains, they were to describe enormous geometric figures, drawn in characters of reflecting luminosity, among which was the proposition regarding the square of the hypotenuse, commonly called the "Ass's Bridge" by the French. "Every intelligent being," said the geometrician, "must understand the scientific meaning of that figure. The Selenites, do they exist, will respond by a similar figure; and, a communication being thus once established, it will be easy to form an alphabet which shall enable us to converse with the inhabitants of the moon." So spoke the German geometrician; but his project was never put into practice, and up to the present day there is no bond in existence between the earth and her satellite. It is reserved for the practical genius of Americans to establish a communication with the sidereal world. The means of arriving thither are simple, easy, certain, infallible—and that is the purpose of my present proposal."

From the Earth to the Moon, direct route in 97 hours. (1865)[2]
Jules Verne (1828 - 1905)

Johannes Kepler

Johannes Kepler (Dec 1571 – Nov 1630) was a German astronomer, physicist and mathematician; a great protagonist in the development of the scientific revolution of the European Renaissance, and one of the founders of modern astronomy; he is universally known for the formulation of physical laws about the movement of celestial bodies. He was the son of the marriage between Heinrich Kepler and Katherina Guldenmann, who would have a total of four children. Similar to Brahe, his interest in astronomy wakes up at a very early age, on 1577 at the age of five he contemplates a comet; while at nine he witnesses the eclipse of the Moon of January 31, 1580, and later he describes it in one of his optic works, reporting that the satellite looked quite red.

The young man finished his basic studies in 1583 and he was then about twelve years old; he then entered the Protestant Seminary of Adelberg in 1584, and two years later to the Upper Seminary of Maulbronn, which was a preparatory school for the University of Tübingen; he graduated in 1588 and the following year he enrolled in this university. There he initially studied philosophy, dialectics, rhetoric and ethics; also Greek and Hebrew in the area of languages; whereas in natural sciences he studied physics and astronomy; the final courses included theology and human sciences. Finally he finished his master's studies in 1591. It was in this university where he met the Copernican heliocentric theory through his professor of astronomy Michael Maestlin; the young Kepler studied it decidedly, welcomed it and soon became its most decided and faithful defender.

Originally he intended to enter the Lutheran Church, but the administrative rigor that prevailed among his ministers was not his complacency, so he ended up accepting in 1594 the position of provincial mathematician of the University of Graz, in the Austrian state of Styria; from this moment he was entirely devoted to science. After a productive professional life, Kepler finally died in 1630 at the age of 58 in Regensburg, in Bavaria, Germany.

The Cosmographic Mystery

Kepler approached with such zeal the astronomical studies that he soon produced his first treatise on the subject. As an astronomer, Kepler tried to understand, describe and explain the movements of the celestial bodies during most of his life. At first he considered using the Pythagorean philosophy of the mathematical harmony of the cosmos, usually known as the *Harmony of the celestial spheres*; but he soon became a strong supporter of the Copernican model. Initially he published his thoughts and theories in a book called *Mysterium Cosmographicum*, or *The Cosmographic Mystery*, published in Tübingen in 1596.

The *Cosmographic Mystery* exposes Kepler's cosmological theory, which is based on the Copernican system, and according to which the five regular Pythagorean polyhedrons dictate the structure of the universe and reflect God's plan through geometry. And although this main idea of the work was erroneous, we owe a great debt of gratitude to that work, since it represents the first firm step for the purification of the Copernican system of the remains of the Ptolemaic Aristotelian theory that were still included in it. The reasons for abandoning the Ptolemaic system in favor of the Copernican are exposed in the first chapter with remarkable clarity.

Kepler's greatest desire was to obtain more correct values about the distances and average eccentricities of the planets, to show that his theory was absolutely true; and the only place in the world where this information could be obtained was in the observatory of Tycho Brahe. As an astronomer, Kepler was very aware of the excellent observation program that the very famous Brahe was carrying out; and when his book *The Cosmographic Mystery* was published in 1596, he sent him a copy attaching a letter of explanation, as well as a request for accessing his data with the purpose of testing his

model more rigorously. This gave rise to a continuous correspondence that culminated a few years later with the invitation of Tycho to Kepler for working with him in the new observatory that was being built in Benatek, in Prague.

Kepler and Brahe work together

Harassed by the religious persecution in Styria, in January 1600 Kepler went to Prague where he was invited by Brahe; who then had the best astronomical observation center in the Benatek castle, had the best data on planetary observations of the time and who had read some of Kepler's works. In February 1600, Tycho Brahe and Johannes Kepler met for the first time at the observatory: Brahe's desires to have Kepler as his assistant, and Kepler's ones to have access to Brahe's data, soon materialized.

The two professionals working together must have been the perfect team: the largest observational astronomer of the moment was joined by someone who would prove to be the best mathematical analyst of the time. But although their knowledge and skills were perfectly complementary, their personalities were not: the association was far from pleasant and harmonious, possibly due to their very different natures and backgrounds. Rather the relationship between both was complex and marked by distrust and professional jealousy; and until the death of Tycho in 1601, Kepler failed to have full access to all the data collected by that, which were considerably more accurate than those previously used by Copernicus.

Kepler had joined Tycho with the primary objective of gaining access to his observational data, which had not yet been published; but he soon discovered that Tycho considered them his private treasure and that he would only release them at his will and in limited quantities. The two were so opposed by nature that only their professional interests kept them together. Kepler eagerly desired Tycho's data, but Tycho considered that his observations represented his life's work and that additionally they contained the secret of the Universe and, rightly, he thought that only Kepler had the intellectual capabilities to extract it. Up to certain point, Brahe's selfishness was justified in the fear that the laws that the German astronomer was going to deduce from his data would carry Kepler's name, and not his.

When Brahe died in October 1601, Kepler replaced him in the positions of mathematician and imperial astrologer of Rudolph II. At that time, and because the scientific method was barely developing, the difference between astronomy and astrology was not yet clearly established; this would definitely occur during the movement of the Enlightenment of the eighteenth century. Then, finally the German astronomer had in his hands the treasure of astronomical data of Tycho, and had to exploit them magnificently to discover his three famous laws that would describe, for the first time in mathematical and geometrical terms, the movements of the stars; a great advance in the cosmology of the moment.

During his routine practices, Brahe had realized the effect of the Earth's atmosphere on astronomical observations, and wrote about the fact that the stars did not appear in the sky precisely where it had been predicted: every time he made an observation, he noticed the difference between the predicted altitude and the actual observed altitude of the body, and thus he built a table of which what we now know as *refraction of light* is deduced. Kepler spent some time looking for an explanation for these discrepancies, and his conclusions he presented in a good work entitled "*A Supplement to Vitello, which explains the optical part of astronomy*", published in 1604 and usually known as *Kepler's Optics*; and in it the author introduced the science of optics within astronomy. The refraction of light consists of the change in the direction of a ray of light as it passes between two media of different physical properties and in a direction that is not perpendicular to the contact surface.

De Stella nova

In a manner very analogous to Brahe's experience thirty-two years earlier, Kepler observed on October 17, 1604, a supernova in the constellation of the Serpentarium, which would later be called Kepler's Star, also known as SN 1604. The phenomenon had been observed by other European astronomers since day 9 and in different cities such as Prague, Verona, Rome and Padua; and it continued to see itself for almost a year. Inspired by Brahe's earlier work, *De Nova stella*, Kepler made a detailed study of his own observations on this new phenomenon, and presented his conclusions in the work *De Stella nova in pede serpentarii*, or *On the new star located at the feet of Serpentarium*, published in Prague in the year 1606. Kepler's observations, combined with data from other European astronomers, make *De Stella Nova* a very important document both for the study of the supernova itself, and for the study of early seventeenth-century astronomy. In this work the astronomer presents the evidence that the Universe is not immutable or static as Aristotle proposed, but it is subjected to important and varied changes. At that time the current constellation of Ophiuchus was called Serpentarium, or The Snake Hunter.

Astronomia Nova: Kepler derives the laws of celestial movement

Kepler was the first important astronomer after Copernicus to conscientiously adopt and defend the heliocentric theory, but Tycho Brahe's data showed him that it could not be entirely correct in the way that the Polish astronomer had developed it. When Kepler joined Tycho at Benatek Castle in February 1600, the planet Mars was in opposition to the Sun and Brahe had prepared a table of its oppositions observed since 1580 and had developed a theory that represented its lengths in opposition very well, with errors only of two minutes of arc. Thus, always based on the informations left by Brahe and especially those related to the opposition and the retrograde movement of the planet Mars, Kepler realized that planetary movements could not be explained by his own model of perfect polyhedron and the so-called *Harmony of spheres*. From this moment on, the German astronomer devoted himself to finding another mathematical geometric model that fitted much better with the Copernican heliocentric system. In the process he managed to deduce his famous *Three laws of planetary movements*; he so promptly determined the first two laws that were published in his work *Astronomia Nova*, or *New Astronomy*, published in Prague in 1609; and which astonished the scientific world of the moment and revealed him as the best astronomer of the time.

In his theoretical development, Kepler began by using the traditional model for the planetary movement, the circumference for being the most perfect of geometric figures, and proposing eccentric trajectories and movements in epicycles for the different planets. But soon he found that the observational data did not adjust perfectly with the theoretical models established by Copernicus and they showed a persistent error of eight arc minutes; which then led him to conclude that the planets did not follow circular trajectories around the Sun. Unmotivated, Kepler understood that he should abandon the circumference and the idea of a perfect world for which reason he resorted and tried other geometric shapes for the celestial orbits. Finally, after considering the oval, he used the concept of the ellipse, a rare geometrical figure originally introduced and described by Apollonius of Perga in the third century BC; so happily he discovered that it fitted very well with Brahe's measurements and data. It was then when Kepler established that the planets describe elliptical orbits with the Sun in one of its foci; and thus he starred in one of the greatest milestones of modern astronomy.

Fortunately, Tycho had concentrated a lot on the planet Mars that exhibits a very pronounced elliptical orbit; otherwise it would have been impossible for the German astronomer to perceive that the orbits of the other stars should follow the same pattern. Thus, Kepler had discovered his *First law of the planetary movement*: The celestial bodies move around the Sun following elliptical trajectories with the King star located in one of the two foci that contains the ellipse. In general terms, the Copernican theory still needed deferents and epicycles circles, although less of those necessary in the Ptolemaic system; and it was not until the German astronomer discovered this

law that the new heliocentric theory acquired its total simplicity and at the same time its great efficacy. In this way, Kepler finally managed to get rid of the ancient Ptolemaic concepts of epicycle, deferent eccentric and equant, and replaced them with the statement that later became known as *Kepler's First Law*, pointing the way to the definitive abandonment of the circumference as an exclusive representative of planetary movements.

The discovery of this first law of celestial movements required a greater intellectual effort for the emancipation of the Ptolemaic tradition than is normally accepted. The only hypothesis on which all previous astronomers had agreed without exception was that all celestial movements were perfectly circular, or composed of uniform circular motions. The replacement of the perfect circles by the "deformed" ellipses involved the definitely abandonment of the supposed aesthetic foundation that had governed astronomy since the Pythagorean and Aristotelian times. For the ancients the sphere and the circumference were perfect, ideal figures; and the stars, divine beings, were perfect bodies: it was mandatory that a perfect body moves according to a perfect figure.

Later on Kepler was dedicated to investigate the speeds of the planets through the respective elliptical orbits already established. He focused again on the planet Mars, and noticed that it moved faster the closer it was to the Sun, and slowest the farther away it was from the Sun, so that the surface described by the straight line connecting the Sun to Mars is always proportional to the time. This allowed him to derive the *Second Law*: The areas swept by the celestial bodies in their elliptical trajectory are proportional to the time used by them to travel the perimeter of such trajectories. This law explains why when the planet is closer to the Sun, in the perihelion, it moves more quickly, and when it is at its farthest distance, the aphelion, it moves more slowly.

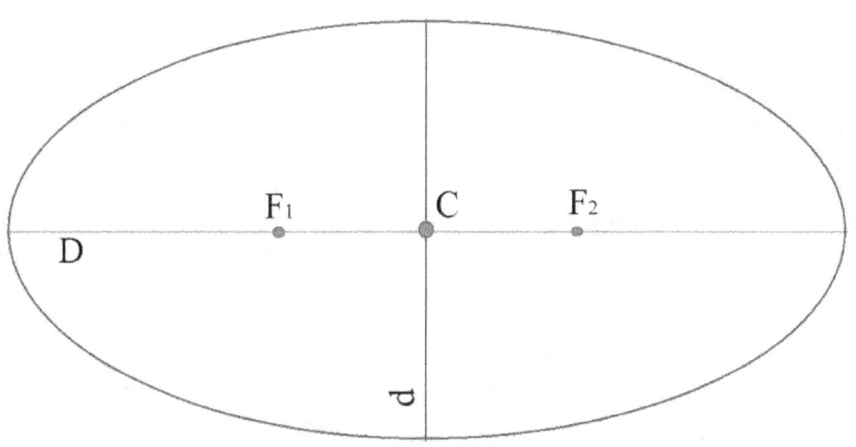

Image 6.1
Elliptical geometry

The geometric concept of the Ellipse was introduced by Apollonius of Perga in the third century BC. It has three main points, the geometric center C and the Foci F_1 and F_2; and two main axes, the greater diameter D and the smaller one d. Johannes Kepler found that this figure represented very well the planetary movements around the Sun, the King star being located in any of the two foci. From then on the deferent and epicycles circles would definitely be useless in astronomy.

All that development took Kepler eight years and almost a thousand pages of carefully elaborated calculations before deciphering the problem and discovering his first two laws of planetary motion; his third law would have to wait another nine years. Such a delay was caused not only by the tedious and laborious nature of the extensive numerical calculations that were needed, but also by the psychological problem of having to relinquish the sacrosanct dogmas of the past, especially those that established that the movements of celestial bodies should be circular and at constant speed.

Kepler's great achievement as an astronomer was the discovery of his *Three laws of planetary motion*. The first two he published in his *Astronomia nova*; which contains the results of researches made by Kepler for about eight years about the movement of the planets, particularly on Mars. Here Kepler's first two laws of the planetary movement are presented, which meant a radical change in astronomy and broke with a tradition of 2000 years. The work is based on the observations of the Danish astronomer Tycho Brahe and has a broad mathematical development. Its full title is "*New Astronomy based on causes, or celestial physics, treated by means of commentaries on the movement of the star Mars, from observations of Tycho Brahe*." This title is interpreted as the first time that the idea of combining physics and astronomy in a sole treaty is expressed, thus giving rise to what we now call *Celestial Mechanics*. In fact, all of Kepler's research was guided by the idea that the Sun is the origin of the forces that move the planets in their orbits and that increase or decrease with the respective Sun-planet distance.

The work is structured in five parts. In the introduction a series of axioms related to a *force of attraction* between the different celestial bodies are enunciated, and there it is affirmed that the Moon's force arrives as far as until the Earth and produces the tides. So likely Kepler was very clear about the attraction's force concept between the stars, but he was wrong in arguing its nature: he understood that force as a mutual attraction similar to the magnetic force, which had been explained by the physicist William Gilbert in the year 1600.

In the first part the cosmological models proposed until then are discussed: the geocentric Ptolemaic, the heliocentric Copernican and the intermediate model proposed by Brahe. In the other parts of the work several detailed models for the orbit of Mars are exposed, demonstrating that simple modifications of traditional models are not enough to explain the observations of the Danish astronomer. He then enunciates and explains his first two laws and presents a fully satisfactory model for the orbit of the planet Mars with an ellipse that has the Sun in one of its foci. Finally, Kepler argues by analogy that the other planets will also conform to the two laws exposed. With this great contribution, Copernican astronomy was definitely placed above all previous cosmological systems.

The *New Astronomy* presented the arguments and evidences on the discovery of his two first laws of planetary motion. Previous attempts to adjust the data by the displacement of the Earth, according to Ptolemy, or the Sun according to Copernicus, were explained because each focus of the elliptical orbit is displaced from the geometric center, which is equidistant from both. The total credit for the discovery of these laws must be given to Kepler; who was not careless or forgot his great debt to Brahe. On the cover of his *New Astronomy* he wrote: *Prepared by Johannes Kepler from the observations of Tycho Brahe*. Nor did he forget about Copernicus: on the back cover he announced that he had proof that the Polish astronomer considered his own theory not simply as a useful hypothesis, as was generally believed, but as a system that actually represented well the Universe.

The Third law of celestial movements

Once the first two laws were established, it still lacked Kepler to quantify and relate the trajectories of the different planets to each other. After a strenuous dedication and after years of research, the astronomer deduced in May of 1618 his famous *Third law of planetary movement*: The square of the orbital periods of

the celestial bodies is proportional to the cube of its respective distance to the King star.

In the *Cosmographic Mystery,* Kepler had supposed the existence of a *Spiritus Motricium* or *Driving Soul* in the Sun, and this idea he now develops it further. The confirmation of his idea of the similarity between the movement of the Earth and those of the other planets, naturally led him to resume the suggestion made in that work that this movement is caused by a force that emanates from the Sun; and as the effect of any force of such kind must necessarily vary in one or another way with the distance to the King star, it occurred to him to speculate about the variation of the velocity of a planet with respect to that distance. According to Kepler, this force emanates from the Sun but, unlike light, does not extend in all directions, but only in the plane near which the respective planes of all planetary orbits are located, so that it simply diminishes as the distance increases. Therefore, the speed of a planet in its orbit varies inversely with the distance to the Sun, which is the central idea of his third law.

Between the years 1617 to 1621 Kepler wrote a work called *Epitome Astronomiae Copernicanae,* or *Epitome of Copernican Astronomy,* whose great novelty was to contain the first printed version of his Third Law of planetary motion. The work was divided into seven books that treat much of Kepler's astronomical knowledge, as well as his thinking about physics and metaphysics; it is in this work that the physical concept of inertia was exposed for the first time.

After making reference to the recent discoveries of Galileo Galilei, In *Epitome of Copernican* astronomy the German astronomer extends its recently deduced laws to explain the movement of the other stars as the natural satellites or moons: he establishes that the fundamental laws discovered for the planet Mars are true and applicable also for the other planets, for our Moon that revolves around the Earth and for the natural satellites around Jupiter discovered by Galileo, or *Medicean Planets* as they were initially called. The theory of the Moon is discussed in detail and treats the inequalities of Evection and Variation. Additionally, he establishes the need to correct the distance from the Earth to the Sun, which since the times of Hipparchus and Ptolemy had always been taken as equal to 600 times the Earth's diameter, which implied a solar parallax of three minutes of arc.

The Third Law refers to the relationship between the revolution periods of the planets and their respective distances from the Sun. The periods were well known precisely for the large number of revolutions that had been observed and registered since antiquity, but their distances were not very well established, to the point that there was no solid basis to determine how the periods varied with the respective planet's distance. A question of fundamental importance that the astronomer had already raised in the *Cosmographic Mystery*. To unlock the subject he would have to tackle one of the central problems of astronomy of all times: the determination of the distances between the celestial bodies, a subject that was practically stagnant since the works of Aristarchus in the third century and of Hipparchus in the second century BC. Then, the German astronomer devoted himself to developing methods to determine the relative distances between the planets, once again exploiting the registers of Tycho Brahe and demonstrating again his brilliant analytical abilities; and finally Kepler was rewarded with the discovery of his third law.

The methods used by Kepler were geometric and consisted in defining a triangle that had the Earth and the studied planet in two of its corners, and another appropriate body in the third. Knowing the length of one of the sides, called the baseline, and measuring two of the angles involved, then the Earth-planet distance could be calculated by elementary geometry; and then the distance from that planet to the Sun would be determined. Kepler selected as a baseline the distance from Earth to the Sun, which we now call the *Astronomical Unit* (AU), which for his time it was extremely difficult to measure with good precision. Therefore, Kepler's measurements were in terms of Astronomical Units, which were very precise in the relative

sense and this was enough for him to establish the relations and the equations he was looking for.

With his measurements of interplanetary distances, Kepler quickly established his *Third Law of Planetary Motion*, which determines the relationship between the period of revolution of each planet around the Sun and its respective distance from the King star. In mathematical terms: the square of the orbital period is proportional to the cube of the distance to the Sun. This law indicates that the most distant planets take longer to complete their orbit, both because they have more road to travel, and for they move more slowly.

These three laws are the great intellectually achievement of Johannes Kepler, and they would finally allow to understand, unify and to predict all celestial movements. He discovered them embedded in the extensive set of observational data compiled by Tycho Brahe, and he extracted them with the most brilliant series of mathematical analyzes that had ever been carried out up to that point. His great interest in astronomy had been inspired by the firm conviction, also felt by Pythagoras, that there should be some sort of order in the movements of the celestial bodies, and that it could be expressed in both geometric and mathematical terms. Initially Kepler had proposed that the planets' orbits could be explained based on the five regular solids of geometry, and from there his work was driven by the desire to prove that this hypothesis was correct. In deducing his *Three Laws*, he showed that the general belief he had held, along with the Pythagoreans, to a certain extent was true: he had revealed the harmony and order of the Universe. Order and harmony totally different from his initial hypothesis, but which he considered as real even though he did not fully understand its causes or nature.

Kepler's explanation of planetary movements was the first serious attempt to understand and interpret the Solar System's mechanism. He did not advance beyond the notions of mechanics of the XVII century; he supposed it was necessary that a force acts constantly to maintain a star in movement, and that this would stop where it was if the force stopped acting. For Kepler this active force was pure and simple magnetism, and he never tired of emphasizing this every time an opportunity presented itself.

Although Kepler deserves much credit for having tried to find the causes of celestial movements, he cannot be called a forerunner of Newton. His strength is not directed to the Sun but it is tangential, that is, it is not attractive but rather a promoter. It is remarkable how firmly Kepler clung to the close analogy between Gravity and Magnetism on the one hand, and between the *Driving force* of the Sun and Magnetism on the other hand. And yet, he could not see the identity between the Gravity and the Force that keeps the planets in their orbits. This is more evident when we appreciate that he, in the notes of his *Somnium* written between 1620 and 1630, expressly attributes the tides in the oceans to the bodies of the Sun and the Moon which attract the waters of the sea with a certain force similar to the magnetic one.

These laws are mixed in Kepler's works with many digressions, and there is even a hint of the *Law of universal gravitation* in his works. In fact, he exposes that in the celestial sphere there must exist an unknown force similar to the attraction of the Earth. For example, he argues that if the Earth and the Moon were not kept in their respective orbits by one *Vital force* or another, they would precipitate against each other! Kepler tries to establish a relation between the terrestrial gravity and the force of attraction that the Sun exerts on the planets, but then he deviates when supposing an analogy between the universal attraction and the magnetic attraction.

The Lunar movement according to Kepler

As explained above, Kepler extended his *Laws of the planetary movement* to sustain the movement of the other celestial bodies, including the natural satellites around their respective planets. Thus, according to Keplerian laws, the lunar globe moves in an elliptical orbit with the Earth located in one of the foci of such

trajectory; and the whole set, or Earth - Moon system, moves around the Sun in another elliptical trajectory.

With respect to the Moon, the German astronomer found that the introduction of elliptical motion in its theory was very problematic due to the variability of the eccentricity of the lunar orbit; and over the years he made many changes in the way he represented the observed lengths. In fact, he must have envied his predecessors who could use a circular epicycle to account for each new inequality.

Regardless of Tycho, Kepler had discovered the *Annual Equation of the Moon*. The lunar eclipse in February and the Full Easter Moon of 1598, as well as the Solar eclipse of March 7, occurred more than an hour after the announcement in the calendar made by Kepler, while the lunar eclipse in August of the same year happened earlier than expected. Therefore, for the 1599 calendar he suggested that the period of the Moon with respect to the Sun is a little longer in winter than in summer; and later in the *Epitome* he proposed two arguments for it. First he stated that the Moon could be delayed in its movement by a force that emanates from the Sun, which would be greater in winter when the Earth and the Moon are closer to the Sun than in summer. Secondly, he argued that the cause of the phenomenon could also be that the earthly rotation speed depends on the distance from Earth to the Sun, and that it is a bit faster in winter, so the Moon takes more time in this epoch than in summer to travel through equal arcs of its trajectory.[3]

Also in the *Epitome* this astronomer argued on the theory of eclipses and explained why the Moon seems to be illuminated by a weak reddish light during total eclipses: he rightly supposed that sunlight, when crossing through the terrestrial atmosphere, it deviates from its path to reach the satellite and then to be reflected. Likewise, Kepler dealt with the total eclipses of the Sun, he trated with the luminous brightness that appears around the King star's globe during these phenomena and cited what was previously stated by Plutarch: *"The Moon sometimes obscures the Sun completely, but always for a short time, and it is never large enough to prevent a certain luminosity from appearing around the circumference of the Sun, which makes the darkness never black and deep nor completely dark."*[4] What is constituted in the first historically recorded considerations on the solar corona.

Other Keplerian Works

Although they were contemporaries, Kepler and Galileo never met personally; but each was well informed about the other's works and they often written themselves. On April 8, 1610, Kepler finally got the edition of Galileo's little book, *Sidereus Nuncius*, or *Sidereal Messenger*, which was sent to him by the own author. In some forty pages Galileo gives a very precise description of his telescopic observations and states that he has seen amazing things in the sky, so he asks the German astronomer for his interpretations. Kepler responded publicly in 1610 with a booklet called *Dissertatio cum Nuncio Sidereo*, or *Dissertation about the Sidereal Messenger*, praising him for his constructed telescopes and for the discoveries made with them, as well as adding own interpretations and hypotheses, all within the framework of the Heliocentric theory.

For great disappointment of the German astronomer, the hypothesis of fixing the orbits of the planets inside perfect polyhedrons, or Platonic Solids, as he had done in his first work *Cosmographic Mystery*, never worked. *Harmonices Mundi*, or *The Harmony of the World* of 1619, is a book written by Kepler where he began exposing this theory, to finally show that it is incompatible with the astronomical observations and with the two laws of planetary movement exposed by him in his *New Astronomy*, so he ended rejecting it. The third law, which indicates that the cube of the average distance from the planet to the Sun is proportional to the square of its orbital period, is newly presented in chapter 5 of *The Harmony of the Worlds*. Additionally, in this work Kepler tried to justify the planetary movements from the causal point of view, postulating a force similar

to magnetism which he thought emanated from the Sun.

The *Rudolphine tables* are planetary and stellar datasets collected by Kepler from the astronomical observations made by Tycho Brahe; they are presented in tabular form and were widely used by astronomers around the world for more than a century to determine celestial movements and so to calculate the positions of planets and stars. On his deathbed Tycho asked Kepler to complete his astronomical tables, this accepted such a request and the *Rudolphine Tables*, so named for being dedicated to the Emperor Rudolph, finally appeared published in 1627; although the observational data are essentially those of Tycho, the theory underlying them is that of Kepler himself.

The publication of these tables in 1627 constitutes the final act of Kepler's fruitful professional life; who died on November 15, 1630 in Regensburg, Germany. Astronomer that managed to completely purify the Copernican heliocentric system from the remnants of the Hellenistic conceptions that Copernicus could not eradicate. The Solar System was completely revealed and understood as a whole, and all its individual components were described for the first time with the same set of physical laws written in mathematical geometrical language.

There is a work by this great astronomer that was published posthumously in 1634 by his son Ludwig with the title: "*Somnium, seu opus posthumum de Astronomia Lunari*", "*The Dream, or posthumous work of Lunar Astronomy*". It concerns a fiction novel written by him towards 1608, and that normally is considered like the first work of Science fiction of history, preceding Jules Verne by more than two centuries a half. It is a strange and passionate adventure both fictional and scientific, framed within the estate of astronomy during the first decade of the seventeenth century, and narrated in the voices of a dreamer, a young man named Duracotus, who most likely represents the same Kepler. Thanks to a magic spell by a *Spirit*, Duracotus, his mother Fiolxhilde and a group of characters of his time make a fantastic and oneiric trip to the Moon.

Although it is dated at 1609, the text was not published before Kepler's death at 1630. Its author was very aware of the conflict developed in Europe between the official teaching of the Catholic Church and the Copernican revolution that was at its height. Kepler was not as trustful person as Giordano Bruno, who was burned at the stake for holding similar beliefs in other worlds; for that reason he tried to hide his strange history as a dream or a fantasy from which he could retract when it would be indispensable. Thus, similar to Copernicus and basically for the same reasons, Kepler was also cautious at the moment to publish some of his scientific thoughts.

The work has been translated from the original Latin and is now available in different languages and on multiple WEB sites, as well as in printed books. The most widespread English translation is that of the Reverend Norman Raymond Falardeau of 1962; and for this text I am guided by its version presented by Les Coleman on the Frosty Drew Observatory & Sky Theater website.[5] From whose work I selected some paragraphs for their important content and meaning for this text; I also attached some illustrative images and made some explanatory comments. Kepler's text is presented in italics, while my own notes are in plain text.

First, there are some specific terms that the reader should know in advance:

Duracotus: Is the protagonist and the narrator of the story, his mother is Fiolxhilde.
Spirit: It should not be interpreted precisely as a evil spirit or devil; but rather as a kind of extraterrestrial Spirit, an alien in modern terms.
Levania: Is the name that Kepler uses to refer to the Moon.
Volva: Is the name for the Earth.
Subvolva: The lunar hemisphere that is always facing Earth.
Privolva: The lunar hemisphere opposite the Subvolva, and which is always hidden from the Earth. Although it can be completely illuminated

by solar rays, it is often and mistakenly known as the *Dark Side of the Moon.*

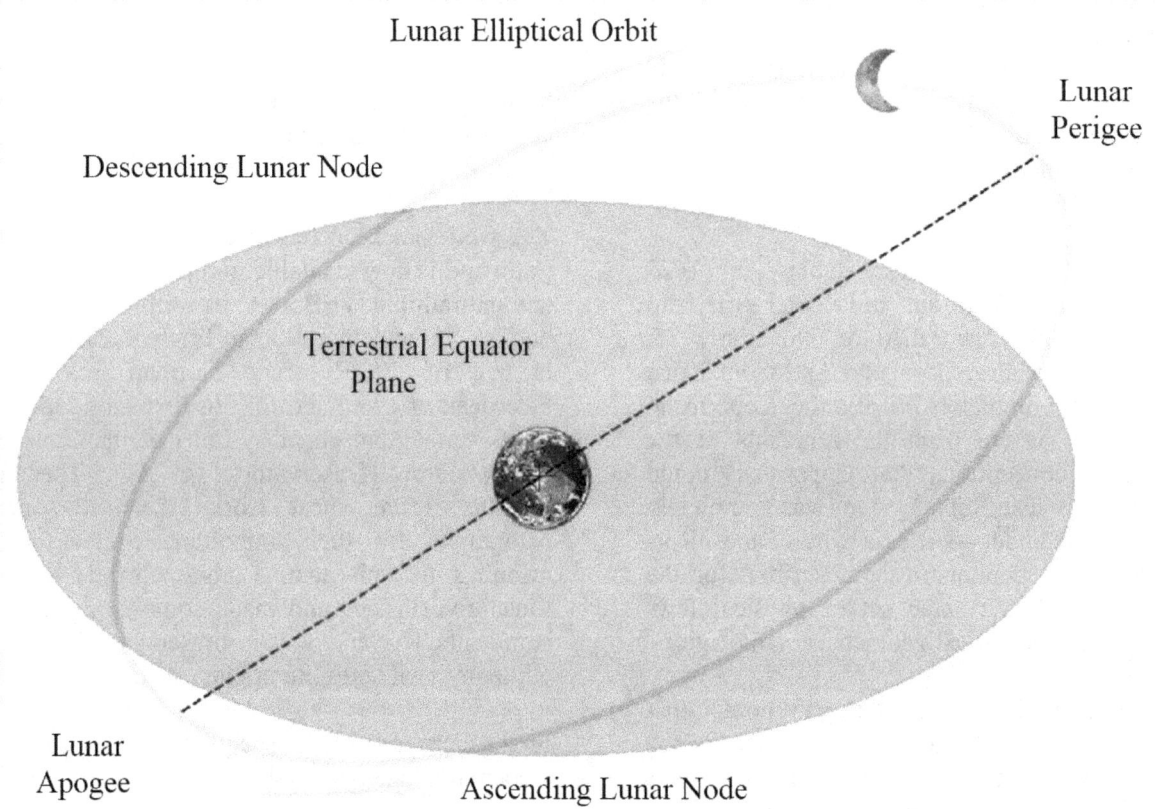

**Image 6.2
Lunar Elliptical Orbit**

Johannes Kepler extended his Laws of Planetary Motion to explain the kinematics of natural satellites around their respective planets.

Somnium, Seu Opus Posthumum de Astronomia Lunari

Joannis Kepleri

The Dream, or Posthumous Work of Lunar Astronomy

Johannes Kepler

Selected, commented and illustrated paragraphs.

In the first paragraphs Kepler made the introduction and contextualized his work. When he wrote that "*there were many malicious usurpers of the arts who, because they didn't understand anything, due to the ignorance of their mind, distorted them and enacted laws harmful to the human race*"; most likely such *malicious usurpers of the arts* refers to the members of the Church that persecute the philosophers of the time that promote theories contrary to the *Holy Scriptures*, and for that reason such clerics *enacted laws harmful to the human race*; such as to bring people to the fire !!.

He also described how *Duracotu*s, the protagonist of the work, through his mother knew one of the *Spirit*s of the region, who c*onjured* Levania and was responsible for directing the respective trip to the Moon. Then the author started the fictional and fantastic part of the text describing the way in which such a trip was made.

In 1608, when discord broke out between the two brothers, Emperor Rudolph and Archduke Matthias, the population scrutinized their actions comparing them with examples taken from the history of Bohemia. At that moment I was driven by the same curiosity to devote myself to the study of the legends of Bohemia; when I came across the legend of the heroine Libussa, so celebrated in the art of magic, something happened. One night, after carefully contemplating the Moon and the stars, I settled peacefully on my couch and fell into a deep sleep. While I slept, it seemed to have taken a book from the bookshelf to read it. The course of the book was as follows.

My name is Duracotus and my homeland is Iceland, called Thule by the ancients. My mother Fiolxhilde died recently, giving me a license to write, something I already ardently wanted to do. While she lived, she diligently made sure that I did not write, for she claimed that there were many malicious usurpers of the arts who, because they didn't understand anything, due to the ignorance of their mind, misrepresented them and enacted laws harmful to the human race. Under these laws, many men would surely have been condemned and devoured in the abysses of Hekla. My mother never told me what my father's name was. But she said he was a fisherman who died at the advanced age of 150, in the seventieth year of his marriage and when I was three years old.

In my childhood my mother, holding my hand or lifting me up in her arms, often took me to the lower ridges of Mount Hekla, especially around the feast of St. John when the Sun is visible for 24 hours and there is no night. She gathered many herbs and at home cooked them with various religious rites. She made small sacks from goat skins and when these were filled with herbal concoctions, she brought them to the neighboring harbor to be sold to placate the captains of the ships. Thus she provided herself with the means of sustenance.

Once, out of curiosity, I opened one of the sacks and took out the herbs and a linen cloth ornamented with needlework and displaying various symbols. My mother, unaware of what had happened, sold it. For I opened this bag, I defrauded her of her earnings. My mother, inflamed with rage, gave me to the skipper as his instead of the sack, so that she could keep the money. The next day, he unexpectedly set sail from the port in a favorable wind bound for Bergen in Norway. After several days under the rising north wind, he was brought between Norway and England. He went through the canal and went to Denmark because he had letters from an Icelandic bishop to give them to Tycho Brahe, the Danish, who lived on the island of Hveen. Then I, a fourteen-year-old boy, was getting very sick from the sudden jolting of the boat and the unusual temperature of the air. When the ship was unloaded along with the letters, the captain left me there in the house of an island fisherman and sailed with the promise of returning.

After the letters were delivered, Brahe, in a very good mood, began to ask me many questions. I did not understand him because I didn't know the language, except for a few words. He devoted all his time to his students, whom he cared for considerably and, because of Brahe's liberality, they could often talk to me. With a few weeks of practice I began to speak Danish in a tolerable way, until I was no less prepared to answer them than they were to interrogate me. I marveled at many unknown objects. I related many recent events in my homeland to my admirers. Finally, when the ship's captain came back to take me, Brahe held me back. This made me extremely happy.

The astronomical exercises pleased me to an extraordinary degree. For entire nights Brahe and his students devoted themselves to the study of the Moon and the stars using wonderful machines. This practice reminded me of my mother, as she also frequently talked with the Moon. For this stream of events, although they considered me a semi-barbarian because of my birthplace and my homeless-boy circumstances, I came to a great knowledge of the most divine of sciences, which prepared my way for greater achievements.

After living on this island of Hveen for several years, I wanted to visit my homeland. Because of the science I had acquired, I assumed that it would not be difficult for me to rise to some degree of honor in my own nation of unskilled men. I requested and achieved permission to depart from my employer; I said goodbye and went to Copenhagen. My fellow travelers took me freely under their protection because of my familiarity with their language and country. I returned home five years after I left.

My first source of joy on my return was discovering that my mother was still living and performing the same services as before. Since I was still alive and had the means to make a living, I ended up with her ongoing bitterness for having abandoned her son in a fit of rage. Autumn was approaching and those long nights of ours were drawing near too. Since during the month in which Christ was born the Sun hardly rises at noon and sets again immediately. My mother stayed close to me now that she was free from her work and didn't leave me no matter where I went. She questioned me about the lands I had visited and even questions about heaven. My mother was pleased to compare the degree of knowledge I had gathered and what she herself discovered to be true. She declared that she was now ready to die, so that she could leave her inheritor son the information that only she possessed.

By nature I had a great thirst to learn new things. Then I asked my mother about her arts and what teachers from her country stood out above the rest. Then, on a certain day, when she had leisure for speaking, she told me in this way everything she knew from the beginning: "Duracotus, my son, knowledge is available not only in other provinces which you visited but also in our own homeland. You have made me realize the charm of other regions. But even if we have coldness, darkness and other discomforts that I now feel oppress us, we still abound in talented people. We have among us highly gifted spirits who detest the bright light of other regions and the chatter of men; they perceive our dark areas and speak to us intimately. Of these spirits nine are important. One of these, by far the kindest and most innocent, was particularly known to me, and he is evoked by twenty-one characters. Often, in a fraction of a second, I was transported by his power to other coasts that I selected for myself. If I had kept away from certain places due to their distance, I now gained ground by asking about those places as if I had been there. He reviewed for me a lot of data about those objects that you examined with your eyes, accepted from a report or obtained from books. I would especially like you to be a spectator, my partner, from that region that he told me. How wonderful were those things that he told me about it. He conjured Levania."

Without delay, I agreed that she should summon her teacher. We sat together, and I prepared to hear both the full purpose of the trip and the description of the area. Spring was upon our region. As soon as the Moon was in the Half Moon, it began to shine once the Sun had hidden under the horizon, joined to the planet Saturn in the sign of Taurus. My mother, retiring from me to the nearest crossroads and throwing a few

words in a loud clamor, exposed her request. After completing the ceremonies she returned and, demanding silence with the palm of his right hand extended, she sat down near me. We had barely covered our heads with a cloth, as was the custom, when it behold there came the whistle of a hoarse and lisping voice, and immediately it began to speak in this manner, but in the islandic language.

The Spirit and the Voyage to Levania

The island of Levania is fifty thousand miles in the air. The journey to and from this island to our Earth is very rarely available; but when it's accessible it's easy for our people. However the transport of men, linked as it is to the greatest danger in life, is more difficult. We do not admit sedentary, very robust or delicate men in this entourage. We choose rather those who spend their time persistently riding fast horses, or who frequently sail to the Indies, accustomed to survive on doubly baked bread, dried fish, garlic and other unpleasant plates. There are dried old women especially suited for our purpose. The reason for this is well known, as from their childhood they are accustomed to riding goats, or on mantles, and to cross narrow corridors and the immense expanse of the Earth. Although the Germans are not suitable, we do not reject the dry bodies of the Spaniards.

The entire trip, much far it may be, is completed in four hours maximum. Because we are always very busy, our departure time is not until the Moon begins its eclipse in its eastern section. If the moon is full while we are still on the way, our return trip will be impossible. The occasion becomes so brief, that we have few humans and no other beings except the most useful towards us. Forming a column, we take advantage of any man of this type and all of us pushing up we elevate him to the heights. The initial stress is the worst part for him, as he turns upward as if by a gunpowder explosion and flies over mountains and seas. For that reason, he must be drugged with narcotics and opiates before the flight. His extremities should be carefully protected so that they don't torn from him, the body from the legs, the head from the body, and so that the recoil doesn't extend to each member of his body. Then he will face new difficulties: intense cold and alteration of breathing. These circumstances which are natural to spirits are applied forces to man. We continue our way placing moistened sponges in our nostrils. With the first section of the trip complete, our transportation becomes easier. Then we expose our bodies freely to the air and retract our hands. All these people gather in a ball within themselves because of the push, a condition that we ourselves produce almost by a mere signal from the head. Finally, upon reaching the Moon, the body goes of its own accord to its intended place. This critical point is of little use to our spirits because it is excessively slow. Therefore, as I said, we accelerate by gravity and we will go to the front of the man's body, lest by a very strong impact into the Moon he may suffer some damage. When the man awakes, he usually complains that all his limbs suffer from an ineffable lassitude, from which, however, he completely recovers when the effect of the drugs disappears, and thus he can walk.

Image 6.3
A Voyage to the Moon

Engraving made in 1868 by the French artist and sculptor Paul Gustave Doré (1832 – 1883)

There we quickly retreated to the caves and dreary places, lest the Sun, for the moment in the open air but about to eclipse itself, a little later expel us from an enjoyable resting place and force us to follow the departing shadow. Our ingenuity is exercised in the moments of decision. We join ourselves to the demons of this province and a society begins when the Sun begins to decay over the locality. Gathered together in crowds we deviate from our course in the shade. And if the shadow hits the Earth with its sharp point, which often happens, we will fall heavily on Earth and with our fellow soldiers, since we are not allowed any other result when men have witnessed the Sun's eclipse. From this it follows that the eclipses of the Sun are feared.

Lunar Geography and Lunar life

Next the author is dedicated to exposing the lunar geography, or selenography, the different climatic cycles, the varied lifestyles that the inhabitants of the satellite would take, and other possible life forms.

The whole of Levania extends no more than 1.400 German miles in circumference, a quarter of our Earth. It possesses very high mountains, very deep and wide valleys and, consequently, it yields a lot to our Earth in perfect roundness. The entire surface is porous, as it is traversed by hollow caverns and continuous caves, especially prolonged through the Privolvans. These hollow places are the main means that Privolvans have to protect themselves from heat and cold.

The Lunar Hemispheres: the Privolvan and the Subvolvan

In the same way that geographers divide the earthly globe into five zones due to celestial phenomena, Levania consists of two hemispheres, that of the Subvolvans and the other of the Privolvans. The circle that divides its hemispheres, similar to the colure of our solstices, passes through the celestial poles and is called a divisor. (Colure is either of the two principal meridians of the celestial sphere.)

In the following paragraphs Kepler exposes the way in which our planet is seen from the Moon, and he argues about the great differences in both Levanian hemispheres depending on the visibility or not of the Earth or Volva. First of all, he explains that our planet would be almost four times bigger than the satellite when it is seen from there.

The most pleasant of all the occupations in Levania is the contemplation of his Volva. The Levanians enjoy the sight of their Volva as we do with our Moon, which the Privolvans lack completely, because they are in the darkest. Due to its perennial presence of Volva, this region is called Subvolva, just as the rest is called Privolva because it has been deprived of the sight of its Volva.

When our Moon rises full and travels to distant homes, the inhabitants of Earth see it as the open circle of a large wooden barrel. When it rises to the center of the sky, the Moon brings to mind something like the shape of a human face. (Equally) The Subvolvans see their Volva in the middle of their own sky, which take this position for those who live in the middle or in the navel of this hemisphere, with a diameter a little less than four times as large as our Moon to us, so that if we establish a comparison of disks, their Volva's surface is fifteen times greater than our Moon.

What belongs separately to each hemisphere is the great diversity among them. Not only the presence and absence of Volva show very different spectacles, but these common phenomena differ so much here and there in their effects, that one could more correctly call the Privolvan hemisphere as intemperate, and the Subvolvan as temperate. In general, the Subvolvan hemisphere compares favorably with our cantons, cities and gardens, while the Privolvan resembles our fields, forests and deserts.

Lunar Life: the Selenites or Levanians

Then the author refers to the inhabitants and the different forms of life in the respective lunar hemispheres.

These are the appearances in both hemispheres of Levania: the Subvolvan and the Privolvan. From these considerations, it is not difficult for me to issue a silent judgment on how much the Subvolvanos differ from the Privolvans in other aspects.

Everything that springs from the ground, or walks on it, is of a monstrous size. The increases on size are very fast and life is short-lived because all living things grow to a bodily mass so enormous. The Privolvans don't have a fixed place of residence; in the space of a single day they traverse their entire world in hordes following the waters in retreat, either in their legs that are longer than those of our camels, on wings or on boats. If a delay of many days is necessary, they crawl through the caves according to each one's nature. There are many divers among them and all their living creatures breathe very slowly. By combining nature with art, they can take refuge at the bottom of deep waters; they say that those who are in the depths of the water withstand the cold, while the upper waves are boiling by the Sun. Those that remain on the surface are boiled by the midday Sun and serve as food for the wandering colonists.

Other creatures that find that breathing is more necessary, retire to caves that are supplied with water through narrow channels so that the water gradually cools down in its long path; but when night comes, they go out to eat.

The plants on the ground, and there are some on the tops of the mountains, sprout and die on the same day, daily making room for new growing things. The bark of trees, the skin of living creatures, or if something else takes their place, occupies most of the corporeal mass because it is spongy and porous.

If any creature is taken by surprise by the heat of the day, its skin becomes hard and scorched, and it falls at night. Others whose spirits have been exhausted by the heat of the day lose their lives, but return through the night due to a paradoxical cause, such as the production of flies here on Earth. Here and there, all over the ground, there are masses scattered in the shape of pineapples. Their husks are burned by the Sun during the day, and they die, but in the night they produce living creatures when the hiding places are opened. In the Subvolvan hemisphere a special means of relief from heat are clouds and uninterrupted storms, which sometimes take hold of half or more than half of the region.

Astronomy from the Moon

In the following paragraph Kepler introduces the reader to one of the main questions in the text: What is it like to practice astronomy from the Moon?, or in other terms, how do the firmament and the Universe look like when looking at them from the satellite?; and on this basis he proposes that studying astronomy from there it would be considerably different from doing it from our planet.

I am going to talk about the form of the province itself, beginning as geographers do with those things that happen to it from above. Even if the whole of Levania has in common with us (on Earth) *the appearances of fixed stars, yet one observes very many movements and numbers of planets different from those which we see from Earth so that all of their astronomy has another meaning.*

Next, Kepler presents his views on the difference in astronomy practiced from our satellite, and about the influence of the celestial movements on the passage of time and climate there. While the Earth rotates on its axis in a daily period of twenty-four hours, causing an apparent daily movement of the stars in the sky, the Moon rotates around its axis in a period of approximately 27,32 earthly days, or sidereal month, and moves around the Earth in the same time lapse in relation to the fixed Stars. These differences in the relative movements of these two bodies determine the distinct aspects in the apparent movements of the other celestial bodies, especially the Sun and the Fixed Stars, and therefore mark different ways to understand astronomy on each place.

The Earth and Sun Viewed from the Moon

Additionally, due to such a lag in the relative movements between the Earth and the Moon, from the same point on our planet we always see the same hemisphere of the satellite: from the same place we always see the same face of the Moon, which is facing the Earth, the other part we never see it and hence the traditional term *Hidden Face of the Moon*. But the fact is quite different for the satellite's inhabitants: Those on a hemisphere, that Kepler denominates Subvolvan, permanently enjoy the presence of their Volva, the Earth; while the inhabitants of the opposite hemisphere, the Privolvan, always lack this spectacle.

In previous chapters we have explained how the Moon and the Sun seem to be of the same size when we contemplate them from the Earth, very much in spite the fact that they actually have markedly different sizes; this phenomenon is due to the great difference between the respective distances: the King star is much bigger but it is also much farther away, and that's why it looks like the same size as our satellite. Let us now pose the question: How do the Earth and the Sun look like when viewed from the Moon? In this work Kepler's answer is that our planet looks like as it was almost four times the apparent size of the King star when contemplated from the lunar globe.

Of these two hemispheres, the Subvolvans always see their Volva, or our Earth, which for them is like our Moon; while the Privolvans are completely deprived of the vision of their Volva.

Since the stars are moving, Levania seems to be no less immobile for its inhabitants than our Earth for us. One of our months is equal to one of their nights and one day. (The Earth turns on itself in an earthly day, therefore) *Just as in one of our years the Sun revolves 365 times and the orbits of fixed stars 366 times; or more precisely, in four years the Sun revolves 1.461 times but the orbits of fixed stars 1465 times for us.* (But, according to Kepler's calculations, the Moon has a rotation speed 30,4 times lower than the Earth's) *So for them in one year the Sun goes around 12 times in a (levanian) year and the orbit of fixed stars 13 times; or more precisely, in 8 years the Sun goes around 99 times and the orbits of fixed stars 107 times. But they are more familiar with a 19 (levanian) year's cycle. In that number of years, the Sun ascends 235 times and the stars set 254 times.* (Therefore and according to Kepler, the Levanian year comprises 12,375 levanian days, and each of these lasts approximately 30,4 terrestrial days.)

The Sun rises in the central or innermost part of the Subvolvans when the last quarter of the Moon is visible to us; then it passes to the innermost parts of the Privolvans when the first quarter appears for us. What I say about the central parts must be understood of all the semicircles conducted through the poles and halves at right angles to the divisor. You can call them the semicircles of the Medivolvans.

A Static Earth, nailed in the sky

The Subvolvans' Volva remains as if it was fixed with a nail to the sky and was immobile in this place. (While) *Other stars and the Sun cross from sunrise to sunset. Nor there is any night when none of the zodiac's fixed stars hide behind the Volva, and emerge once more in the opposite region. Although the same fixed stars don't achieve this every night, they still all change completely between them; that is, those that move up to 6 or 7 degrees from the ecliptic. In 19 years the entire circuit is made to return exactly to their original positions.*

The Eclipses Seen from the Moon

As it has been said, due to the sizes and relative distances between the Sun and the satellite, it is coincidental that these stars, when viewed from the Earth, cover very similar angles in the sky and appear with a size apparently equal. But a very different situation manifests itself when the Sun and the Earth are looked at from the Moon; also because of the relative sizes and distances between these bodies, the Earth seen from the satellite looks like almost four times the size of both the King star and the Moon. And this marks a considerable difference in the way of perceiving the eclipses, both those of Sun and

those of Volva, which is the central idea on the next paragraphs in Kepler's text:

The most diligent observers see that this Volva does not remain the same size. During the hours of the day when the stars move rapidly, the diameter of the Volva is much larger, so it is clearly four times larger than our Moon.

Now, what should I say about the Sun's and the Volva's eclipses occurring in Levania at the same time that the eclipses of the Sun and the Moon occur here on the earthly globe, but evidently for different reasons? When we see the total eclipse of the Sun, their Volva is eclipsed, whereas when our Moon eclipses, the Sun is eclipsed for them. However, not in all these things I agree exactly. They themselves often see partial eclipses of the Sun when none of the Moon fails to us. On the contrary, they are often exempt from eclipses of their Volva when we have partial eclipses of the Sun. They have eclipses of their Volva in Full Volva just as we have those of the Moon in Full Moon; and they have eclipses of the Sun in the New Volva as we have them in New Moon. Because they have long days and nights, they experience the eclipses of both celestial bodies more frequently. A large number of our eclipses cross over to our antipodes, and theirs to their antipodes. The Privolvans do not see any of this, but the Subvolvanos do see everything.

The Subvolvanos never see a total eclipse of their Volva, but through the body of the Volva they see crossing a small spot, reddish on its edges and dark in the center. This small spot makes its entrance from the eastern section of the Volva, and exits the western edge; the same applies to the natural sites of the Volva, quickly anticipating them. The duration extends to one sixth of their hour, or four of ours.

The cause of the Subvolvans' solar eclipse is the Volva, as our Moon causes ours. This cannot occur without the Sun having to cross from the south-east through behind the immovable Volva into the west; because their Volva measures four times greater than the Sun. The Sun would then disappear very close behind the Volva with the result that part or the Sun's whole body would be hidden by it. There is often a great eclipse of the Sun's whole body because it lasts for several of our hours, when the light of both the Sun and the Volva is eclipsed at the same time. This is an important experience for the Subvolvanos who have other nights not as dark as these days (of eclipse) *due to the brilliance and magnitude of their omnipresent Volva. In the Sun's eclipse, both celestial bodies, the Sun and the Volva, are hidden from the Subvolvans* (so the darkness is total, ¡overwhelming!).

Image 6.4

Earth and Moon in space

Drawing from the book *Recreations in Astronomy*, written in 1879 by the American bishop and episcopal author Henry White Warren (1831 - 1912)

The drawing illustrates very well Kepler's statement referring to the fact that, looking from the Moon, the Earth exhibits phases similar to those of our satellite.

With regard to the Subvolvanos, the eclipses of the Sun have this point in common. Quite often it happens that the brightness rises on the opposite side when the Sun has barely hidden behind the body of the Volva, as if the Sun had expanded and encompassed the entire body of the Volva; however at other times and in many sections the Sun appears less than the Volva. Complete darkness does not always occur, unless the centers of the bodies coincide closely and, by the regular arrangement of the diaphanous centers, come together. The Volva suddenly disappears so that it cannot be discerned at all, for the Sun is completely hidden behind the Volva.

The Planetary Movement Seen from the Moon

Then Kepler describes how the course of the planets would look on the background of the Levanian sky:

In addition to the many inequalities (in the movements) *of the six planets, Saturn, Jupiter, Mars, the Sun, Venus and Mercury which we perceive, they see three others; namely: two in the length, one daily and the other during the cycle of 8,5 years; the third is in latitude during a circuit of 19 years.*

The Levanian Time and Climate

The differences in the relative movements of the Earth and Moon determine the differences in the apparent movements of the other celestial bodies, especially the Sun and the Fixed Stars, and therefore mark different rhythms of the passage of time and climate on each place: A complete sunny day in the Moon is equivalent to almost fifteen days of ours; or in other words, the same site of the satellite remains about fifteen earthly days under the suffocating heat of the Sun, from when this rises until it sets for that place.

It occurs one (sunny) *day* (equivalent to) *14 of our days or a little less in which the Sun seems larger. The Sun is slow under fixed stars and there are no winds. Then, everything becomes intolerable. Therefore, during the space of one of our months or of a Levanian day and in the same place, the heat* (during the day) *becomes fifteen times hotter than our Africa, and the cold* (during the night), *unbearable.*

Kepler then argues about the way in which the satellite's inhabitants would view the different portions of the Earth's illuminated hemisphere, and states that they would perceive a kind of *Earth phases* in a similar way as we see those respective *Moon phases*. He also explains how such terrestrial phases determine a special way of measuring the passing of time by the Levanian inhabitants.

The Volva of the Subvolvanos does not grow nor wane less than our Moon. This is due both for the presence of the Sun and for the digression of the Volva. If you study its nature the time is the same; but Subvolvans measure it by one method, and we do it by another. The Subvolvanos think that a day and a night is the space of time during which all the increases and decreases (phases) *of this Volva* (Earth) *are completed.* (On Earth) *We call this space of time a* (synodic) *month. The Volva rarely hides itself from the Subvolvans, even in new-Volva, on account of its size and brightness; especially for the Subvolvan polar inhabitants who lack the Sun at the time. On one Subvolvan night, even if it is 14 of our nights, the presence of their Volva illuminates the terrain and protects it from the cold; such a large mass and so much shine can only keep it warm. In general, for those who live between the Volva and the poles under the Medivolvan circle, the New Volva is the sign of noon and the first quarter of the afternoon. The Full Volva separates equal parts of the night, and the Last Quarter brings the Sun back.*

The night of the Privolvans lasts 15 or 16 of ours, terrible and with endless darkness, as are our nights without Moon. The rays of Volva never shine on them and for this reason everything becomes rigid by the ice, the frost and from the savagest and most powerful winds.

From these observations we can draw conclusions about those who inhabit the places described above.

I will explain first what is common to both hemispheres. All of Levania undergoes the same alternations on day and night as we do (on Earth), *but during the year there are no other annual changes. Throughout Levania their days are almost equal to their nights, except for the fact that for the Privolvans each day is regularly shorter than their own night, while the day of the Subvolvans is regularly longer. What is altered in a cycle of eight years should be mentioned later. Below both poles half of the Sun is hidden to equal the night, the other half shines forming a circle around the mountains.*

Image 6.5
Lunar day

Drawing in the book *Recreations in Astronomy* written in 1879 by the American bishop and episcopal author Henry White Warren (1831 – 1912)

The Subvolvanos differentiate the hours of the day by means of these and other phases of their Volva, so that the closer the Sun and the Volva are, the closer is the midday for the Subvolvanos and the afternoon or the sunset for the Medivolvans. The Subvolvanos are much better equipped than us to measure the night periods that regularly last 14 of our hours. We said that outside that sequence of Volva's phases, whose Full Volva marks the middle of the Medivolvan night itself, the Volva already determines their hours. Although the Volva does not seem to change its place in any way, our Moon, on the contrary, turns (around our planet) *within its place and adequately explains the surprising number of marks* (on Earth) *which persistently change from its rising to its setting. When the marks return after one of those revolutions* (a terrestrial rotation on its axis in twenty-four terrestrial hours), *the Subvolvans have an hour in time equal to a little more than one of our days and nights. This is then the only uniform measure of time* (that selenites have).

It is not enough that the Volva distinguishes the Subvolvans hours of the day in this way, but it also give clear indications of the parts of the year if anyone pays attention to it, or if the utility of fixed stars escapes to anyone.

Considering everything that Kepler has explained up to this point, we can establish the following conclusions regarding the way in which the Levanians measure their time: a Levanian hour is equivalent to a 24-hour earthly day; one Levanian day with its respective night lasts approximately one terrestrial month of 28 or 29 earthly days. Finally, a Levanian year has a duration of twelve Levanian days, equivalent to twelve terrestrial months and therefore it is also equal to one earthly year. But it should be noted that while on Earth we contemplate 365 Sun's raises, in Levania it only perceive twelve raises of the King star.

Also, due to the different course or movement of the stars on the Levania's firmament, especially the Sun, climate changes take place very differently than they do on our planet:

On Levania there are some variations of summer and winter, but they must not be compared with our own nor as we have in the same places at the same time of the year. In a period of ten years their summer changes from one part of the star-year to the opposite part, from the same intended place. In a cycle of 19 stars or in 235 (Levanians) *days summer occurs 20 times and winter with the same frequency towards the poles, and at the equator 40 times. As we have our months* (of summer), *they have 6 days in total during the summer, the rest belongs to the winter. The same alternation* (of climate) *is hardly felt around the equator because the Sun does not digress to the sides beyond 50 forward and backward from those places. But it feels closer to the poles and those places that have or do not have the Sun alternately at six-month intervals, as well as on Earth those of us who live below one of the two poles. The globe of Levania is also divided into five zones that correspond in a certain way to our terrestrial zones; that is, the Torrid and Frigid Zones have only 10 degrees each; all the rest falls in proportion to our Temperate Zone. The Torrid Zone passes through the middle parts of the hemisphere, half of its length through the Subvolvans, the other through the Privolvans.*

Ending the Dream

When I reached this part of my dream, the wind rose with a torrential rain that disturbed my sleep and ended one of the last books I brought from Frankfurt. While the Spirit, the orator, Duracoto with his mother Fiolxhilde and the listeners had been left behind, just as they had been with their heads covered, I returned to normal and found that my head was on a cushion and my body wrapped up in a blanket.

This work shows us the genially of the German astronomer. Kepler traveled to the Moon, imaginatively of course, and from there he practiced his profession, Astronomy. Immediately the astronomer realized the great difference between looking at the sky from the Earth or from the lunar globe. He devoted himself to studying, mentally, the celestial movements watched from the Moon; he also took care of the geography, the climate and the lifestyles of the imaginaries *Selenites*. His conclusions he left us in this exciting work.

Undoubtedly Kepler was the best astronomer of his time. He unveiled the most profound secret of celestial movements. As for the lunar globe, he probably did not intend to write a fictional text, that was not his style. But he was forced to do so because, as the sad experience of Giordano Bruno showed, in his time it was very dangerous to promulgate some ideas.

Galileo Galilei

Galileo Galilei (1564 – 1642) was a philosopher, mathematician, astronomer, physicist and Italian engineer, a great protagonist of the Renaissance and the Scientific Revolution. Galileo is the largest of the foundational period of modern science, excepting Newton. In astronomy his great achievements include the redesign and introduction of the telescope as a research tool in this science, his decided support to the Copernican Heliocentric Theory and a great amount of astronomical observations with transcendental discoveries. He is important as an astronomer, but perhaps even more as a physicist, for laying the foundations of dynamics as a science of body motion, discovering for the first time the great importance of acceleration in such movements.

Galileo was born in Pisa in 1564, seven years before Kepler, and died eleven years after him in 1642, the same year that Newton was born. He was the oldest of six sons of the marriage formed by Mrs. Giulia Ammannati and the mathematician and musician Vincenzo Galilei. The family was in charge of Galileo's education until the boy was ten years old, and then he was sent to the convent of *Santa Maria de Vallombrosa* in Florence, where he was introduced to religious life. His father wanted him to be a merchant, but when his intellectual abilities were very remarkable, he decided to send him to study medicine at the local University of Pisa, which Galileo joined at his age of 17 and where he studied medicine, philosophy and mathematics.

Although his father wanted the young to devote himself completely to medicine, he was definitely inclined towards mathematics, physics and astronomy. While in Pisa, the first signs of his scientific curiosity emerged stimulated by his observations of the cathedral's candelabras rhythmically swinging. Then he made some of the first real scientific experiments, and established that a pendulum of a specific length would oscillate with exactly the same period of time, whatever its amplitude and whatever the weight or nature of the suspended body: still a student, he discovered the *Law of the isochrony* in the movement of pendulums. He also devoted efforts to understand the movement of free fall of bodies, and established the bases for the development of a new branch of physics: *Mechanics.* His father was not rich and had to support other five children; with high college fees and since he could not get a scholarship for Galileo, the young man had to leave the university in 1.585 without any diploma.

Galileo continued demonstrating his inventive ingenuity by developing other instruments; and to financially sustain himself the young man seeks a job as professor in different universities. Finally Guidobaldo del Monte interceded for him before Ferdinando I de' Medici, Grand Duke of Tuscany, who hired him in November of 1589 for the chair of mathematics on University of Pisa, four years after he had been denied a scholarship and without a degree in something. He worked in this institution until 1592, year in which he joined the University of Padua to be professor of geometry, astronomy and mechanics; functions he performed up to 1610. In July of this year he accepted the positions of First Mathematician at the University of Pisa again, and First Mathematician and First Philosopher of the Grand Duke of Tuscany; for that reason he left Venice for moving to Florence. The material advantage of a new position as mathematician and philosopher of the Grand Duke of Tuscany was undoubtedly the main reason for his return to Florence in the summer of 1610, and there he would spend the eighteen most fruitful years of his life.

When he was a professor at the Pisa and Padua universities, he taught, as usual, about the Ptolemaic theory of the planets, adhering more or less faithfully to that system, which hardly coincided with the observed facts and did not fully convinced Galileo; who on the contrary was increasingly attracted to the Copernican system. He had written to Kepler in Graz claiming that he had converted to the Copernican theory many years before. These were Galileo's first steps in the astronomical field.

In the physics area Galileo was the first to establish the law of bodies that freely fall. He argued that each body, if left free, would continue to move in a straight line with uniform velocity, and that any change, whether in the speed or in the direction of the movement, must be explained as due to the action of some force. This principle was later enunciated by Newton as the first law of motion, and it is usually known as the *Law of Inertia*. This law explains an enigma that, before Galileo, the Copernican system had been unable to explain and that is related to the terrestrial rotation: If a body is released from the top of a tower, it will fall at the foot of the tower and will not deviate anything to the west. But, as the Earth is spinning, it should have moved some distance to the West during the downfall. The reason which this does not happen is for this object, precisely due to that Law of Inertia, preserves the speed of rotation that before being released shared with everything else on the surface of the planet.

Galileo had established the basic and unique property of gravity: all objects, whatever material they are made of and whatever their size and mass, fall identically under the influence of a gravitational field.

The Telescope and the first astronomical observations

The German manufacturer of ophthalmological lenses Hans Lippershey (1570 – 1619) is usually cited as the first person to request, in 1608, the patent for an artifact capable of making very far away objects appear as if they were really close to the observer. The news about this novelty would have spread quickly through Europe to reach the ears and the hands more indicated. In May of 1609 Galileo receives from Paris a letter from a former pupil, the French Jacques Badovere, who confirmed a persistent rumor: the existence of an instrument that allows seeing so amplified the objects that are very far away. Galileo was very enthusiastic about Lippershey's striking invention and hastened to redesign it, built his own lenses and thus made it a true astronomical observation instrument: a telescope ten times more powerful than the original design. In November of the same year, he manufactured an instrument with a resolution power, or *Magnification*, of about twenty times. Later he built a much more powerful and improved telescope, with an opening of almost five centimeters and which had a magnification of 33.

The magnifying power of a telescope does not refer to how many times the celestial body studied looks like bigger, but to how many times the instrument makes the same star seem more close to us; thus, a telescope with a magnification of 30 times makes us to see the object in question as if it was really 30 times closer to us.

When he directed his telescopes to the sky at the beginning of 1610, Galileo made multiple discoveries that seriously questioned the old cosmological theories: sunspots, mountains on the Moon, the phases of Venus, etc. But his most transcendental discovery was that of the four moons orbiting the planet Jupiter, since this allowed him to conclude that not all celestial objects revolved around the Earth and that our planet was not necessarily the exclusive center of the Universe, as Aristotle maintained.[6]

Apart from this effective reinvention of the telescope, the authentic Galileo's merit consisted on understanding its great utility as a scientific instrument in the study of the starry sky. The new artifact, initially conceived and used for common practical purposes as in war and navigation, became in their hands a powerful medium for studying the Universe. His great enthusiasm in his first telescopic examinations of the sky with such a modest and elementary telescope, but which revealed so many wonders of the cosmos, is quite remarkable in his notes and letters to colleagues, which were later collected and edited with more calm and scientific criteria in his famous text *Sidereus Nuncius*.

The *Sidereus nuncius*

On the cosmological level, Galileo decidedly adopted the Copernican heliocentric system; he corresponded with Kepler and accepted his novel laws of heavenly movement. In one of his

letters, Galileo wrote to Kepler wishing they could laugh together in the face of the stupidity of *the gang*. The rest of his letter makes it clear that the gang refers to professors of philosophy who attempted to discredit Jupiter's newly discovered moons, using *arguments that cut logic as if they were magical enchantments*. Galileo's telescopic observations and Keplerian laws of celestial movements combined to destroy forever the millennial and radical Aristotelian Ptolemaic belief in the fixed Earth and the perfection of the heavens, and for putting in its place the Copernican cosmological theory.

In his best known work, *Sidereus nuncius* or *Sidereal Messenger,* originally written in Latin and published in Venice in March 1610, Galileo presented the discoveries of his first telescopic observations. This work became the first astronomical treatise based on scientific observations made with a telescope, caused the final collapse of the Ptolemaic Aristotelian cosmology and marked the birth of modern astronomy. In it the Italian astronomer considered that Jupiter and its satellites are a strong proof that not all celestial bodies revolve around the Earth, and with it he tried to dissuade the followers of Ptolemy from their cosmological thoughts. And by the way, he wanted to correct some heliocentrists who blindly and persistently maintained that all celestial bodies revolved around the Sun.

The *Sidereal Messenger* is a short text in which Galileo presented in a very detailed way, resorting even to different graphical representations, his observations and discoveries made with his newly built telescopes. The astronomer paid special attention to our Moon, the planet Jupiter and its satellites, to Venus and Saturn, the Sun and many stars of the deep sky. Galileo was very insistent that his findings constitute very strong evidences against the traditional Ptolemaic Aristotelian theory, and, on the contrary, very much in favor of the novelty heliocentric model proposed by the Polish astronomer Nicolaus Copernicus.

The Moon as it was viewed by Galileo

The Aristotelian cosmology presupposed two great regions in the Universe: the *Sublunary world* that comprises the Earth and everything that lies between it and the edge of the lunar sphere, and has the characteristic of being an imperfect world always subjected to change; the other is the *Supralunary region* which includes the Moon and extends to the sphere of the fixed stars. This Supralunary region is ideal in nature and there are no more than perfect geometric forms represented by the spheres, and eternal and immutable celestial movements, that is, circular movements. But when Galileo directed his telescope to the Moon he found multiple arguments to contradict or refute that Aristotelian cosmology.

Image 6.6

Galileo teaching to Doge of Venice the use of the telescope

Fresco made in 1858 by the Italian painter Giuseppe Bertini (1825 – 1898).

Galileo always based on data provided by his telescopic observations that demonstrated the validity of his arguments. He observed a transient zone between the illuminated and dark regions of our satellite and delimited by a curve he called *Terminator*, and whose contour was not perfect, regular or continuous, but very irregular, sinuous or zigzagging; which allowed him to propose the existence of mountains on the Moon and thus invalidate Aristotelian hypotheses about a smooth, ideal lunar sphere. From these observations he deduced that the brightest regions of the lunar globe would correspond to geographic irregularities and equated them to terrestrial mountain ranges; while the dark regions would represent valleys, plains and depressions that had not yet been illuminated by sunlight. With which this astronomer was able to contradict the Aristotelian cosmology according to which the supralunar skies were perfect and their celestial bodies perfect spheres.

Among all the works of Galileo, the *Sidereal Messenger* was undoubtedly the one that caused the greatest sensation and aroused the greatest curiosity in the entire scientific world of the time. The text begins with the telescope's invention story and then addresses the description of the Moon's physical surface observations, the first celestial body that this great astronomer studied. The first discovery of Galileo with the help of his telescope and published in the *Sidereus nuncius* in 1610 is related to the mountains on the Moon. He described the lunar mountain ranges as very tall and that look very bright on the side facing the Sun, but which are very dark on the other side, and he presented detailed sketches and explanations for this. He noted that the summits of the highest mountains are illuminated at a considerable distance from the edge of the crescent moon; so, with simple geometric reasoning and taking the earth's diameter as a reference, he concluded that the lunar mountains are at least four times higher than those of our planet. His method of calculation was so brilliant but it was based on erroneous hypotheses that led him to distorted results.

In order to explain well his exposition, Galileo included in the *Sidereus nuncius* five detailed diagrams or sketches on the general physical appearance of the satellite's surface. He presented a sketch for each Quarter of the lunar phases, and then a repetition of the fourth to show that the features were not a coincidence but those were repetitive.

By demonstrating that the Moon was not a smooth and perfect sphere he refuted the Aristotelian thesis that the Supralunar world was perfect, immutable. In this regard, Galileo went further and took the trouble to elaborate and to present various drawings of the lunar surface as it was revealed by his telescope, and to make estimates on the height of its mountains, although them were wrong for using incorrect data on the distance of our satellite.

Apart from studying in detail the lunar surface with his telescope, Galileo also worried about the orbital movement of the Moon. He argued that the angular velocity of the satellite must be greater in the New Moon than in the Full Moon, because when it is closer to the Sun it describes a smaller orbit with reference to the King star. He compared the Sun with the suspension point of a pendulum, and the Earth and the Moon with two weights attached to the respective strings and each one forming a pendulum. If the Moon is placed at different distances from the point of suspension, the oscillation period of that pendulum is altered and this allowed him to conclude the satellite moves more quickly at the time of the New Moon than in that of Full Moon.

After having discussed the way in which sunlight reflects from the lunar surface towards the Earth and arguing that the same should happen in the case of our planet, he stated that this is a more blunt proof against those who sustain that the Earth must be excluded from the family of the planets claiming that it lacks both movement and light and that it doesn't shine like those. Instead, he confirmed, always by means of those demonstrations and natural observations, that the Earth moves and surpasses the satellite in light and brightness.

Jupiter and its moons revealed by the telescope

On January 7th, 1610, this astronomer made a transcendental discovery: he caught four small stars very close and revolving around the planet Jupiter. Galileo initially called these bodies the *Medicean Stars* in honor of Cosimo II de' Medici, Grand Duke of Tuscany, his former student and his great patron. Those were the moons of Jupiter today called *Galilean Satellites*: Io, Ganymede, Callisto and Europe. This fact constituted the most important and conclusive proof that not all celestial bodies revolved around the Earth, as erroneously established by the ancient geocentric Aristotelian Ptolemaic model.

The text reflects the emotion and amazement of Galileo when he observed the rapid movement, the appearance and the disappearance first of three, after four stars around Jupiter. On the first night of observation, on January 7th, Jupiter was accompanied by three neighbors that he considered as fixed stars, small but much brighter than other similar stars, and they appeared arranged as in a straight line parallel to the planet's equator. He continued seeing them on the nights of 8th and 9th but in different positions with respect to Jupiter, and for the night of the 10th he concluded that such changes on position was not due to the movement of Jupiter, but to that of the stars themselves. The above, along with the fact that they occasionally disappeared on one edge of the planet, hiding behind it and then reappearing at the other end, led him to the accurate conclusion that the stars were actually orbiting Jupiter in a plan close to its equator.

In the *Sidereus nuncius* Galileo presented many and detailed graphs on the relative positions of such stars according to his observations between January and March of 1610. The astronomer considering that these stars changed their relative position night after night, but always remaining around the same straight line, deduced that they were satellites of Jupiter. He had realized the biggest discovery in his live: four small stars moved around a larger planet, in a plane very similar to the plane of the ecliptic, in circular orbits of different amplitude and with a higher velocity as smaller was the distance from the satellite to the central planet.

Due to his persistent observations, he succeeded in following the four *Medicean Stars* in their respective movements, their relative positions with each other and with the planet itself, their occultations or eclipses and their transits in front of Jupiter. In this way he was able to quickly calculate the various elements or parameters of the orbital movements of such satellites: he found that the periods of revolution of each of the four stars varied from a little less than two days for the most closest, up to almost seventeen days for the outermost.

This Galileo's great discovery evidenced the existence of bodies that revolved around other larger bodies and, therefore, all together followed a single trajectory around the Sun: he determined that Jupiter with its family of four satellites employs a period of twelve years in his great revolution around the King star, similar to the Earth with his Moon in its interval of a year. The astronomer soon realized that such occurrence would provide a formidable argument to eliminate the doubts of those who opposed the heliocentric system: That Jupiter would have satellites was very problematic, because it suggested that there were at least other centers of rotation in the universe, apart from the earth itself.

Galileo looks to Venus

In September of 1610 the astronomer discovered that the planet Venus manifests cyclic phases similar to those of our Moon; and he considered this fact as an additional test in favor of the Copernican system, which manages to describe this phenomenon thanks to the heliocentric hypothesis. Although he delayed its publication until 1623 when *The Assayer* (*Il Saggiatore* in Italian) was published, he made the observation on the phases of Venus in 1610. Galileo contemplated the phases and the variation in size of the illuminated part of the planet; which constitutes a proof that Venus revolves around the Sun: the largest size is manifested when the planet is in the full phase, while the smaller size occurs in the new phase.

Galileo observed a complete set of phases of Venus with the help of his telescopes, and considered what he saw as an emphatic proof of the veracity and superiority of the Copernican system. For his time, the phases of Venus were very disconcerting: according to the three main planetary systems of the moment, the Ptolemaic, the Copernican and the Tychonian, the three stars could be aligned with Venus between the Earth and the Sun, that is to say in the EVS order, situation in which such planet would not be perceptible to the naked eye, since the sun's own glow would hide it and the situation would be very analogous to that of the New Moon. On the other hand, the ESV alignment gives rise to a fully illuminated Venus, which is similar to the case of a Full Moon. As conclusion: all systems predicted a set of phases for this mysterious planet, but the Ptolemaic does not predicted the whole set.

With its appearances similar to those of our Moon, Venus attracted Galileo from the beginning of his observations, and this circumstance convinced him of the movement of the planet around the Sun. On the basis of his discoveries and reasonings, Galileo now decidedly and openly supported the Copernican doctrine with all the known problems that were to arise.

Other observations: the Milky Way, the Sun, Fixed Stars and Saturn

Later the Italian astronomer dedicated to contemplate the Milky Way, he focused on the constellation of Orion and verified that certain stars that at first glance seem only one, actually they are sets or clusters of stars.

Thus, the number of stars visible with his telescope increased; but unlike the Moon, the planets and the Sun, such stars did not increase in size; which proved the Copernican hypothesis about the considerable distance that the fixed stars are and the existence of a huge gap between Saturn and those stars.

Late on 1610 in Rome, the astronomer made another discovery that refuted the idea on perfection of the heavens: sunspots. Galileo argued that they are on the surface of the Sun and having a rotational movement; reason why the astronomer determined that the King star is rotating on itself, which can be used as an argument in favor of earth's rotation. Later on, in the *Dialogue on the systems of the world*, Galileo took up that theory of sunspots turning it into a powerful argument against the cosmologic system of Tycho Brahe, the only and last refuge left to geocentrists.

Galileo also devoted much time to the planet Saturn and detected its rings but he couldn't identify them as such, but as strange appendages or as two handles, one on each side of the star. He thought that the planet didn't have a spherical shape and that better it seemed as a globe with handles. Galileo wrote that the body was not one alone, but that it was composed of three stars that almost touched each one, and that never moved or changed positions with respect to each other. He also added that they were arranged in a line parallel to the zodiac, that the one in the center was about three times the size of those on the sides and that they had a very similar shape to this: ºOº.

Dialogues about the systems of the world

Later, and protected by Pope Urban VIII and the Duke of Tuscany Fernando II de Médici, Galileo published in Florence on February 21, 1632, his *Dialogo sopra i due massimi sistemi del mondo tolemaico e copernicano*, or *Dialogue Concerning the Two Chief World Systems Ptolemaic and Copernican*, where he openly and forcefully ridiculed the ancient Aristotelian ptolemaic geocentric system. The book is definitely pro-Copernican and the dialogues take place over four days in Venice and include three interlocutors: *Simplicio*, a radical and obtuse Aristotelian defender of Ptolemaic cosmology; *Salviati*, a Florentine defender of Copernican heliocentrism and who probably represents the author himself; and *Sagredo*, a neutral Venetian but very enlightened and eloquent. The discussions of these protagonists go around the

two current conceptions of the Universe: the Aristotelian-Ptolemaic and the Copernican one.

His theory of ocean tides, presented on the fourth day of the *Dialogue Concerning the Two Chief World Systems*, is brilliant and characteristic of the genius of the Italian astronomer; pity that it is the only one he presented in this work that is wrong. According to Galileo the two movements of the Earth, the one of rotation on itself and that of translation around the Sun, cause that the points located in the terrestrial surface undergo accelerations and decelerations in cycles of twelve hours, which would originate the tides observed in the oceans. Although he was wrong in his approach, Galileo completely discredited, in his time, the theory about the lunar origin of the causing forces of such phenomena. We know today that the complicated phenomenon of the tides is caused by the gravitational attraction of the Moon and the Sun on the huge bodies of water on Earth. Actually, the effect of the Moon is stronger because it's closer to Earth than the Sun.

Unlike Copernicus and Kepler, and because he had friends in the Church, Galileo was not fearful when publishing his scientific thoughts; and this he would pay expensive. In the *Dialogue* the astronomer criticized the Aristotelian cosmological principles and defended his own with such vehemence, that he achieved enemies in some ecclesiastical spheres. After its publication in 1632 in Florence, Galileo was ordered to go Rome for giving account of his actions before the Inquisition, where he was judged as a heretic and for contradicting the cosmological theological dogmas of the time. He was convicted to life imprisonment and the sentence was signed by seven cardinals, although it was not ratified by Pope Urban VIII who probably considered Galileo not so much a heretic but rather a superb and thoughtless. According to the legend, when the astronomer was leaving the courtroom he was stomping the floor and murmuring: "*Eppur si mouve*", "*And yet it moves*", (the Earth).

Galileo was condemned initially by the Inquisition in privately session in 1616, and then publicly in 1633, this last occasion he had to retract his cosmological theories and to promise that never more would sustain the idea that the Earth has any kind of movement. But promptly the initial sentence of life imprisonment was changed by house arrest; Galileo was approaching seventy years old and he would live another nine, the last four he was blind. After his trial Galileo remained confined in his residence in Arcetri, Florence, from December 1633 to 1638. There he lived in acceptable comfort, he resumed his studies on the movement of the bodies he had begun forty years before in Pisa and compiled them into a masterly text: *Discourses on two new sciences*, published on 1636 and which became the last work written by Galileo; in it he established the fundamentals of mechanical physics and introduced the concept of inertia The work marked the end of the validity of Aristotelian physics and became the basis for the modern science of the *Dynamics of movement*.

While living in Arcetri Galileo received some visits from his friends, which allowed that some of his works in course of writing could cross the border. In January 1638 Galileo definitively lost his sight and then received authorization to settle in his house in San Giorgio, near the sea, there he remained some years surrounded by several collaborators, working in astronomy and other sciences. Finally, on January 8, 1642, Galileo died in Arcetri at age 78, and the funeral honors were made in Florence on January 9.

Similar to Kepler's *Somnium*, Galileo's *Sidereus nuncius* has been translated into multiple languages, and now the work is available online on different WEB sites and in printed books too; for this text, I was based on the English translation made by Edward Stafford Carlos which is on the *Wikisource* and *Project Gutenberg* websites.[7] From there I selected some paragraphs according to their importance for our theme, I included some modern reproductions of the original images of the text; and finally, I annexed some of my own comments.

Image 6.7

Cover of the *Sidereus nuncius*, Venice edition of 1610.

Sidereal Messenger

*Unfolding great and marvelous sights,
and proposing them to the attention of every one,
but especially philosophers and astronomers,
being such as have been observed by
Galileo Galilei
a gentleman of Florence, professor of mathematics
in the university of Padua, with the aid of a
telescope lately invented by him,*

*Respecting the Moon's Surface, an innumerable number
of Fixed Stars, the Milky Way, and Nebulous Stars,
but especially respecting Four Planets which revolve
round the Planet Jupiter at
different distances and in different periodic times,
with amazing velocity, and which, after remaining
unknown to every one up to this day, the
Author recently discovered, and
determined to name the
Medicean Stars.*

VENICE, 1610

Selected, illustrated and commented paragraphs.

Very beautiful and admirably nice is to see the body of the Moon, away from us almost sixty terrestrial radiuses, as close as only two of these dimensions; so that the same diameter of the Moon appears almost thirty times larger, its surface almost nine hundred, and the volume almost twenty-seven thousand times greater than when viewed with the naked eye: and therefore, with the certainty of sensible experience, anyone can understand that the Moon is not covered by a smooth and polished surface, but rough and irregular, and, like the face of the Earth, it is everywhere full of large protuberances, ravines and deep cavities.

But what surpasses all the wonders, and mainly led us to warn all astronomers and philosophers, is to have discovered four wandering stars, not known or observed by any of them before us; which, like Venus and Mercury around the Sun, have their revolutions around a certain prominent star among known ones, and sometimes are in front of it, sometimes behind it, though they never depart from it beyond certain limits. And all these things were discovered and observed a few days ago with the help of a pair of glasses that I invented after receiving the illumination of Divine Grace.

First we will say of the hemisphere of the Moon that turns towards us. For clarity, I divide the hemisphere into two parts, one clearer and the other darker: the lighter one seems to surround and fill the entire hemisphere, while the darker one clouds the face itself and makes it appear to be full of scattered spots. Now these spots, as they are somewhat dark and of considerable size, are plain to everyone, and every age has seen them, wherefore I shall call them great or ancient spots, to distinguish them from other spots, smaller in size, but so thickly scattered that they sprinkle the whole surface of the Moon, but especially the brighter portion of it.

And these spots were not seen by anyone before me. From repeated observations of these points, I came to the conviction that the surface of the Moon isn't perfectly smooth, uniform and exactly spherical, as many philosophers believe about it and other celestial bodies, but uneven, rough and with many cavities and protuberances, not unlike the surface of the Earth diversified by mountain ranges and deep valleys. The things that I saw and from which I could draw these conclusions are the following:

On the fourth or fifth day after the New Moon, when the Moon shows its radiant horns, the dividing line between the part in shadow from the shining part does not spread uniformly according to an elliptical line, as would occur in a perfectly spherical solid, but it's drew by an uneven, rough and very sinuous line. In fact, much luminosities such as excrescences extend beyond the limits between the light and dark, and on the contrary some dark particles are introduced into the illuminated part. I also noted that the small points mentioned above coincide, all and always, in this: in having the black part directed towards the Sun, while those on the opposite side of the Sun are crowned by bright contours, almost as burning mountains. A spectacle similar to the one we have on Earth during the sunrise, when the valleys are not yet illuminated but we see the surrounding mountains shining on the opposite side of the Sun. Likewise, the same as the shadows of the terrestrial cavities diminish as the sun rises, these lunar spots also lose their darkness as the luminous part grows.

Not only it can be seeing the limits between light and dark in the uneven and winding Moon, but, what awakens more astonishment, in the dark part of the Moon there are many bright cusps, completely separated and differentiated from the illuminated part and away from it a distance not small. After a certain time, they gradually increase in size and brightness, and after two or three hours they join the illuminated part that has already become larger. Meanwhile, more and more peaks like sprouting here and there are illuminated in the dark part, they enlarge and finally they also join the luminous part that has been widening more and more. The previous figure also gives us an example of this phenomenon. And on Earth, before the Sun rises and while the shadow still occupies the plains, ¿Are not the tops of the highest mountains

illuminated by the Sun's rays? ¿Perhaps is it not true that, as time goes by, the sunlight is scattered until the lower and wider parts of those mountains get illuminated; and that in the final, after the Sun has risen, the illuminated parts of the plains and mountains come together? The diversities in such protuberances and cavities of the Moon seem to exceed the roughness of the Earth's surface, as I will show later.

Meantime, I will not overlook a fact worthy of attention that I observed while the Moon began its first quarter, as shown in the above drawing. In the luminous part a large dark protuberance penetrates direct towards the lower horn, which I observed long and discerned as completely dark, and that after about two hours began to show, just below half the sinuosity, a kind of luminous vertex. This was growing little by little, took a triangular shape and remained completely separate from the luminous part. A little later, three small points began to shine around it, until, turning the Moon already at sunset, the triangular figure, extended, widened and as big as a great promontory, joined the remaining luminous part, still surrounded by the three mentioned points. However, as I also said above, the blackish part of the spot is directed towards the solar irradiation, while a more resplendent strip that surrounds the spot on the opposite side of the Sun, turns towards the dark region of the Moon.

In fact, the large spots of the Moon do not look like broken or full of depressions and protuberances, but more equal and uniform; in fact, only small bright areas appear here and there. So if someone wishes to unearth the ancient opinion of the Pythagoreans, that is, that the Moon is almost a second Earth, the brightest part would best represent the solid surface, while the darkest would be the water; and I never doubted that, looking from distance the earthly globe illuminated by the Sun, the terrestrial surface appeared lighter, and the watery part darker.

Also, on the Moon, the larger spots are more depressed than the brighter parts; in fact, whether in a Crescent or Waning Moon, always on the borderline between light and darkness the contours of the brightest part around the large spots stand out, as I observed when making the figures. The brightest part stands out especially in the vicinity of the spots and around a certain point located in the upper or northern part of the Moon, so that before the first quarter, and probably also in the second, it rises considerably above and underneath the large protuberances, as shown in the figures.

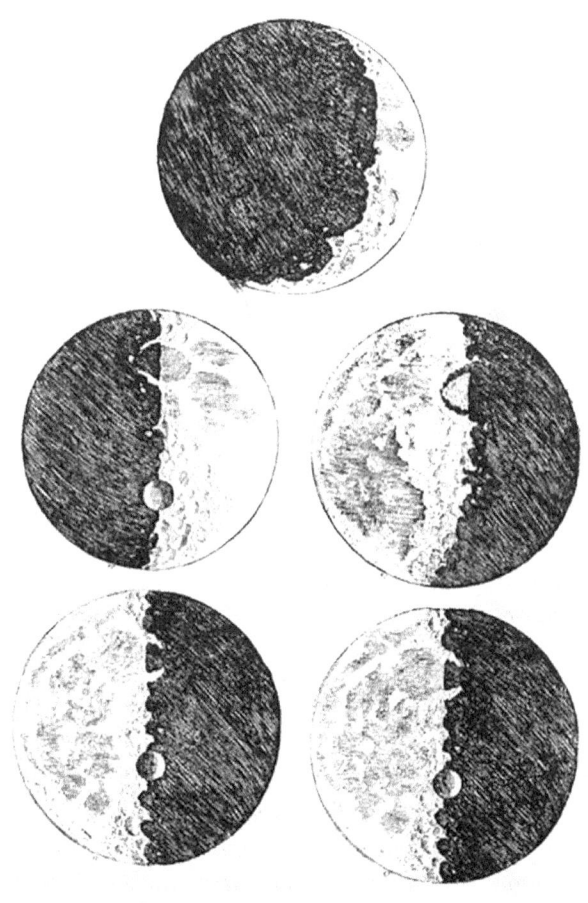

Image 6.8
The Moon as Galileo saw it

Modern reproduction of the graphic representations of our satellite as Galileo Galilei saw it with his telescope in 1610.

The original images, drowns by himself, were included in his greatest work *Sidereus Nuncios*.

And I also want to remember something else that I noticed not without a certain wonder: almost in the middle of the Moon there is a larger cavity than all the others and perfectly round in shape. Which was close to the two Quarters, so it was possible to reproduce it in the two previous figures; as far as lighting is concerned, it offers the same appearance that on Earth would offer the similar region of Bohemia if it were surrounded by very high mountains and arranged in perfect circle. The one of the Moon is surrounded by mountains so high that the extreme region bordering the dark part is illuminated by the solar ray before the boundary between light and shadow reaches the diameter of the figure itself. As in other points, the shaded portion of this too faces the Sun while the bright part is towards the dark side of the Moon. For the third time I call attention on this: between these places always the darkest ones are close to the boundary between light and darkness; the farthest, on the other hand, now seems smaller and now less dark; so that when the Moon, in the opposition, is full, there is very little difference between the darkness of the valleys and the splendor of the peaks. As in an incontrovertible testimony of the asperities and inequalities that are found in the lightest part of the Moon.

Since in fact on the Moon and around its perimeter there are many dispositions of prominences and depressions, the eye that looks from afar is almost on the same level as the summits of those prominences: no one should be surprised that the look that touches them is presented according to a uniform line and not steep at all. Another reason can be added to this explanation: there is around the body of the Moon, just as around the Earth, an envelope of some substance denser than the rest of the ether, which is sufficient to receive and reflect the Sun's rays, although it does not possess so much opaqueness as to be able to prevent our seeing to penetrate through it, especially when it is not illuminated. Such a cover, when illuminated by the Sun's rays, renders the body of the Moon apparently larger than it really is, and, if its thickness were greater, it would be able to stop our sight from penetrating to the Moon's solid body. Now well, it is of greater thickness about

the circumference of the Moon, greater, I mean, not in actual thickness but with reference to our sight-rays which cut it obliquely; and so it may stop our vision, especially when it's in a state of brightness, and may conceal the true circumference of the Moon on the side towards the Sun.

That the lighter surface of the Moon is dotted everywhere with swellings and depressions, I believe that it is sufficiently manifest by the phenomena already explained. It remains to be said of its magnitudes, which show how the asperities of the Earth are much smaller than those of the Moon's; of minor importance, I say, but also speaking in an absolute sense, not in relation only to the dimensions of the terrestrial and lunar globes, and this is clearly demonstrated in this way.

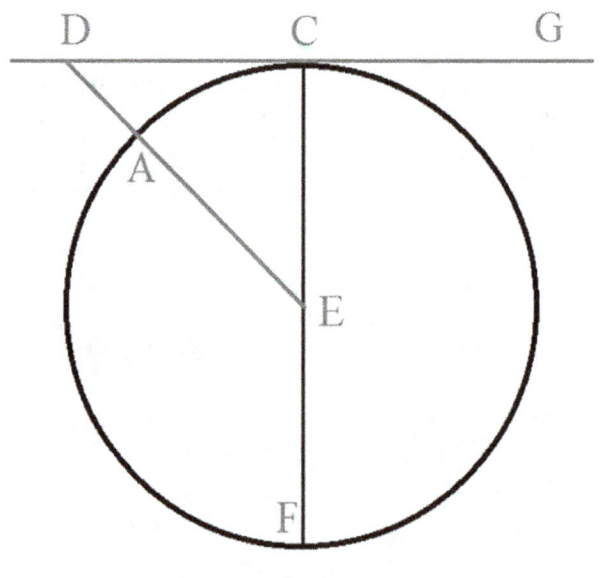

Image 6.9

Calculation of the height of the lunar mountains

Reproduction of a graph for the determination of the lunar mountains' height, which appeared on the *Sidereus nuncios*.

Having observed the Moon several times over the Sun, I detected that in the dark part of the Moon some vertices, although quite far from the limit of light (the terminator), seemed saturated; and by comparing their distances with the total diameter of the Moon, I verified that this distance sometimes exceeds one twentieth of the diameter. Suppose the distance to be exactly 1/20th part of the diameter and let us the diagram (in Image 6.9) represent the Moon's orb, whose maximum circle is CAF, E its center and CF a diameter, which consequently bears to the diameter of the Earth the ratio 2:7; and since the diameter of the Earth, according to the most exact observations, contains 7000 Italian miles, CF will be 2000, and CE 1000; and the 1/20th part of the whole CF will be 100 miles.

Let now CF be the diameter of the maximum circle that separates the luminous part of the Moon from the dark one, which does not differ significantly from that other maximum due to the great distance of the Sun from the Moon, and let the distance of A from the point C to be 1/20th part of that diameter. Let the radius EA be drawn and let it be extended to cut the tangent line GCD, which represents the ray that illumines the summit, in the point D; then the arc CA or the straight line CD will be 100 of such units. As CE contains 1000, the sum of the squares of DC and CE is then 1010000, which is equal to the square of DE; therefore the complete DE will be more than 1004, and AD more than 4 since EA is1000. Therefore the height AD, that on the Moon indicates any projection that reaches the Sun's ray GCD and separated from the limit C by a distance CD, is then greater than 4 Italian miles. But in the Earth there are no hills which reach to the perpendicular height even of one mile. We are therefore left to conclude that it is clear that the prominences of the Moon are loftier than those of the Earth.

At this moment I am pleased to clarify the cause of another lunar phenomenon worthy of admiration that I observed. When the Moon, both before and after the conjunction, is not far from the Sun, its sphere not only appears in our sight in the part adorned with the bright horns, but also in the dark part, naturally the opposite to the Sun, a certain faint peripheral glow seems to draw a circle and differentiate it from the darker background of the ether itself. But if someone looks for something to hide those horns, like a roof, a chimney, or any other obstacle between the view and the Moon, but distant from the eye, and in such a way that the remaining part of the lunar globe would be visible to him, then he would be surprised to see that this region of the Moon also shines with no little light, even though it is deprived of sunlight, something more visible when the frost of the night has grown due to the lack of Sun.

Therefore, this second glow is not congenital and proper to the Moon, and is not provided by any star or the Sun, since there is no other body in the vastness of the universe than Earth. So, What solution should be proposed? Isn't it that the body of the Moon and any other opaque and dark body are illuminated by the Earth? What disturbs us? Here's the reason: The Earth, with a fair and grateful exchange, pays back to the Moon an illumination like that which it receives from the Moon nearly the whole time during the darkest gloom of night. Let's see it more clearly.

The Moon in conjunctions, when it is between the Sun and the Earth, is illuminated by the solar rays in its upper hemisphere opposite to the Earth; while the lower hemisphere that looks at the Earth is wrapped in darkness and, therefore, does not illuminate the Earth at all. The Moon, continuously moving away from the Sun and gradually illuminating in some part of the hemisphere in front of us, shows us the white horns, but still thin, and lightly illuminates the Earth. The Moon grows and approaches the quarters, the illumination of the Sun increases the reflection of its light on the Earth; the light of the Moon extends then along a semicircle and our nights are more illuminated. Finally (in opposition) the entire lunar hemisphere in front us and the Sun is shining with very bright rays: (during the night) the entire surface of the Earth is illuminated, perfumed by the moon's (reflected) light. Then, waning, the Moon sends us fainter rays and the Earth is weaker illuminated; the Moon begins (again) the conjunction and the dark nights fill then the

157

Earth. But with equal measure, the Earth corresponds. In fact, while the Moon is in conjunction with the Sun it faces the entire surface of the Earth's hemisphere exposed to the Sun and vividly illuminated, and it receives by reflection that earthly light; therefore, the lower hemisphere of the Moon, devoid of sunlight, as a result of this reflection (it is somewhat illuminated and) looks not very bright. The Moon, moving away from the Sun's quadrant, sees only half of the illuminated hemisphere of the Earth, the West, because the eastern half is shrouded in darkness: therefore, the Moon itself is less illuminated by the Earth and its secondary light appears to us darker.

If the Moon is placed in opposition to the Sun, it will see the hemisphere of the Earth that is between it and the Sun completely dark and impregnated with black night; if then this opposition occurs in the ecliptic, the Moon will not receive any illumination, deprived of both the illumination of the Sun and that of Earth (and then it will be totally eclipsed). In its different positions between the Earth and the Sun, the Moon receives more or less light reflected by the Earth, depending on whether it looks at a larger or smaller part of the illuminated terrestrial hemisphere. Because this relationship occurs between the two globes: when the Earth is more illuminated by the Moon, the Moon receives less light from Earth, and vice versa.

Therefore, we have a valid and excellent argument to remove any doubt from those who, even silently accepting the revolution of the planets around the Sun in the Copernicus system, are so disturbed by the movement of the Moon around the Earth, while each year they together make their revolution around the Sun, to believe that this structure of the universe should be rejected as impossible. Now, in fact, we do not have a single planet that revolves around another, while both travel in the great orbit around the Sun; but the sensitive experience shows us four stars which wander around Jupiter, in the same way as the Moon turns on the Earth, while all together with Jupiter rotate in a wide orbit around the King star with a period of twelve years.

And finally we should not omit the reason for which the Medicean stars appear sometimes to be twice as large as at other times, though their orbits about Jupiter are very restricted. We cannot search for the cause in terrestrial vapors, because they seem larger and smaller while Jupiter and the nearby fixed stars are unchanged.

That they approach and recede from the Earth at the points of their revolutions nearest to and furthest from the Earth to such an extent as to account for so great changes seems altogether untenable, for a strict circular motion can by no means show those phenomena; and an elliptical motion (which in this case would be nearly rectilinear) seems to be both untenable and by no means in harmony with the phenomena observed. The solutions that come to mind in this respect, I expose them in good taste, and offer them to the judgment and criticism of the philosophers. It is known that due to the interposition of terrestrial vapors, the Sun and the Moon seem larger, while the fixed stars and planets are smaller: therefore, the two luminaries near the horizon seem larger, while the stars, smaller and just visible, they become smaller even more if those vapors are illuminated; for this reason the stars during the day and in the twilight seem very dim, but not the Moon, as we previously noted. And that not only the Earth, but also the Moon, is surrounded by vapors, it results both from what we said above and from what we will say more extensively in our System. Therefore, we can correctly believe the same for the other planets, and it does not seem altogether improbable that there is a denser envelope than the ether around Jupiter as well, around which, like the Moon around the sphere of the elements, the Medicean planets rotate. And because of the interposition of this ethereal sphere, they will be smaller when they are in the apogee, but larger when they are in the perigee, according to the disappearance or attenuation of that same sphere.

The *Sidereus nuncios* constitutes the first astronomical treaty based on tests and on data provided by systematic observations made with a telescope. Unlike Kepler, which relied on purely theoretical and rational arguments,

Galileo is based on undeniable experimental evidence to describe both the Moon and other bodies and other celestial phenomena. For this reason, the *Celestial Messenger* becomes the first scientifically elaborated description of our satellite.

This great Italian astronomer redesigned the apparatus now known as telescope and turned it into a true scientific research instrument in the field of astronomy. Apart from his great achievements in the celestial science, Galileo also devoted himself to physics and discovered natural laws, those related to the movement of bodies in the terrestrial sphere, such as pendulums and other bodies that freely fall. So, alongside Copernicus and Kepler, Galileo Galilei is one of the titanic characters founders of modern science.

Image 6.10

The Moon photographed by the astronauts of Apollo 17 mission in December 1972

This image is included here because it is very representative and illustrates very well what was exposed by both Kepler and Galileo in their respective works.

Bibliographic Citations

[1] Gutenberg Project. George Tucker. *A Voyage to the Moon*. http://www.gutenberg.org/ebooks/10005

[2] Internet Archive. Verne, Jules. *From the earth to the moon: direct in ninety-seven hours*. https://archive.org/details/cu31924086273913/page/n6

[3] Dreyer, John L.E. *A History of Astronomy from Thales to Kepler*. Cambridge: Dover Publications, Inc. 1953.

[4] Abetti, Giorgio. *The History of Astronomy*. London: Sidgwick and Jackson, 1954.

[5] Frosty Drew Observatory & Sky Theatre. Les Coleman. Somnium - A Dream. By Johannes Kepler https://frostydrew.org/papers.dc/papers/paper-somnium/

Anecdotario. https://anecdotariohistoricoactual.wordpress.com/2014/09/06/somnium-kepler/

Creighton University. The Somnium Astronomicum by Johann Kepler Translated, with Some Observations on Various Sources. https://dspace2.creighton.edu/xmlui/handle/10504/109241

The Somnium Project. https://somniumproject.wordpress.com/somnium/

[6] John D. North, *Fontana History of Astronomy and Cosmology*. Mexico: Economic Culture Fund, 2005.

Philip's Astronomy Encyclopedia, London: Philip's, 2002. http://www.sozvezdiya.ru/Philip's_Astronomy_Encyclopedia_2002.pdf

[7] Wikisource. The Sidereal Messenger. https://en.wikisource.org/wiki/The_Sidereal_Messenger

Project Gutenberg. The Sidereal Messenger of Galileo Galilei. https://www.gutenberg.org/files/46036/46036-h/46036-h.htm

Liber Liber. Galileo Galilei Sidereus Nuncius. https://web.archive.org/web/20110629154736/http://www.liberliber.it/biblioteca/g/galilei/sidereus_nuncius/html/index.htm

University of Oklahoma University Libraries. Galileo Galilei, Siderevs nuncius (Venice, 1610). https://digital.libraries.ou.edu/histsci/books/1466.pdf

Chapter 7

Modern Astronomy And the True Lunar Movement

The lunar globe as it was visible on Friday 09, February 2018 at 06: 00 TUC, 23 days 3 hours 30 minutes after New Moon, it achieving to see the 35.0% of its hemisphere illuminated by the Sun; this lunar phase is called Waning Quarter.

Courtesy of NASA's Scientific Visualization Studio: https://svs.gsfc.nasa.gov/4604

"SUPPLY OF THE LUNÍCOLAS TO THE SUN

SOLAR MAGEST:

The lunícolas, so subjugated and prostrate in front of the immense circumference of your excessive grandeur, and assuring you perpetuity of eternal light, ardor and attraction, implore the effects of your Providence to remedy promptly the bewilderments that in this lunar world cause the doctrine of the inhabitants of the nearby terrestrial globe, which, having been a comet in the past, by the grace of your solar attraction stopped in our vicinity, and still remains little by little becoming a planet."

Viaje estático al mundo planetario (1793)[1]

Lorenzo Hervás and Panduro (1735 – 1809)
Spanish polygraph, linguist and philologist.
(In Hispanic popular culture the *Lunícolas* are the imaginary inhabitants of the Moon.)

"Others again, belonging to the doubting class, expressed certain fears as to the position of the moon. They had heard it said that, according to observations made in the time of the Caliphs, her revolution had become accelerated in a certain degree. Hence they concluded, logically enough, that an acceleration of motion ought to be accompanied by a corresponding diminution in the distance separating the two bodies; and that, supposing the double effect to be continued to infinity, the moon would end by one day falling into the earth. However, they became reassured as to the fate of future generations on being apprised that, according to the calculations of Laplace, this acceleration of motion is confined within very restricted limits, and that a proportional diminution of speed will be certain to succeed it. So, then, the stability of the solar system would not be deranged in ages to come."

Jules Verne (1828 – 1905)

The Successors of Galileo

After the transcendental discoveries of Galileo about the Moon, the planets, the Sun and the stars, the telescope enjoyed a great welcome as an instrument of scientific investigation by class of the astronomers during XVII and XVIII centuries, who constantly improved it both in size as in quality. Finally, during the nineteenth century two other developments of great importance for the time were added to it: the first was the invention of photography that, with its ability to integrate light and time, it managed to overcome human vision and, besides, allowed to prepare a permanent record of such heavenly observations. The second development was the incorporation of the spectroscope to the telescope, which allowed analyzing in detail the light coming from the celestial bodies and in this way to determine its chemical composition, turning it into one of the most powerful tools of astronomy.

On the course of seventeenth and eighteenth centuries, and from the multiple technical developments, especially in optics, and of the new mathematical and physical theories, a great impetus was given to the sciences and very especially to astronomy, in which the systematic observation of the stars played a fundamental role. From the beginning of the 17th century, great driving men of what we know today as *Modern Astronomy* emerged: Christian Huygens, Giovanni Domenico Cassini, Isaac Newton, Ole Rømer, Edmund Halley and a long list. Of this early period, the most famous telescope was the one built by the Polish astronomer Johannes Hevelius in the 1670s, which was 45 meters long.

The eighteenth centenary is usually known as *Century of Lights* due to the great cultural and intellectual movement that occurred in the Europe of the moment, and which is now recognized as *The Enlightenment*. The leading exponents of this movement argued that human reason and knowledge could combat the superstition, ignorance and tyranny of the rulers, all with the purpose of building a much better world. The Enlightenment had great repercussions in the scientific, political, economic and social fields of the time. In the field of the sciences, astronomy had considerable advances from the hand of great and consecrated physicists, mathematicians and astronomers. The most precise, realistic, determination of the lunar movement was a central theme in the scientific development of the time.

Godefroy Wendelin

Godefroy Wendelin (1580 – 1667), also known as Govaert Wendelen, was a Belgian cleric and astronomer to whom astronomy owes his great step toward determining the extent of the Solar System; which he achieved in 1630 by taking up the method originally proposed by Aristarchus of Samos to calculate the distance to the Sun from Earth, but using the help of the newly invented telescope and better instruments for measuring angles. After all the considerations and necessary practices, he determined that the angular distance of the Moon to the Sun at the moment in which the satellite is in its First quarter phase and looking like exactly half illuminated, it was equal to 89° 45'. Therefore Wendelin, using elementary geometrical formulas, could determine that the distance of the Sun from the Earth was equal to 6876 times the diameter of the Earth. Recall that Aristarchus had obtained 87° for that angle, and a value of 176,1 times the Earth's diameter for the distance from the Sun to our planet.

With regard to the Sun, Wendelin arrived at a value that is too small but of an approximation in truth far superior to any which has been done before, and as close as an observation of that kind would allow; the accepted value at present is that the King star is distant at about 11740 times the Earth's diameter. But for the case of the satellite, the results are more positive because continuing with the trigonometric calculations according to the values reported by Wendelin, we obtain a distance from the Moon to the Earth of 30 times the diameter of the planet; a value that turns out to be surprisingly close to the modernly accepted 30,16 times the terrestrial diameter for such distance.

Christiaan Huygens and his telescopes

Christiaan Huygens (1629 – 1695) was born in The Hague, The Netherlands, and was famous not only in astronomy but also in related sciences, such as physics and mathematics. One of his greatest passions was optics, both on a theoretical level and in the practical field of telescopes' lens manufacturing. He soon realized that an improvement of the lenses was a necessary condition for the progress of astronomy, and he discovered a new method to give them the required curvature with considerable precision. Therefore, he was able to build telescopes with greater magnification power and much superior than those of Galileo. Towards 1655 Huygens finished a telescope of a great quality for his time: with a diameter of 5 cm and a length of three and a half meters, which provided about fifty magnifications. Later he built telescopes with increasing focal lengths: five, ten, twenty and finally 37 meters of focal distance. These long instruments had to be installed on heavy wooden structures and fastened by strong ropes.

By directing his powerful telescopes to the sky, he soon obtained concrete results for the case of the planet Saturn: he was able to elucidate the true form of the mysterious planet of three bodies; that is, he could come to the conclusion that it was not three different objects but a central planet, Saturn, surrounded by a concentric ring made up of a great variety of smaller bodies. Fact observed previously by Galileo, but who could not fully understand it. *Huygens* also performed in March, 1655, the discovery of Saturn's first natural satellite, later called Titan. He announced such discoveries in his texts *De Saturni Luna Observatio Nova* of 1656, and *Systema Saturnium* in 1659. In the latter he clearly explained the different appearances of the planet as due to the profiles of its ring inclined 20º with respect to the ecliptic's plane; and he also determines that the newly discovered moon takes just under 16 days to orbit the planet.

Hi made other great contribution in the field of optical physics by formulating the undulatory theory of light; in his work *Traité de la lumière*, or *Treatise on Light* of 1690, he argued about reflection, refraction and the double refraction of light.

Giovanni Domenico Cassini, his moons and the size of the Solar System

Giovanni Domenico Cassini (1625 – 1712) was born in *Perinaldo*, Italy, was geodesist, engineer and astronomer; the initiator of four generations of astronomers. He studied initially at the Jesuit College of *Via Balbi* in Genoa, and later entered the seminary of the Abbey of *San Fruttuoso* in *Camogli*. At twenty-five years of age he was called from Genoa to occupy a position as professor of astronomy at the University of Bologna, and during the nineteen years in which he remained there, Cassini gave considerable impetus to the studies of astronomy with the modest means at his disposition. At the same time, he was an astronomer at the Panzano Observatory from 1648 to 1669. In this last year he settled in France, and in 1671 he was appointed member of the French Academy of Sciences, founded in 1.666, and first director of the Paris Observatory, created in 1667. Subsequently, in 1673, he acquired French citizenship and remained definitely in that country. Cassini would remain as director of that observatory for the rest of his life; after forty years of extensive astronomical observations, and like Galileo, he would be completely blind in 1711 and finally he died in September 1712.

In the cosmological aspects, *Cassini* initially adhered to the Ptolemaic geocentric system, but his later astronomical works led him to accept rather the Copernican heliocentric system model. Although sometimes he rather considered the intermediate cosmological theories of Tycho Brahe.

Cassini, while still in Italy, observed the characteristics on the surfaces of the planets Mars, Venus and Jupiter: he discovering their rotation movements on their axes, calculated their periods and appreciated their seasonal changes. Also, when studying Saturn and its ring, he discovered a division or band that separates the ring into two unequal parts; which now bears the name of *Cassini Division*. This

band gave him the idea that in reality the ring is made up of a swarm of very small satellites, which revolve around the planet with different speeds and that from the Earth cannot be seen separately. Likewise, he spent a lot of time studying the planet Jupiter with his system of moons, and in his work *Ephemerides Bononienses Mediceorum Siderum*, or Ephemerides Bolognese of the Medicean Stars of 1668, the astronomer sustained the configurations of the four natural satellites of Jupiter previously discovered by Galileo.

Shortly after arriving in Paris and working at the Observatory, Cassini discovered in 1671 Saturn's second moon, Iapetus. The following year he discovered the third moon, Rhea; and finally in 1.684 two others, Tethys and Dione. Apart from these satellites of Saturn, Cassini also studied our Moon in detail, and in 1679 the French Academy of Sciences published his graphic representation of the satellite surface known as *Carte de la Lune*, which was widely used until the arrival of photography in the astronomical field.

The attempts to measure the distance of the celestial bodies and the size of the Universe as a whole are now, after more than 2000 years, about to be completed and it is appropriate to summarize them here. In 1672 Cassini sent his colleague Jean Richer to Cayenne, in French Guiana, South America, while he stayed in Paris. From these cities the two astronomers made simultaneous observations of the planet Mars in its opposition to the Sun: they calculated the parallax of the planet, determined its distance from Earth and, by triangulation, they also calculated the distance from Earth to the Sun finding a value of 139 million of kilometers, a figure only 7,1% lower than accepted value today that is 149,6 million kilometers, distance today known as *Astronomical Unit*. This allowed for the first time an estimate of the size for the solar system as a whole based on direct measurements.

Image 7.1
Paris Observatory

Engraving of the Paris Observatory during the 18th century, where Giovanni Domenico Cassini made his astronomical observations. Wolf, Charles J. E. (1902): *Histoire de l'Observatoire de Paris de sa fondation a 1793*.

Isaac Newton: his Laws of movement and the Universal Gravity

Isaac Newton (1643 – 1727) was born in *Woolsthorpe* in Lincolnshire, England, at the beginning of 1643, one year after the death of Galileo. His parents were Isaac Newton and Hannah Ayscough, although he didn't get to know his father because he died a few months before his birth. The young Newton made his first studies at King's School of the population of Grantham, in Lincolnshire between 1655 and 1660. Then he entered in 1661 to the Trinity College in Cambridge University where he graduated in 1665; year in which the university was closed because of the bubonic plague. Newton then retired himself to his hometown Woolsthorpe to continue on his own studies on mathematics, mechanics, optics and gravitation: he devoted entirely to the study of the foundations of natural philosophy and caused a total revolution in the fields of physics and astronomy with his discoveries about gravitation, with his developments on optics and the theory of colors, and with the invention of the reflecting telescope, which is based not on lenses but on mirrors and is also known as Newtonian telescope. Later he returned to the University of *Cambridge*, where he managed to be a Lucasian Professor of Mathematics in 1669; field of knowledge in which Newton made great contributions with the development of the infinitesimal calculus exposed in his text *Method of fluxions and infinite series* of 1736.

Regarding that obligatory withdrawal of the physicist and his stay in his native town, one of his manuscripts is available referring to his thoughts and the issues he addressed while he was there at his twenty-two:

"In the same year I began to think about the gravity that extends to the orb of the Moon, and I discovered how to estimate the force with which a globe that rotates inside a sphere presses the surface of the sphere, according to Kepler's rule that the periods of the planets are in a sesquialterated proportion of their distances from the centers of their orbs (sesquialterated means one and a half time, or, as we say, the square of the years are as the cubes of the orbits). I deduced that the forces which keep the planets in their orbits must be reciprocal as the squares of their distances from the centers around which revolve: and so I compared the force required to keep the Moon in her orb with the force of gravity on the surface of the Earth, and I found the answers quite approximate. All this was in the two years of plague of 1665 and 1666, because in those days I was in the prime of my age for invention, and I cared about mathematics and philosophy more than at any other time since then."[3]

Later and continuing with some ideas of Galileo, from the year 1684 Newton established the modern science of *Dynamics* upon formulating his three laws of motion of bodies. *Mechanics* is the branch of the physical sciences that studies and analyzes the states of motion and rest of material bodies; whereas *Dynamics* is the branch of physics that studies the evolution in time of physical systems, mainly in relation to the movement of bodies when they are subjected to the action of forces; that is, it deals with the relations between forces exerted on the bodies and the movements that they induce in them. Newton published his theories in his greatest work *Philosophiae Naturalis Principia Mathematica*, or *Mathematical Principles of Natural Philosophy*, firstly edited in London in 1687; which marked a remarkable turning point in the history of science and knowledge; and it is usually considered the most influential work in the physical sciences. In this work the English physicist established, after a series of definitions and propositions, the three axioms or *Laws of the movement of bodies*.

Mathematical Principles of Natural Philosophy

Compiling the theoretical discoveries of Kepler and Galileo in the fields of physics and astronomy, Isaac Newton managed to construct a general mathematical and geometric model that explains both the movement of the celestial bodies as well as that of terrestrial objects for the first time in history. In this work, originally written in Latin, Newton presented the fundamentals of physics and astronomy

expressed in a purely mathematical and geometric language. It is a deductive work where the mechanical and dynamical properties of matter are demonstrated by means of theorems deduced from very general propositions. The work consists of three books, preceded by two preliminary chapters dedicated one to general definitions and the other to the axioms or *Laws of movement*. The first two books are entitled *The movement of bodies*. In Book I he presented the general theory of the movement of bodies in ideal conditions: bodies with mass but without form or volume, without problems of elasticity or viscosity, and moving in a vacuum, that is to say in media without physical resistance. Book II is much more concrete in nature, involving more specific problems in which he introduced the properties of bodies, as the mass, density and the elasticity; and additionally the complications due to the viscosity and physical resistance of the medium in which they move are involved.

Image 7.2
Carte de la Lune

Prepared by the astronomer Jean-Dominique Cassini in 1.679 while working at the Paris Observatory

In the third book of the *Mathematical Principles*, entitled *The World System*, Newton took up, completed and expanded the concepts previously expressed by Kepler and Galileo, and placed the principles of dynamics on solid mathematical foundations, with its general laws of motion and with his theory of universal gravitation. He addressed the concepts of inertia, strength and mass; and established the *Law of Universal Gravitation* with its multiple consequences for the orbits of the planets, natural satellites, and of the other stars; with which he laid the foundations for the future theory of celestial movements. He then deduced again the laws of Kepler and Galileo as a logical consequence of his ones, which are additionally true for the gravitational effects on the Earth's surface. He also found, as a consequence of his formulations, the explanation of the mutual perturbations between the different celestial bodies. With all the above he sustained and supported the Copernican heliocentric system, according to the laws of Kepler and Galileo, the new mathematical methods developed and based on his new *Law of Universal Gravity*.

The Newtonian Laws of motion and his Theory of gravity underlie much of modern physics and engineering, and signal the beginning of the modern scientific era. The first is the *Law of Inertia* that states that: if none force acts on a body, this will maintain its rest or rectilinear and uniform movement state indefinitely. This is what is observed, for example, in the case of the Earth-Moon system: it is the Earth's gravity force that keeps the satellite in an orbit around the earthly globe. If the Moon was not subjected to any force, it would follow a uniform rectilinear motion, or it would always remain at rest; the combination of straight-line movement and the gravitational force of terrestrial attraction is what gives shape to the satellite's orbit. The second is known as the *Fundamental Law of Dynamics*: a force applied to a body induces to it an acceleration that is directly proportional to the magnitude of the force and inversely proportional to the mass of such body. Finally, the *Law of Action and Reaction* states that if a body exerts a force on another, known as Action, the latter exerts exactly the same force on the first, but in the opposite direction, constituting then the Reaction.

When Newton combined his laws of dynamics with Kepler's laws about the orbital motion of the stars, he deduced the *Law of Universal Gravitation*; which states that any two bodies attract each other with a force directly proportional to the product of their masses, and inversely proportional to the square of the distance that separates them. And additionally he ratified the Keplerian theory of planetary elliptical motion: the path followed by a body moving under the action of a force which is inversely proportional to the square of the distance, has an elliptical shape, rather than a spiral as many believed at first.

In 1684, shortly after observing the comet that now bears his name, Edmond Halley visited Newton at Cambridge to inquire him about the orbit of a star subjected to a gravity force that diminishes with the square of distance; the response of the genius was immediate: I have already calculated it, it is an ellipse. Unable to find his original calculations, Newton promised to make new ones and send them to him. The English physicist returned then to the subject, applying it again to the specific case of the Moon, an idea born twenty years before, but now considering the most accurate measurement of the terrestrial diameter made by the French astronomer Jean Picard in 1670; and after checking the validity of his law of gravitational attraction, he considered solved the problem in February of 1685.

Newton is consecrated as the greatest genius of classical physics in 1687 with his work *Mathematical Principles*, in which he exposed his three *Laws of the movement of bodies* as well as the *Law of universal gravitation*, the latter being the one that explains why such elliptical orbit of the planets around the Sun: it is the gravity of the King star which holds the planets in their respective elliptical trajectories. With Newton, the theoretical, physical and mathematical frameworks for a *Heliocentric Solar System* were clearly established, within the scope of classical physics, non-relativistic. Additionally and by extension, Newtonian physics provides an excellent description and justification of the Moon's movement around the Earth: leaving aside the influences of other stars, it's the force of gravity of our planet that holds the satellite in its respective elliptical orbit.

In the middle of the Renaissance, in the first half of the 16th century, the astronomer Nicolaus Copernicus reintroduced in Europe the theory of a *Heliocentric Universe*, which would be sufficiently demonstrated with the theoretical and experimental works of the seventeenth century by the astronomers Tycho Brahe, Johannes Kepler and Galileo Galilei. Finally, in 1687 the English physicist Isaac Newton in his work *Mathematical Principles of Natural Philosophy* laid the theoretical, physical and mathematical basis for the support of the modern Heliocentric Theory, in which the Earth is only one of the planets that revolve around the Sun; and where a mathematical system is presented that makes it possible the calculation and the most precise and long-range prediction of the movements of the celestial bodies, such as the planets, the moons and their respective eclipses. With all above, the geocentric worldview of Eudoxus, Aristotle and Ptolemy, which was in force for almost two thousand years, came totally discredited: if Ptolemy's Mathematical Syntaxis and Geography opened the history of Renaissance cosmology, Newton's Mathematical Principles signaled its closure and inaugurated the new era of modern science and astronomy.

From the second edition, the text is finalized with a *General Scholium* that infers a rational explanation of the existence of a *Higher Being* organizer of the world, and it's well known for the sentence *Hypotheses non fingo*, or *I don't imagine hypothesis*, in reference to the rational methodology of this genius. Newton came to know three editions of his Principia: the first in 1687 with a circulation of one 400 copies; the next two were corrected and expanded by himself, one in 1713 and then the one in 1726. The English translation of Andrew Motte appeared in 1729, after the death of Newton; while the French edition was published in 1756.[4]

The Three-body problem and the Common center of mass, or Barycenter

The Newton's laws make it possible to describe, calculate and predict with sufficient accuracy the orbital movements of any star, being it a planet, a natural satellite, a comet or an asteroid; or also the trajectories of artifacts manufactured by man, such as an artificial satellite, a rocket or a spaceship. But his theories and equations work easily and perfectly when dealing with only two bodies, such as the Earth-Moon system. Because for the case of three or more bodies the situation becomes quite complex given the amount of forces and movements involved; which gives origin to what is normally known as the *Three-body problem*, and which kept the astronomers, mathematicians and physicists of the time well occupied.

When two bodies of unequal magnitude, such as the Earth and the Moon, interact gravitationally with each other, the system's common gravity center, known as the *Barycenter*, is at a distance of both inversely proportional to its respective masses, and the attraction of each one makes the other revolve around this center of gravity. So the Moon, which is the smallest body and is at the largest distance from the Earth-Moon system barycenter, describes a larger orbit and that necessarily includes the one described by the Earth around the same point. For a similar reason the Sun, being immensely greater than all the planets that surround it, makes a movement around the common center of mass of the Solar System as a whole, a movement that is considerably small when compared to the movement that perform any of the planets.

Newton's investigations had clearly shown that it was a mandatory result of his law of gravitation that, excluding the attraction of the other bodies of the system, the orbit of any star that revolves around the barycenter should be an ellipse in both physical and mathematical terms. But repeated and accurate observations have shown that this is not the figure of the lunar orbit, nor that of any planet, nor that of the common center of gravity of a planet and its satellites. In fact, the so-called *Three-body problem,* when considered in its entirety, presents difficulties that all the mathematical tools of modern analysis are not able to overcome. But these difficulties can be minimized considerably by assuming that one of the three bodies involved has a much greater mass than any of the others, as it's the case of the Sun compared to the planets; or that it is very remote compared to the distance of the others, as is the distance from the Sun compared to the distance from the Moon to the Earth.

The movements of the Moon according to Newtonian physics

When Newton studied Kepler's third law of celestial motion, he was able to demonstrate that the planetary movements obeying it could be explained by an action of the Sun assuming that the King star attracts the planets with a force inversely proportional to the square of the respective distances. Then it was natural to wonder if the Earth would exert a similar force on our satellite, and investigate if that force was of the magnitude necessary to keep the Moon in its orbit: it occurred to the great physicist that a force similar to that what makes a body to fall on the Earth, by continually diverting our satellite from its natural lineal trajectory, could force it to turn around the planet. In this way the physicist discovered that the Moon during its movement around the Earth, to stay in its orbit, had to move for every minute of time through a distance of 13 feet. Then, reasoning about the distance that a body travels when it freely falls for a minute on the earthly surface, and applying the inverse square rule, he discovered that for this reason the Moon would deviate 15 feet. A less rigorous scientist would have been satisfied with the results of the analysis, but Newton instead believed that the discrepancy between the two numbers was too great, and then abandoned the investigation for a time. As a unit of measure for longitude, the foot equals 30,48 cm.

And meanwhile the French astronomer and priest *Jean Picard* (1620 – 1682) performed in 1.670 a much more accurate measurement of a degree of latitude, providing a very accurate measure for the diameter of the Earth: 12658 km, with a margin of error of only 0,66%; and at the same time he would facilitate an

experimental confirmation of the Newtonian theory of Universal Gravitation. Shortly after Newton engaged in a discussion with the encyclopedic Secretary of the Royal Society Robert Hooke, his great rival, about the curve that a body that falls from a certain height describes; which induced Newton to reexamine that problem of the Moon that he previously abandoned without satisfactorily resolving it. As he now had a much more accurate value of the Earth's diameter than that he used in his first calculations, then he established the perfect equality of the two results, proving then that his hypothesis was valid: it is the Earth's gravity force that sustains the Moon in its respective elliptical orbit.

But perhaps the greatest importance in the applications of Newtonian law of gravitation is related to certain irregularities in the movements of the celestial bodies, and more specifically to the Moon. In its trajectory around the planet, our satellite is mainly directed by the great gravitational attraction of our terrestrial globe. If there were no other body in the universe, then the center of the Moon should necessarily describe an ellipse, the center of our planet being located in a focus of it. But if we consider now the King star and its gravitational influence on the Moon, we would see that the movement of the lunar globe would not have such simplicity. With its great gravitational power, the Sun attracts both the Moon and the Earth albeit in different proportions, because the same force influences more drastically a small body than a great mass one. The ultimate result is that, given its smaller mass, the movement of the satellite is more drastically affected by the gravitational influence of the King star in comparison with our planet: the shape of the lunar trajectory is not exactly an ellipse, nor is the Earth exactly in any focus. This is the central theme of the *Three-body problem,* and was one of the greatest contributions that Newton made to the definitive understanding of the lunar movement.

Considering the great solar gravitational perturbation on the Moon's trajectory, Newton proceeded to estimate the important inequalities of the lunar movement; he showed that its apses advance and its nodes recede with reference to its orbit; facts already known and established by historical records and by direct observation, although they had not yet been satisfactorily explained. With his analyzes Newton established that the part of the gravitational force of the Sun, F_g, that exerts disturbing effects and leads to the first two inequalities of the lunar movement, the *Equation of the center,* discovered by Hipparchus, and *Evection*, discovered by Ptolemy, is the radial component F_{gr}; although he did not enter to make any calculation on these effects.

With respect to the inequality of the lunar movement that is called *Variation,* and which was discovered by Tycho Brahe, it is described by Newton in the third book as dependent on that portion of the solar attractive force acting over the Moon in a tangent direction on its way around the Earth, F_{gt}. This force is maximum and alternately accelerates and retards the lunar movement in the different quadrants of its orbit, and it reduces to zero in the conjunctions and oppositions because the tangential component is zero and then all the Sun attraction acts perpendicularly, in radial direction over the lunar globe. It is also zero in the quadratures, since there the tangential component of the solar attraction is equal to the attraction exerted on the Earth by the Sun; consequently, there is no difference in the disturbance of the movement of the Moon in relation to that of the Earth. But the influence of that tangential force is greater when the Moon is just in the octants; it's at 45 degrees of the points of syzygy and quadrature, where the variation of the lunar movement rises to almost 35' 42''. Newton evaluates it at 85' 10'', but he was wrong by supposing that the Earth is in the center of the elliptical orbit of the Moon, and by not considering the consequence that arises from the ellipticity of the Earth's orbit.

A fourth inequality in the movement of the Moon, which was discovered by Kepler and is called *Annual Equation,* Newton explained it arguing that arises from changes in the moon's orbit size caused by different degrees of solar attraction over the Earth in different positions, conformable the planet is closer or farther from the Sun according to the ellipticity of its orbit.

Thus, when the Earth is in perihelion and the Moon in syzygy, the greater attraction of the Sun causes a corresponding increase in the size of the main axis of the lunar globe's elliptical orbit; and in aphelion, an opposite effect occurs due to the smaller solar attraction. But the satellite moves more slowly when its orbit expands and then moves away from the Earth, and faster when such orbit contracts and the Moon approaches the planet.

These *four inequalities* of the lunar movement were those discovered way of the observations; but the theory of gravitation has made us know other more, which probably would never have existed without the force of gravity; of these Newton mentioned one that he called the *Semestrial Equation*, and showed that it arises from the different positions assumed by the line of the lunar nodes.

Lunar Gravitational force and the Terrestrial oceanic tides

Newton also gave an explanation of the oceanic tides, already conjectured but not demonstrated or evaluated by his predecessors, and showed that they are caused by the solar and lunar gravitational attractions on the liquid masses of the planet. Even in ancient times there was agreement that the tides were related to the position of the satellite; it was known that the tides were especially high during the Full Moon or during the New Moon, and this circumstance obviously pointed to the existence of some connection between the lunar globe and these terrestrial water movements; although as to the nature of that connection no one had any precise conception until Newton announced his *Laws of motion* and his *Law of Universal Gravitation*. Just as the Earth exerts a force of gravitational attraction on the lunar globe, this in turn exerts via the principle of reaction a force of attraction over our planet, which accounts for that reflux of water that wanders freely in the oceans: water is attracted from the side of the Earth that faces the Moon; which is called High tide.

Image 7.3
The Three-body problem and the Sun-Earth-Moon System

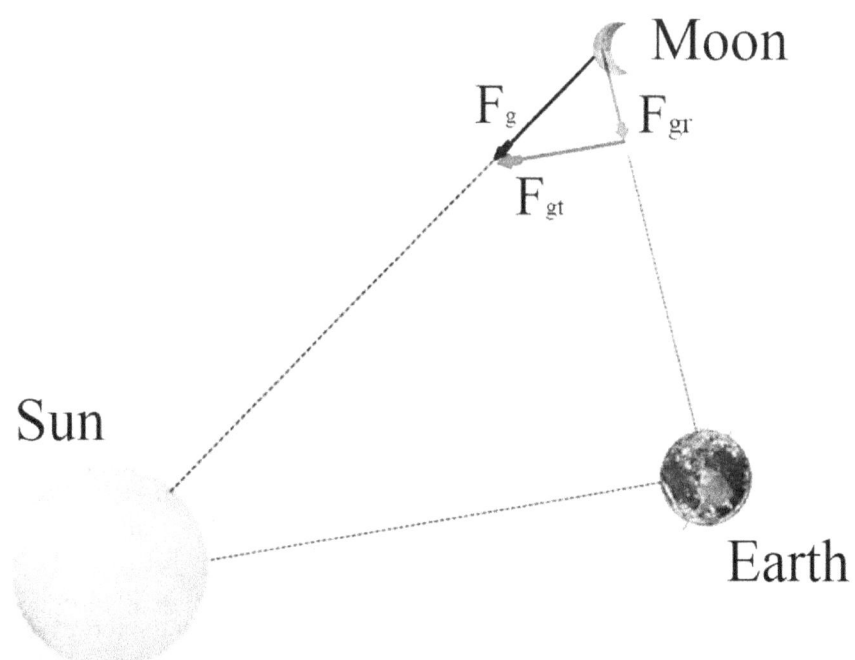

Isaac Newton established that the inequalities in the lunar movement known as the *Equation of the center* and the *Evection* are caused by the radial component of the solar gravitational force F_{gr}; while the *Variation* is caused by the tangential component F_{gt}.

173

The genius made it perfectly clear that the rise and fall of the oceanic liquid were simply a consequence of the attractive power exerted by the Moon on the oceans that lie on our globe. He further argued that to some extent the Sun also produces tides; and he was able to explain how it is that when the King star and the satellite conspire the combined result is to produce especially high tides, known as *Spring tides*, or *Living tides*; whereas if the solar action is weak and at the same time the lunar force is strong, then the phenomenon of *Neap tides*, or *Dead tides*, occurs.

The English physicist established that due to the orbital motion of the satellite around the Earth, the two tidal flows take place with an average delay of approximately 50 minutes per day, a result that agrees well with the observation; he also concluded that the lunar gravitational effect was the largest, since the Moon is closer to Earth; and additionally he determined that when solar and lunar attractions add up, the tides are the highest. Comparing the magnitudes of elevation of the tides due to the Sun and the Moon, the physicist proceeded to derive the mass of the satellite as a function of that of the King star and, consequently, in function of the mass of the Earth; but, although novel and remarkable, the method provided a value approximately twice as large as that accepted today; which is because his theory of the tides was based on certain premises that then had to be modified. After this great discovery of Newton, it was possible for the first time to determine the masses of certain celestial bodies, comparing the gravitational attractions in other bodies with the attraction between the Earth and the Moon.

Lunar Gravitational force and the Earth's axis movement

During his theoretical developments, Newton perceived that each star should gravitationally disturb the movement of others, and thus he could satisfactorily account for certain phenomena that had maintained perplexed all previous researchers.

The phenomenon of the *Precession of the equinoxes*, so mysterious until that moment, was explained by Newton as due to the attractive forces of the Sun and the Moon over the widened Earth's sphere at the equator: with a wonderful perception the physicist realized that the gravitational attraction of both the King star and the satellite on the bulky matter around the Earth's equator would force the Earth's axis to vary its orientation and to have a slow conical movement in space, which is known as *Precession of the Earth's axis*, and which has the same general characteristics as the *Precession of the equinoxes*. For this movement Newton obtained a theoretical datum that agrees well with the observed value; but this agreement is due to the accidental compensation of some errors derived from the imperfect knowledge of our planet, the distance from the Sun and the mass of the Moon.

Such phenomenon can be explained in this way: Given that the Earth is not a perfect sphere, but is elongated and has protuberances in its equatorial zone, the gravitational forces of the satellite and the Sun will exert on that protuberances some effects of attraction that change continuously the direction of the Earth's axis and, consequently, the point of the space to which the axis is directed must be constantly changing.

Newton theoretically determined the average value for the angular variation on the orientation of the Earth's axis in space, and stated that the part due to solar attraction is approximately 9'' of arc per year, while that part due to general attraction of the Moon is almost 41'', therefore the joint action becomes approximately equal to 50'' per year, that is 1° every 72 years. With which the Earth's axis describes a full circle around the pole of the ecliptic, and it takes approximately 25000 years to complete its revolution. For the moment, the Earth's axis points towards what we know as *Polar star*, or *Polaris*, but as time passes it will point towards different stars; for example, in about 12000 years it will be directed towards the bright star Vega.

On the other hand, the gravitational action of the luminaries over the equatorial regions of the Earth produces, in addition, changes in the obliquity or degree of inclination of the terrestrial axis with respect to the ecliptic; and the intensity of the lunar attraction is, again, stronger than that of the Sun. That mysterious movement by which the Earth's axis swings among the fixed stars had long been an unsolved enigma. But Newton showed that the gravitational influence of the Moon on Earth would be more felt in its equatorial prominences, thus causing an oscillation, a cyclic variation of the angle of inclination of the Earth's axis, which is known as *Nutation*, and which is responsible for this apparent displacement of the remote stars.

Ole Christensen Rømer: the moons and the Speed of light

Another brilliant astronomer of this era was the Danish Ole Christensen Rømer (1644 – 1710), who entered the Academy of Sciences in Paris in 1672, and spent seven years in the Observatory after its foundation. It was in Paris in 1.676 that he made his great contribution not only to physics but also to natural philosophy, for being the first person to determine the *Speed of light* with a very close value of 210606 kilometers per second. The work of Rømer described below is a good indication of the important role played by the moons and their eclipses in the development of modern cosmology.

By examining the extensive observations and records of *Cassini* on the planet Jupiter, Rømer noted that the duration of the eclipses of its moons, that is the time elapsed between the disappearance of the satellite behind the disk of the planet and its subsequent resurgence at the other end, varied in such a way that it was shorter when the Earth was closest to Jupiter, and it lasted longer when our planet also retreated to the maximum of that planet. Indeed, he realized that this happened because the speed of light was finite, and not, as previously thought, infinite: the light coming from Jupiter has a shorter distance to travel when the Earth is closer to that planet that when it is more retired, and for that reason the duration of the eclipses of the *Jovian Moons*, when they are contemplated from the ends of the Earth's orbit, seem to vary. Rømer concentrating on the duration of the eclipses of the *Io* satellite according to the previous conditions, one night in 1676 he determined that the light took about 11 minutes to cross the Sun-Earth distance, which for its time was taken as 139 million kilometers according to what was established by Cassini four years before; although modern estimates are closer to 8,33 minutes. Thus, Rømer obtained a value of 210606 kilometers per second for the speed of light; by comparison the accepted modern value is 300000 kilometers per second.

Edmund Halley

Once the scientific theory of the heliocentric universe was established and accepted during the seventeenth century, astronomers continued to give special attention to the Moon and its eclipses with the purpose of understanding and predicting such phenomena with greater precision. The English astronomer, mathematician and physicist *Edmond Halley* (1656 – 1742) was sixty-four years old when in 1720 he became director of the Royal Observatory of *Greenwich* in England, and even so he immediately undertook a series of lunar observations to cover a whole revolution of the its nodes, more than eighteen years, which led him to an amazing conclusion.

The period of the lunar orbit is now cognized with good precision, and by means of ancient records on eclipses it can also be determined for some two thousand years behind. By studding the satellite's movement, *Halley* discovered that the period that the Moon requires to carry out each of its revolutions around the Earth, the synodic month, has been steadily decreasing, although certainly very slowly too. The variations so produced aren't appreciable when only small time intervals are considered, but they becomes very evident when we have to deal with intervals of thousands of years. This phenomenon is called *Secular acceleration of the lunar movement*, and the effect it produces

can be visualized as follows: If we assume that over all time the Moon has rotated around the Earth in exactly the same period it has at this time, and if based on this assumption we calculate by regression where the satellite should have been about two thousand years ago, we will get a position that the records on ancient eclipses show as different from that in which the Moon was really located in that time. The difference between the position in which the satellite would have been found two thousand years ago if there was no secular acceleration and the position in which the Moon really was according to those records, ascends approximately to an arc in the sky equivalent to one degree, this is twice the apparent diameter of the Moon: a very large discrepancy. The term secular applies to processes timed in centuries; that is, very long-lasting phenomena.

By 1695 Halley was the first to realize that the times and places reported by records from ancient eclipses do not correlate well with the retrospective calculations of his time. Comparing the positions of the Moon deduced from the observations of the Babylonians, that of Hipparchus and Ptolemy ones with those obtained from modern astronomy, Dr. Halley discovered that the average motion of the satellite was constantly accelerating at a rate of approximately one degree in two thousand years. In his time it was generally believed that this acceleration, and the consequent decrease of the orbital period and the average distance from the Moon to the planet, would continue until the satellite approached enough and finally hit the Earth. In some social circles it was believed that the "Moon was falling", and then it was thought that humanity and its creations would be destroyed and that the planet would rotate around the King star without any satellite.

Additionally, in the midst of his observations in 1731, *Edmund Halley* recognized the potential of the lunar eclipses method to provide a solution for the problem of measuring geographic longitude at the Earth's surface. But he is more widely remembered by his application for the first time in 1705 of the Newtonian theories to the comet observed in 1682, which now bears his name, obtaining an elliptical orbit and predicting its return around 1757; failing for a few months because it was really contemplated again at the end of 1758.

James Bradley

The English clergyman and astronomer *James Bradley* (1693 – 1762) was professor of astronomy at *Oxford* and later at the *Greenwich Observatory*; and in 1742 he succeeded Edmund Halley and thus he became the third Royal astronomer. From observations that covered a complete revolution of the lunar orbit's nodes, made between the years 1727 and 1748, he determined that the positions of certain stars seemed to change with time. Five years later he found an explanation based on principles already established in Newtonian physics: the Earth's axis is swaying as a result of the gravitational attraction of the Moon on the equatorial protuberance of the planet. To this swaying, now called *Nutation*, is due the apparent displacement of the stars, which seem to describe a tiny ellipse along its average position in a period of approximately 18,6 years, the same period of the lunar nodes' movement. This great discovery of Bradley on the apparent displacement of the stars constitutes good evidence on the validity of the Newtonian gravitational theory.

Richard Dunthorne

Richard Dunthorne (1711 – 1775) was a British astronomer and surveyor; famous for making important contributions to the procedures for the calculation of geographical longitude on the high seas using tables of lunar positions in the sky; tables traditionally known as *Ephemerides*.

In 1739 Dunthorne published some ephemeris under the title *Practical Astronomy of the Moon*. The work was derived from the lunar theory of Isaac Newton, and thus could elaborate a confirmation of the Newtonian theory with respect to the experimental records; and based on his conclusions he proposed some adjustments in the numerical terms of such theory.

One of the most difficult problems faced by astronomers of the eighteenth century, and which acted as a constant stimulus for the progress of astronomy, had to do with an inequality in the movement of the Moon. The average movement of the satellite seems invariable when it is averaged over a reasonably long period of time, a century for example, but it is not constant when considering much longer periods, but it seems to accelerate. This was first suspected by *Halley* about 1695 based on a comparison he made between some old records of eclipses and the figures that the best modern tables gave for the same eclipses. Subsequently, in 1749 *Dunthorne* took up the subject and added additional ancient data to confirm Halley's suspicions; and resorting to refinements in the accuracy of astronomical measurements, he elaborated a quantitative estimate for the absolute value of the apparent acceleration of the lunar movement and obtained an extremely small value, eighteen times lower than that previously found by *Halley*: only ten seconds of arc per century. Later, other astronomers such as *Mayer* and *Lalande* calculated for it figures between 7'' and 10''.

Other works published by Dunthorne were: *On the movement of the Moon* in 1746, *On the acceleration of the Moon* in 1749, *Regarding comets* of 1751; and finally in 1762 he published an ephemeris of the satellites of Jupiter.

Euler, Clairaut, D'Alembert and the Acceleration of the lunar movement

Leonhard Paul Euler

Leonhard Paul Euler (1707 – 1783) was a Swiss mathematician, physicist and philosopher. He is considered the main mathematician of the eighteenth century and is well recognized for the number of *Euler*, *e*, involved in many formulas of calculation and physics. He applied with great success his analytical tools to the problems of classical mechanics, as well as to the movements of the stars or celestial mechanics. His work in astronomy included questions such as the determination with great accuracy of the orbits of comets and other celestial bodies, and the calculation of the solar parallax.

It seems that Euler was the first to notice that, as a consequence of the Law of universal gravitation, not only the planets describe an ellipse around the Sun, but the King star and its planets orbit all them joint with an elliptical trajectory around their *Common Center of Mass*, or Barycenter; and that this principle is also valid for a planet and its respective satellites. According to this, the common center of mass of the Earth-Moon system describes an elliptical orbit around the Sun and in the same plane of the ecliptic; therefore, the center of the Earth will be above or below that plane depending on whether the center of the satellite, whose orbit in turn is inclined with respect to the ecliptic, is below or above such plane. Additionally, this great mathematician tackled the difficult subject of the lunar movement focusing it from the theory of the Three-body problem. His results and conclusions he presented in his work *Theory of the movement of the Moon that exhibits all its irregularities*, published in St. Petersburg in 1753.

Euler devoted much work to the Three-body problem, as well as to the problem of the perturbation of the planetary and lunar orbits, which is essentially the same. He wrote several texts regarding the theory of the lunar movement: *Considerations to complete the theory of the movement of the Moon and especially its variation*, of 1763; his greatest work on the satellite appeared in 1772 under the title *Theory of the movement of the Moon*; while *On the Theory of the Moon that will take a higher level of perfection* is from 1775; and his last work in this subject was *The improvement of the lunar tables by means of observations of the lunar eclipse*, of 1862.

In 1770 the French Academy offered a prize to whoever would find the physical cause for the phenomenon of the *Secular Acceleration of the Moon*; which was won by Euler and his son *Johan Albrecht*. However, both had shown that the secular acceleration of the satellite could not be explained by Newton's gravitational forces. And here once again there was a crisis around the Newtonian science, so it was proposed that this was the same theme of the 1772 award of the French Academy. The award went back to *Euler*, but now shared with the astronomer *Lagrange*.

Finally Euler made important contributions in the area of optics; and his works developed during the 1740s helped the new theory proposed by Christian Huygens on the undulatory nature of light to become the predominant one, until the later advent of the modern *Quantum theory of light*.

Alexis Claude Clairaut

Alexis Claude Clairaut (1713 – 1765), was a Parisian mathematician and astronomer. Son of a math teacher who would be his tutor; very early Clairaut came to be considered a child prodigy: at thirteen he exhibited before the French Academy of Sciences a summary of the properties of four curves he had discovered. Subsequently, and after the publication in 1731 of a treatise on figures with double curvature, he was admitted to the French Academy.

Trying to understand and explain the complicated lunar movements, Clairaut managed to develop an ingenious approximate solution for the Three-body problem; and in 1750 he was rewarded with the prize of the Russian Academy of Sciences for his essay *Théorie de la Lune*, an essentially Newtonian work in its nature and content.

Jean le Rond D'Alembert

Jean le Rond D'Alembert (1717 – 1783) was an encyclopaedist, philosopher, physicist and mathematician; like Clairaut equally Parisian. He studied the Three-body problem concentrating on the instability of the system and the difficulty to find the equations of the trajectories. He was also dedicated to the problem of the precession of the equinoxes with its displacement of the seasons, and that of the nutation of the Earth's axis; which he addressed from the mathematical and physical perspective.

In the *Three-body problem,* and in particular in the theory of the Moon, the investigations of *D'Alembert* led him to hardly oppose to *Euler* and *Clairaut*, and the intense discussion served to improve the work of all these scientists. They worked at the same time on these questions, albeit almost independently, applying them in particular to the Sun-Earth-Moon trio and encountering the same difficulties that Newton had faced. Finally, *D'Alember* obtained even more precise results and published a complete theory for the satellite, provided with tables, in the first volume of his *Investigations on different important points of the world system*, of 1754.

Following Newtonian physics, *Clairaut* and *D'Alembert* theoretically determined a value of about 18 years for the period of revolution of the lunar perigee, basically double what was obtained from observational data, and for a long time *Euler* and other scientists thought that the only solution was to make adjustments to the law of gravitation established by Newton. But *Clairaut* found in 1749 an error in the approach method that everyone had been using; a fact that was comforting because *Clairau*t and *D'Alembert* could demonstrate that theories of Newtonian dynamics and universal gravitation had successfully passed a rigorous test. Euler did not agree at the beginning, and for that reason he wrote a treatise on the lunar theory: *Theory of the movement of the Moon that shows all its inequalities*, published in 1753, and that included a method for an approximate solution of the Three-body problem, in this case applied to the Sun-Earth-Moon system.

Euler, Clairaut and *D'Alembert* were the first mathematicians and astronomers to advance the *Lunar theory* beyond the point reached by Newton, and the three independently sent memoirs of their work to the Paris Academy of Sciences.

Tobias Mayer and the Lunar Libration

Tobias Mayer (1723 - 1762) was a self-taught German cartographer and astronomer; he was director of the Göttingen Observatory and a professor of mathematics at the University of Göttingen since 1750. *Mayer* made theoretical and practical studies on the Moon to help solve the problem of geographical longitude, and also to analize the physical surface of the satellite. From the theoretical works of Euler, he compared the lunar tables of the latter with his own observations and published *New Lunar Tables* or *Ephemerides* in 1752, together with instructions for their use in determining the *Geographical length* at sea; which made him creditor of the award given by the London Board of Length of 1755. He also published a first *Lunar map* with coordinates that indicated the respective positions on the Moon through the lines of Latitude, Parallels, and Longitude, Meridians; with which he made obsolete all previous maps of our satellite.

Mayer, in his work *Theoria lunae juxta systema Newtonianum,* from 1767, also investigated the elements of the lunar orbit through the Newtonian theory of gravitation, and managed to obtain correct values of its inclination to the ecliptic and for the movements of the nodes and apogee. Finally, *Mayer* developed an algebraic procedure to calculate the phenomenon of Lunar Libration that provided great precision.

Lunar Libration

Since the rotation movement of the satellite on its own axis is synchronized with that of its translation motion around the earthly globe, we have that the hemisphere or the side of the Moon that we can see from the Earth is always the same. This implies that a terrestrial observer could only contemplate a constant 50% of the lunar surface. But on the other hand, since the lunar orbit is eccentric respect the Earth and the satellite's rotation axis is inclined with respect to its orbital plane, a phenomenon known as *Lunar Libration* is produced, and it consists of the variation of the lunar area covered by the angle of visibility between the two stars, both in length and in latitude. This fact allows us to see more than half of the lunar surface looking from our planet; in fact, a terrestrial observer will be able to contemplate up to 59% of the satellite's surface after multiple successive observations.

The total libration of the Moon is a sum of three components; the most important is the libration in longitude caused by the elliptical, and therefore eccentric, orbit of the Moon around the Earth. On the other hand, the libration in latitude is due to the small inclination of the Moon's rotation axis with respect to the plane of its orbit around the planet; and finally, the diurnal libration manifests itself due to the rotation of the Earth, which displaces the observer's angle of visibility, the perspective, with respect to the satellite.

Joseph-Louis Lagrange

Joseph-Louis Lagrange (1736 - 1813) was an Italian mathematician, astronomer and physicist, nationalized in France in 1787. He made great advances in mathematics, was the author of novel works on astronomy and he contributed with some two hundred writings to the Academies of Turin, Berlin, and Paris.

Most of his works sent to Paris were related to astronomical matters, they were duly rewarded, and among them we can mention his essay on the Three-body problem of 1772, his work on the secular equation of the Moon of 1773, his text of 1764 about the libration of the Moon, and an explanation about why the satellite always offers the same face towards the Earth. Additionally, he published an article on the system of the planet Jupiter presented in 1766; and finally a treatise on the cometary perturbations of 1778.

As already mentioned, in 1772 *Lagrange* shared with *Euler* the prize of the French Academy for a work about the problem of the lunar movement. In his text, *Euler* argued that the gravitational theory could not offer an explanation for the secular acceleration of the lunar globe, but that there must be some kind of etheric fluid in space that offered resistance to

both the movement of the satellite and the planet. On the other hand, *Lagrange* proposed a new solution for the Three-body problem, but not for the phenomenon of the secular acceleration of the satellite. The French Academy anew offered a prize in 1774 for the solution of that lunar phenomenon, which was won again by the mathematician *Lagrange* for his argument about the way in which the shape of the Moon determines its own movement. But in any case, his work did not explain the secular acceleration.

Pierre-Simon Laplace and Celestial Mechanics

Pierre-Simon Laplace (1749 - 1827) was a physicist, mathematician and French astronomer. A great continuator of classical Newtonian physics, of celestial mechanics and also of the *Nebular Theory* on the formation of the Solar System, initially proposed by *Descartes* in 1644.

In 1786 *Laplace* elaborated a theoretical analysis based on perturbations of the eccentricity of the Earth's orbit and proposing that the average movement of the Moon should accelerate according to those perturbations. His initial calculations represented the whole effect and therefore he seemed to have united in his theory both the observations of his time and other older ones.

The most influential work of the French astronomer *Pierre-Simon Laplace* was written between 1799 and 1825, in 5 volumes and called *Traité de mécanique céleste*, o *Treatise of Celestial Mechanics*; and in it he resumed some phenomena that are still to be explained by Newtonian physics, such as the anomalous movements that remained unsolved: the planet Saturn seemed to slow down little by little, approaching a larger orbit, more away from the Sun; whereas Jupiter and our Moon showed an accelerated movement, approaching to a more smaller internal orbits. If these tendencies continued indefinitely, as indicated by the millennial records, Saturn would finally escape from the Solar System, Jupiter would fall on the Sun and the Moon over the Earth. But in 1785 *Laplace* demonstrated that the acceleration of Jupiter and the braking of Saturn were not linear, continuous or absolute movements, but oscillatory and therefore also apparent ones; and that those phenomena were due to the variable relative position of such planets with respect to the King star.

Other discoveries related to gravitational theory are also due to Laplace. For example, he found a relationship between the shape of the Earth and certain irregularities in the Moon's movements; as well as he introduced the Earth's rotation in the theory of tides, which had previously been ignored in this regard, and finally, he obtained a complete set of differential equations for the resolution of this problem.

The Celestial Mechanics, the lunar movement and an astronomical deceit

Laplace's fundamental work is his *Treatise of Celestial Mechanics*, where he presents mathematical expressions for the calculation of the movements of celestial bodies. In general, the work contains theories and methods for calculating the translation and rotation movements of the planets, in particular Jupiter and Saturn, and their moons. The work also deals with the main discrepancies in the movements of the planets that seemed to contradict the Newtonian law of gravity; and about the shape and movement of Saturn's rings and their permanence in the plane of the planet's equator. Finally, it also exposes the theory of tides, the libration of the Moon and the precession of the equinoxes. Currently, *Celestial Mechanics* is understood as that branch of Astronomy that aims to study and determine the movements of celestial bodies considering the gravitational interactions between them.

If excepting the Earth and its satellite no other body were present in the universe, surely the movement of the lunar globe would never have exhibited the observed phenomenon of its secular acceleration, and the lunar orbit would have remained forever unaltered. But it is well known that the presence of the King star exerts a

disturbing gravitational influence on the movements of the satellite. In each revolution the Moon is continually removed by the action of the Sun from that place which it would otherwise have occupied; such irregularities are known as *Inequalities* or *disturbances of the lunar orbit* and they have been studied since ancient times. But those who investigated for the first time the phenomenon of lunar acceleration thought it could not be explained as a consequence of the solar gravitational disturbance, and since astronomers did not find any other competent agent to produce such effects, the phenomenon continued to be an enigma without resolution for a long time. Then, at the end of the 19th century the illustrious French mathematician *Laplace* undertook a new investigation into the famous problem of lunar acceleration, and he was rewarded with a success that for a long time seemed to be quite complete and deserved.

When *Laplace* began to study the supposed acceleration in the movement of the Moon, the first thing he did was to discard the affirmations of the skeptics who held that the historical evidence was unreliable, and that other argument that the phenomenon was only an illusion caused by the deceleration of the Earth's rotation because of friction. If this were the case, he wondered, why did the average movements of the planets not increase as well then? Euler's proposal for an ethereal fluid neither did he accept it, since it lacked evidence. In conclusion, he undertook the problem of lunar movement as it was three generations earlier.

Of the numerous disturbances that affect the movement of our satellite, the apparent acceleration of its average movement and the consequent decrease in the duration of the lunar month were still to be explained. *Halley* had first suspected it from a comparison of ancient Babylonian observations, from those recorded by Hipparchus, from observations reported by *Al-Battani* and from other modern ones. As determined by some astronomers, the angular velocity of the satellite increases ten seconds of arc, 10'', every century. Since at that time the phenomenon could not be explained by Newton's gravitational theories, *Laplace* supposed first that gravity was not transmitted from one body to another instantaneously, but as sound and light did: in a finite time. He then showed that this hypothesis could result in a secular acceleration of the Moon, but only if the transmission speed of gravity were eight million times greater than the speed of light.

Laplace did not feel very satisfied with that solution because, like that of the ether, he had no physical evidence to support it. But during his observations and studies on the Jupiter's satellites, he determined that the secular variations in the eccentricity of the orbit of this planet cause a secular variation in the average movements of its satellites. Additionally, in 1787 he found that the shape of Earth's orbit was changing in a similar way to the planet Jupiter: the eccentricity of its elliptical orbit was decreasing and this probably had a connection with the gradual shortening of the duration of the lunar month. He applied all these results to the movement of the Moon and could discover that the lunar acceleration observed by astronomers was actually a mere apparent phenomenon that had as real cause the secular variation of the Earth's orbit eccentricity. He developed the theoretical expression for the secular acceleration of the lunar movement and found a figure of approximately 10.2''; and additionally he demonstrated that after about 24000 years the secular change on Earth's orbit eccentricity would reverse, the satellite would apparently slow down and the duration of the lunar month would then begin to grow again.

Therefore, the hypotheses that he had used in the theory of Jupiter's satellites were applied to our and, since the causes are the combined attractions of the Sun and the planets over the Moon, he discovered that it was a secular inequality not only of the movement of the satellite, but also of the movement of the nodes and the perigee of its orbit; so now the first one increases and the other two decrease. By proving that the real effect of the variation of the eccentricity of the Earth's orbit was exactly the acceleration that had been observed in the average movement of the Moon, he overthrew what was at that time the last barrier to the universal application of Newtonian theories as

the physical explanation of all celestial movements.

In his theoretical development Laplace considered that the lunar orbital period is directly related to the solar gravitational influence; and then, if there is any continuous alteration in the magnitude of the power of the solar disturbing effect, there would be a corresponding alteration also continuous in the lunar orbital period. Considering this, in his time Laplace understood that if he could discover any continuous change in the King star's ability to disturb the satellite, then he would find a causal justification for a continuous change in the period of satellite's movement.

The capacity of the Sun to disturb the Earth-Moon system is obviously related to the distance of the same system to the Sun. If our planet moved in a trajectory of permanently constant dimensions, the disturbing power of the King star would not show any variation of the expected type. But, on the contrary, if the Earth's orbit had some alteration in its shape or size, this would induce changes in the distance between the Earth and the Sun, which would be the desired agent to produce the variation observed in the lunar movement. Now, it is known that the Earth translates in an orbit that is strictly an ellipse, which would remain eternally unchanged if the Earth were the only planet that rotates around the Sun. However, our planet is only one of a good number of stars that circulate around the King star. All these celestial bodies are attracted to each other by gravitational attraction and, as a consequence, their orbits are also mutually affected and so modify the simple elliptical form that they would otherwise have. In conclusion, the movement of the stars is not strictly speaking in an elliptical orbit. But we can assume that it is so, as long as we admit that the ellipse is transforming and gradually changing its dimensions: The ellipticity of the planetary orbits is variable.

It is a remarkable characteristic of the disturbing effects caused by the gravitational attraction of the planets that the ellipse in which the Earth is in motion always keeps its major axis invariable at all moment. In all other aspects the ellipse changes continuously: it alters its position and changes its eccentricity, its shape. Therefore, in the course of time the shape of the path that describes our planet may at some point be closer to a circle, while at another time it may be differentiating more from that circular form. These alterations occur very slowly and are very small in quantity, but they are in incessant progress and can be evaluated accurately. During the past millennia, at present as well as in the millennia to come, the eccentricity of Earth's orbit is decreasing, and for this reason the orbit described each year by our planet is increasingly circular. But we must clarify that the length of the major axis of this elliptical path is not altered under any circumstances and, consequently, the size of the trajectory that our planet describes around the star King is gradually increasing. As conclusion: we have that in the present the average distance between the Earth and the Sun is increasing as a consequence of the perturbations that our planet experiences due to the gravitational attraction of the other planets of the system.

The efficiency of the solar gravitational force to disturb the movement of the lunar globe depends on the distance of the Earth-Moon system to the Sun, and since the value of that distance is gradually increasing, as the millennial records show, it is necessarily deduced that the capacity of the King star to disturb the satellite's movement should also gradually decrease. So Laplace deduced that the satellite's orbit should also be gradually decreasing: the Moon would be moving closer to Earth due to alterations in the eccentricity of the Earth's orbit produced by the attraction of the other planets. Although the variation in the position of the Moon is extremely small and the consequent effect in accelerating the satellite movement is very slight, this is not the case when it comes to large periods of time, many centuries. Laplace knew how strong the efficiency of the planets was in altering the dimensions of the Earth's orbit, and with it he could determine the changes that would propagate in the movement of the satellite. Thus he was convinced that the acceleration of the lunar movement, as it was indicated by the records on the observations of

the ancient conserved eclipses, could be completely explained as a consequence of planetary gravitational disturbance. In his time this was considered a great achievement of the scientific method: unless there was some rational explanation for the *Secular Lunar Acceleration*, the widespread acceptance of the validity of the *Law of universal gravitation* would have been seriously blocked. As of this moment, and during almost seventy years, nobody questioned the veracity of the Laplace's arguments.

Laplace found that the values of these inequalities determined by the theory corresponded well with those derived from the observations of the ancient eclipses recorded by Hipparchus and Ptolemy. According to him, the periods in which these inequalities compensate themselves are immense, but the opinion that the Moon will at some point come into contact with the Earth should be now abandoned; since by a change in the configurations of the disturbing bodies, directly and consequently the effects would happen conversely to what is now observed and the Moon would then move away from the planet; and it would happen like this until a new change reverses again the order of the movements, and there would be no reason to believe that this cycle would end someday.

Then, the French astronomer established in 1787 that the supposed abnormal motion of our satellite, its *Secular acceleration*, was also cyclical and apparent. The millenary observations and their respective registers had deceived and made astronomers believe until then that such variations of the celestial movements were linear, continuous and infinite. Laplace's was an excellent deduction in his time, although later researches modified the results and contributed new elements to the full explanation of these inequalities of the lunar movement.

John Couch Adams corrects Laplace

When Lagrange read the Laplace's document announcing his discoveries, he resumed his own work of 1783 and found some errors, he solved them and this led to his calculations matching the results of Laplace almost exactly. Much later, the English mathematician and astronomer John Couch Adams (1819 - 1892), co-discoverer in September 1846 of the planet Neptune, proved that Laplace's theory could not fully explain the phenomenon it intended to solve; but for a long time the feat of Laplace was considered as a maximum achievement of the dynamic astronomy, of the Celestial mechanics.

When undertaking a new study for the same problem, the astronomer John C. Adams discovered that Laplace had not been sufficiently rigorous in his developments and that consequently there were considerable errors in the result of his analysis: a value two times greater than the real one had been assigned to the planetary influence on the satellite's movement. The problem was purely mathematical and the calculations were repeated by a good number of professionals, and finally it was universally admitted that Couch Adams had corrected the French astronomer in a very fundamental theme of Celestial mechanics.

So, an issue still to be clarified is why the lunar globe's revolution period is now seemingly shorter than a few centuries ago and its movement seems to be accelerating. Then, if we were to express the duration of such a lunar period in terms of the rotations of the Earth around its axis, that is in terms of days and hours, then we would find that the Moon currently demands a smaller number of terrestrial rotations to complete each of its respective revolutions around the planet than those previously required for the same. Obviously, this can be explained by arguing that the satellite moves now faster than millennia ago, which is what the theory of secular acceleration supposes. But it is evident that it can be conceived an explanation of a completely different kind to establish another cause for the same phenomenon: If the period of the Earth's rotation, the day, for some reason would be increasing, the Moon still moving at the same pace as always, without modify its orbital period, it will require a smaller number of terrestrial days to realize each one of its revolutions.

It is interesting to note that there is a physical justification for this increase in the Earth's rotation period and the consequent decrease in the rotation rate of our planet: the tides that flow, rise and fall on the Earth exert an action similar to that of a "brake" on this revolving globe; and there is no doubt that slowly and gradually they are reducing its rotation speed and thus lengthening the duration of Earth's day. Consequently, it is this action of the tides that produces the additional effect necessary to complete the physical sustentation of the lunar acceleration.

As a conclusion, we can argue that the phenomenon known as *Secular acceleration of the satellite movement* is the result of both causes acting together. The first of these is the one discovered by the mathematical astronomer Laplace and according to which the disturbances caused by the other planets on the elliptical trajectory of ours, indirectly affect the lunar movement, although he overestimated the magnitude of its influence. The other cause for the acceleration of the lunar globe movement is more apparent than real: it is not that the satellite moves more quickly or that the size of its orbit is decreasing and getting closer every day to the Earth; but that our clock, the terrestrial revolution, now takes a slower rhythm and, thus, is wasting time. ¡It may appear to us that now the Moon is moving more quickly, when in fact it is the Earth that is slowing down!

William Herschel, discoverer of a planet and many moons

Friedrich Wilhelm Herschel (1738 - 1822) was born in *Hannover*, Sacrum Germanic Roman Empire, currently Germany, where by a deep-rooted family tradition he studied music. But being very young he had to leave his homeland in 1757 because of the French occupation of *Hannover* in the framework of the Seven Years' War; he then emigrated to England where he deepened his musical studies: first he became a professor, then an organist in *Halifax* and later an orchestra conductor of *Bath* in 1766.

His interest in astronomy began at a relatively late age, at 35 years old in 1773; but he did it with a great passion and in the modality of autodidact: buying and reading books on the subject. Very soon he understood the enormous advantage of reflector telescopes, based on mirrors, over the refractors based on lenses. Combining his mathematical and manual skills, he learned to calculate, fabricate and polish the most perfect and powerful metal mirrors of the time; with which he designed, calculated and built his own telescopes that became the best in the world at the time. And with them he would make transcendental astronomical discoveries: he discovered a little more than 2500 new objects from deep space, counting globular clusters, nebulae and galaxies; and as for our Solar System, he understood that it was moving towards the constellation of Hercules; he also discovered a planet and many moons and comets.

On March 13, 1781 Herschel gained great reputation due to the first discovery of a planet since antiquity: *Uranus*. Although it had been observed many times since 1690, it was generally confused and treated like a star. Even Herschel himself considered it at first as a comet, but very soon all the astronomers who systematically evaluated the discovery agreed that there was a new planet in the Solar System and located at twice the distance of Saturn. For which Herschel was awarded by the Royal Society of London with the *Copley* medal, and in addition King George III of Great Britain appointed him Royal Astronomer in 1782. These events led the astronomer to settle permanently in England and acquire that nationality.

Six years later, on January 11, 1787 *Herschel* discovered the first two and largest moons of the planet Uranus: Titania and Oberon. Subsequently, in 1789, this astronomer managed to complete the construction of his largest and most powerful telescope that by the length of its tube became known as the *40-foot telescope*, which was the largest in the world at the time. The instrument consisted of a giant metal mirror 1,2 meters in diameter and 12,2 meters focal length. With it he discovered the 28th of August of 1789 the sixth moon of Saturn,

later called Enceladus. A few days later, on September 17 he visualized for the first time the seventh moon of that planet, today known as Mimas. For some sixty years the 40-foot Herschel Telescope maintained the mark of being the largest telescope in the world, until 1848 when it was surpassed by *William Parsons's Leviathan*.

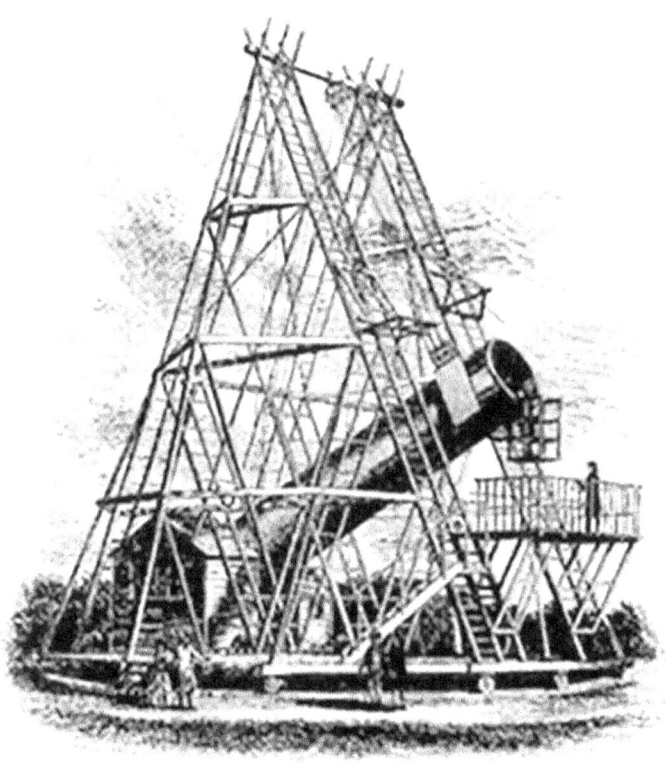

Image 7.4

The 40-foot telescope by *William Herschel* in 1789

With the telescopes built by himself, Herschel discovered in March of 1781 the planet Uranus; and later on he also discovered several moons of Uranus and Saturn.

Asaph Hall

Asaph Hall (1829 - 1907) was an American astronomer from Connecticut. For him being the son of a poor family, at the death of his father he had to work initially as a carpenter. Later he was the assistant of the astronomer *William C. Bond* in the Harvard College Observatory in Cambridge, Massachusetts, from 1857; and continued the studies of astronomy by his own account. He was subsequently appointed director of the Washington Naval Observatory in 1863, where with the help of the large 66 cm in diameter refractor telescope, he discovered in August 1877 the two tiny natural satellites of the planet Mars, called *Deimos* and *Phobos*. His fame as an astronomer grew and finally in 1895 he became professor of astronomy at Harvard University.

William Parsons and his *Leviathan*

William Parsons (1800 - 1867), also known as *Lord Rosse III*, was an Anglo-Irish astronomer who built several telescopes. He was born in York, England; he studied initially at Trinity College in Dublin, Ireland, and then in the Magdalen College of the University of Oxford where he graduated in mathematics in 1822; but he soon became a remarkable amateur astronomer.

Interested in astronomy and inspired by the metallic mirrors developed by *Herschel*, *Parsons* began a series of experiments in casting and polishing the mirror metal, which consist on an alloy of copper and tin. Initially he built telescopes of 48 and 90 cm in diameter, with which he was able to make abundant observations of enough quality.

During the 1840s he began to build at Birr Castle in Parsonstown, Ireland, the famous *Parsonstown Leviathan*: a gigantic reflecting telescope that had a metal mirror 183 cm in diameter and 13 cm thick, with a focal length of 16 m and of three tons in weight, which was assembled in a tube of 16,5 m in length for 12-ton total weight of the finished telescope. The construction of the Leviathan began in 1842 and was used for the first time in 1845. With it, *Parsons* contemplated and cataloged a large number of nebulae, some of which would later be recognized as galaxies; and took some of the first pictures of the Moon. One of the last applications of the instrument in astronomy was in August 1877 when the existences of the two small satellites of the planet Mars, initially discovered by the astronomer Asaph Hall, were verified. Until the beginning of the 20th century, when it was replaced in 1917 by the *Hooker telescope* with a diameter of 254 cm at *Mount Wilson Observatory*, it was the largest telescope in the world in terms of opening size. The *Leviathan* is mentioned in the science fiction novel titled *From the Earth to the Moon*, written by Jules Verne in 1865, who describes it as the largest telescope in the world at the time.

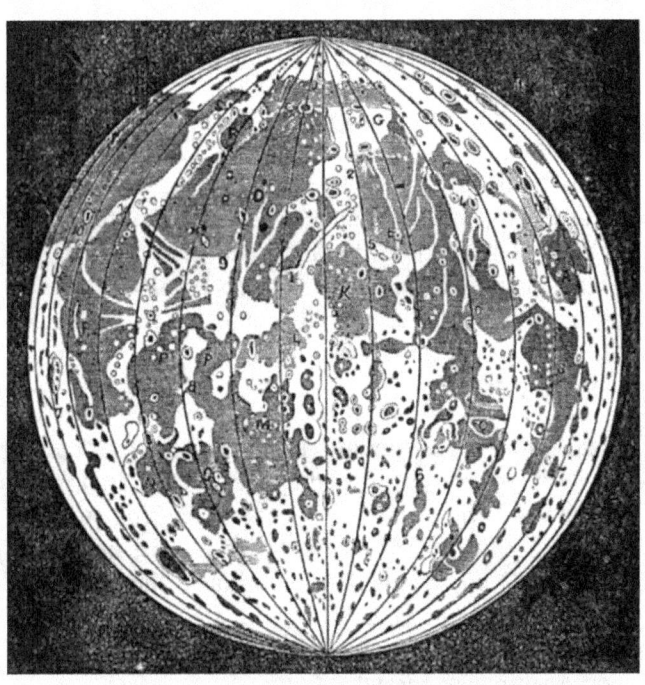

**Image 7.5
The Moon as seen by
Lord Rosse's telescope, 1856**

Representation of the Moon as seen through one of the *Lord Rosse III* telescopes, and presented on the book: *The Moon hoax: or A discovery that the moon has a vast population of human beings*, from 1859.

William Lassell, Neptune and the new moons

The existence of the eighth planet, Neptune, was predicted independently, theoretically and scientifically by mathematicians and astronomers John Couch Adams, English, and *Urbain Le Verrier*, French; who were trying to explain the perturbations observed in the orbit of the planet Uranus. Following the indications of *Le Verrier*, Neptune was finally discovered on September 23, 1846 by the German astronomer *Johann Galle* and his student *Heinrich D'arrest*.

William Lassell (1799 - 1880) was a self-taught amateur astronomer who, as *Herschel* and *Parsons*, made his own mirrors and telescopes and with them he made many important discoveries. Very interested in the newly discovered planet Neptune, on the night of October 10, 1846, he directed one of his telescopes to the sky and soon discovered the largest moon of that planet, later called Triton, and he devoted himself to studying both its orbit and its appearance and brightness.

Later he concentrated on Saturn and on the night of September 18, 1848 he visualized a small moon orbiting it. By coincidence, the satellite had been observed simultaneously by the American astronomer William C. Bond, so the merits of the discovery had to be shared; some time later the moon was baptized as Hyperion. These discoveries excited him and induced him to fully dedicate himself to astronomy. Finally, on October 24, 1851, he discovered simultaneously two new moons of Uranus: Ariel and Umbriel.

Simon Newcomb

Simon Newcomb[5] (1835 - 1909) was a Canadian mathematician and astronomer nationalized in the USA. He basically had no formal education due to the lack of economic resources; and as a young man he had several jobs and became an autodidact in astronomy. Later he applied for a job in the *Nautical Almanac Office of America*, based in *Cambridge, Massachusetts*, where he became practically a *human computer* in 1857. The main function he had to perform there was to prepare new astronomical tables, or ephemerides, to be used in maritime navigation. In parallel he enrolled in the *Lawrence Scientific School* of the University of Harvard and he graduated in 1858.

Because of the American Civil War that began in April 1861, several professors of mathematics in the United States Navy resigned; and in that same year *Newcomb* was appointed professor of mathematics and astronomer at the *Washington Naval Observatory* to fill one such vacancies. He spent more than 10 years there determining the positions of the celestial objects with the modest existing instruments and for two more years with a new 26-inch refractor telescope. He was particularly interested in the theory involved in the orbits of the planets and the Moon , and, with his enormous faculties for computation, he sought to improve their theoretical positions by calculating the perturbations in their orbits, caused by the gravitational attraction of other bodies.

He became a great specialist in celestial mechanics, performed the precise calculations of a large number of orbital parameters concerning almost all the bodies of the Solar System, and he became the principal compiler of one of the most relevant astronomical ephemeris calendars. Additionally, he used some data on eclipses from the astronomer *Ibn Yūnus* to determine the secular acceleration of the Moon.

The tables for the movements of the Moon that existed in his time had been compiled in 1857 by the Danish astronomer *Peter A. Hansen* (1795 - 1874), a specialist in the calculation of the lunar positions based on the gravitational theory; and it was very clear that the satellite was deviating from the position predicted in those tables. *Hansen* had used observations dating back to 1750 to compile his tables, but Newcomb considered that the use of older observations would be very valuable. When Simon Newcomb studied observational data prior to 1750, he discovered that *Hansen's* tables were very erroneous for periods prior to that date. For this reason he devoted himself to making more precise measurements on the Moon to determine well its trajectory, he used data on ancient lunar eclipses to determine its secular acceleration, by which he obtained the perturbations experienced by our satellite over the centuries. For his work during this period on the movement of the satellite, and also on Uranus and Neptune, he was awarded the *Gold Medal* of the Royal Astronomical Society in 1874; he acquired a great reputation as an astronomer and was offered the position of Director of the Observatory of Harvard University in 1875; but he rejected it because his great passion was really in the calculation and development of mathematical theories to explain the observational data.

Simon Newcomb, the self-taught, was later appointed professor of mathematics and astronomy at Johns Hopkins University in 1884, holding this position until 1893; and in parallel he was editor of the American Journal of Mathematics for most of the period from 1885 to 1900. He also served as president of the American Mathematical Society from 1897 to 1898; and was a founding member and first president between 1899 and 1905 of the American Astronomical Society.

Modern Lunar theory: Gravity in action

The Ocean tides and the true lunar movement

The theoretical advances of the 1780s supported the idea that the Solar System was quite stable, that it had evolved to its current equilibrated state, and that it was subject only to self-compensating oscillations. Towards the end of 1787 Laplace announced the missing resolution for the last major anomaly in theoretical astronomy of the time: the *Apparent Secular Acceleration in the lunar motion*. This, he argued, was the result of an indirect perturbation originated in the continuous decrease of the Earth's orbital eccentricity and that leads to a small decrease in the radial component of the gravitational force the Sun exerts on our satellite. But, as the theory of planetary perturbations predicts, these phenomena are oscillatory, cyclical, and in the future the effect would be reversed and the orbital eccentricity of the Earth would begin to increase again and the

Moon would decelerate. But in 1854 John C. Adams proved that *Laplace's* derivation was partially erroneous: for he had only succeeded in explaining half of the apparent secular acceleration of the Moon. The rest would eventually be attributed to the deceleration of the Earth's rotation rate by friction due to the action of ocean tides; which is the effect that *Laplace* had previously ruled out as insignificant. In fact, the slowing down of the Earth's rotation is such that it leads to a considerably greater apparent acceleration of the Moon, but it is counteracted by the satellite's ascent towards an increasingly higher orbit, also due to the tides.

Considering again the general disposition and the gravitational interaction of the Earth-Moon system, we have already described the phenomenon of Oceanic tides as the elevation of the water masses due mainly to the *Lunar gravitational attraction*; and remembering that our planet is elongated, stretched or bulging in its equatorial region, we have that the gravitational force exerted by the satellite is felt most intensely in these bulky areas. Now, due to the inertia generated by the rotation movement of our planet, we have that these deformations or bulges of the terrestrial matter are not aligned with the line that joins the centers of the two stars, but they are deviated some distance to it. This brings a very important consequence for the dynamics of the system: Between the Earth and the Moon acts a *Torque* or *Moment of force* that is given by the intensity of the lunar gravitational force and the distance that the equatorial bulges deviate with respect to the Earth-Moon central line; and that is felt differently by the two stars. (See Image 7.6.)

For our planet the direct effect of such torque is to retard or slow down its rotation movement on its axis, and consequently the duration of the rotation period is increased: the terrestrial day becomes longer, although in fractions of a second. But the effect of this torque is quite different on our satellite since what it does is to propel it, throw it, or rather, to expel it into an increasingly larger and distant orbit from Earth; and consequently its orbital period increases, its angular velocity decreases and the satellite slows down in its orbital motion.

The above complicated phenomena are much more easily understood if they are described in energetic terms. Although both bodies are interacting gravitationally and are exchanging energy, as in any closed physical system, both total energy and angular momentum are conserved for the entire system. But it must be considered that a large part of the energy lost by the Earth dissipates in the form of heat generated by friction between the water and the earthly crust. In this way, the Earth transfers only a part of its rotational kinetic energy and its angular momentum to the orbital motion of the Moon, with which the satellite increases its potential energy by moving to a higher orbit, moving away from its planet! And therefore, by applying *Kepler's Third Law*, its orbital period and its angular velocity decrease. As a final result, both bodies decelerate, the Earth in its rotation movement on its axis and the Moon in its movement of translation around the planet. The kinetic energy and the real speed of translation of the satellite diminish, and as a counterpart its potential energy, its distance to the Earth, increases.

In addition to the changes in the eccentricity Earth's orbit caused by planetary gravitational disturbances, as initially established by *Laplace* and corrected by *John C. Adams* later, there are two tidal effects that influence the satellite motion. First, there is a real decrease in the angular velocity of the Moon's orbital motion, due to the exchange of energy and angular momentum between the Earth and our satellite caused by the tides. This increases the angular momentum of the Moon around the Earth and the satellite moves to a higher orbit, with a shorter period as already explained. Secondly, there is an apparent increase in the angular velocity of orbital motion of the Moon, when measured in terms of *average solar time*, and that arises from the loss of angular momentum of the Earth and the consequent increase in the duration of the day; this last constitute the widely observed phenomenon, which greatly challenged physicists and

mathematicians, and which so disconcerted the astronomers of the past.

All the above phenomenon is known as *Tidal acceleration* and is one of the examples in the dynamics of the Solar System of the so-called *Secular perturbation of an orbit*: a real disturbance that grows continuously over time, which is not cyclical but linear. In the Earth-Moon case it forces the system to self-regulate to remain stable, thus giving rise to two major consequences: the Earth's rotation slows progressively and the Moon also moves away from the Earth progressively. The *Tidal acceleration* is caused by the gravitational force that is exerted by a natural satellite in orbit around a primary planet and that manifests itself as a force of tidal effect.

This acceleration is commonly negative because of the gradual decrease in the rotation speed of the primary object, and the progressive distancing of the satellite with the consequent decrease in its orbital velocity. The overall process eventually leads to the phenomenon known as *Tidal Coupling* or *Tidal Anchorage*, and according to which a small celestial body that orbits around a larger one always has the same face directed towards the body it is orbiting; which is exactly what is observed in the Moon - Earth system.

Most of the improvements in the Newtonian theories of celestial movements, and especially that of the Moon, were made in the form of very laborious and voluminous infinitesimal, algebraic and trigonometric calculations; and also supported by voluminous observational measurements.

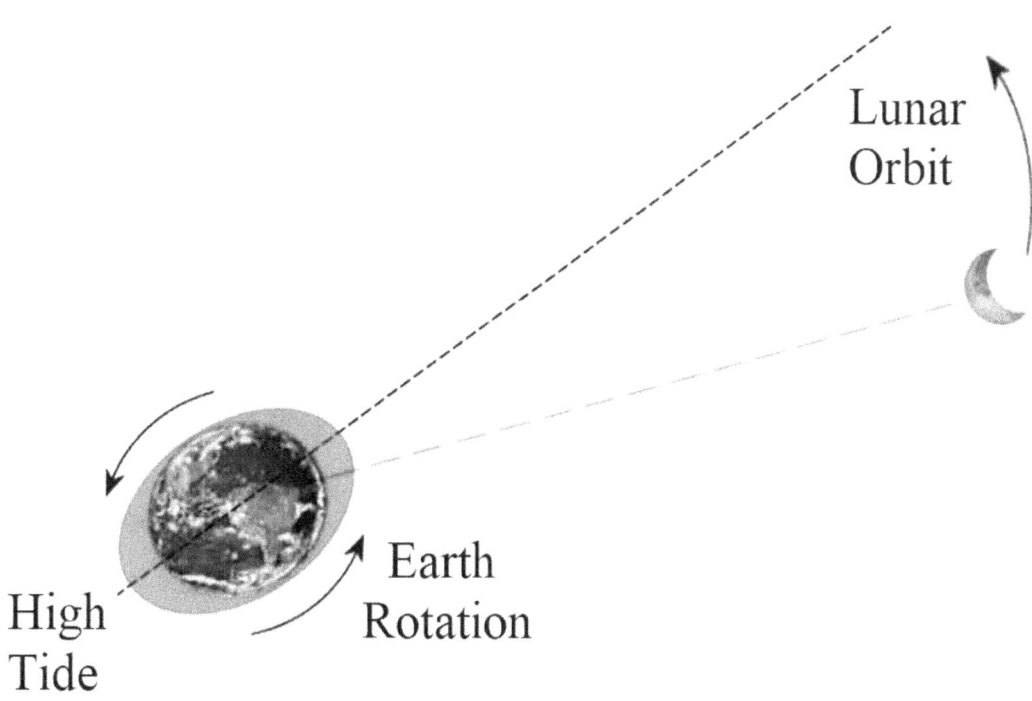

Image 7.6
Earth - Moon System: Tides and Torque

Due to the inertia generated by the terrestrial rotation, the axis of the high tides is deviated ahead of the line joining the centers of the two stars, and this causes a torque that alters the dynamics of the system.

Theorists of the satellite movement until the mid-eighteenth century expressed the perturbations on the position of the Moon using about 25 to 30 mathematical and trigonometric terms. But developments during the nineteenth and twentieth centuries gave rise to very different formulations for the problem, and the number of terms necessary to express the position of the Moon with the levels of precision desired at the beginning of the twentieth century reached up to 1400.

From the second half of the twentieth century *Modern lunar theory* has developed further and in a different way, due mainly to three elements: the development of new numerical methods of integration, the use of automatic digital computing for such calculations and finally due to the modern types of instruments, experimental methods and observational data, which provide a precision level never before reached. Thus, after the *Manned Missions to the Moon* the number of mathematical terms necessary to describe the satellite movement through the methods of modern numerical integrations, using digital computers and according to the precision of the results of the experiments carried out on the surface of the satellite, especially the *Lunar Laser Ranging Experiment*, is several thousand.

The Moon in the popular scope

But Selene has not only captured the interest of astronomers, physicists and mathematicians of all times; but it has always captivated, and with much magic, intellectuals such as poets, novelists, photographers and film producers; and also to all the general public.

Since immemorial times the Moon has always been present in literature in general and in *Science fiction* in particular. If we remember Johannes Kepler, literature, that field within which everything becomes possible, was the one that first raised the possibilities of life, travel and human presence in that Selenite world.

John Wilkins

John Wilkins (1614 - 1672) was an English religious, naturalist and essayist. In his work *The Discovery of a World on the Moon*, published in London in 1638, he proposed the possibility of construction of a ship with the ability to go out into outer space. It is a small book through which Wilkins helped to popularize in England the new science of Astronomy.

In the work, based particularly on the recent discoveries of Galileo in 1610 with his telescopes and in Johann Kepler's *Somnium* published in 1634, *Wilkins* covers a wide range of speculations about the nature of the satellite and the visible lunar characteristics, revealed with amazing details through the new telescopes; he also speculates about the possible inhabitants of the Moon and the potential means to travel there.

Jules Gabriel Verne

Jules Gabriel Verne (1828 - 1905) was a French writer, poet and dramatist; very famous for being the precursor of the modern *Novels of adventures* and the genre of modern *Science fiction*.

His work *De la Terre à la Lune Trajet direct in 97 heures*, or *From the Earth to the Moon direct in 97 hours*, October 1865, is a novel written in an exciting mix of science, fiction and satire. In its background the work is a kind of satire to the American stereotype of the time; but in its most important content constitutes a first serious attempt to describe with scientific rigor the problems that must be considered and solved with the purpose of successfully sending a manned ship to the Moon.

His other novel entitled *Autour de la Lune*, or *Around the Moon*, was published in January 1870. In this work the author narrates with great detail the experiences and adventures of the three explorers installed inside a ship in the form of hollow and giant cannonball, which transports them first to the Moon and then back to Earth. The two novels were subsequently edited

together in September 1872. Undoubtedly, the most outstanding stories about the satellite are these two works of universal literature belonging to Jules Verne and published practically a century before the real trip to the Moon.

Herbert George Wells (1866 - 1946) was a British philosopher, historian, writer and novelist. He was the author of the fictional novel *The first men on the Moon* written in 1901. The text recounts the trip to the lunar globe made in virtue of an anti-gravitational substance named *Cavorite*, compound based on Helium and molten metals which was invented by the brilliant and eccentric scientist *Dr. Cavor*. His adventure partner is the ruined businessman *Mr. Bedford*; and with the newly invented substance they cover a rudimentary spacecraft, which thus loses its weight and then quickly ascends towards the lunar globe. Upon arriving at their destination, travelers discover that the satellite is inhabited by an extraterrestrial civilization, which lives in the caverns of the subsoil and whom he give the name of *Selenites*.

Cinematography is a relatively new art. In February of 1895 the French brothers *Auguste* and *Louis Lumière* successfully achieved to patent the apparatus now recognized as Cinematograph; and they immediately started shooting their first films. And just beginning the new twentieth century, Selene captured the enthusiasm and imagination of film producers.

Georges Méliès (1861 - 1938) was an illusionist and a prolific French filmmaker, very famous for introducing many innovations at the dawn of cinematography.

Le Voyage dans la Lune, or *Journey to the Moon*, is a French science fiction film of 1902, voiceless and in black and white, with a script written and directed by *Georges Méliès*. It was in a sense the first science fiction film in the history of cinema and it is still quite popular: The scene of the face of the Moon receiving the impact of a space rocket fired from Earth is one of the best known of the history of cinema. The script is based on two famous novels: *From Earth to the Moon* by Jules Verne, and *The First Men on the Moon* by *HG Wells*. Subsequently, in 1907 *Méliès* presented *L'éclipse du soleil en pleine lune*, or *The eclipse of the Sun in full Moon*, also voiceless and in black and white.

With all that has been said so far in these seven chapters, humanity has already clearly established the distance, the direction and the speed of our satellite in its orbital movement. To really travel to it, it only remains to build the rockets and space ships, and find the volunteers for the trip. Which constitutes the central theme of the next chapter.

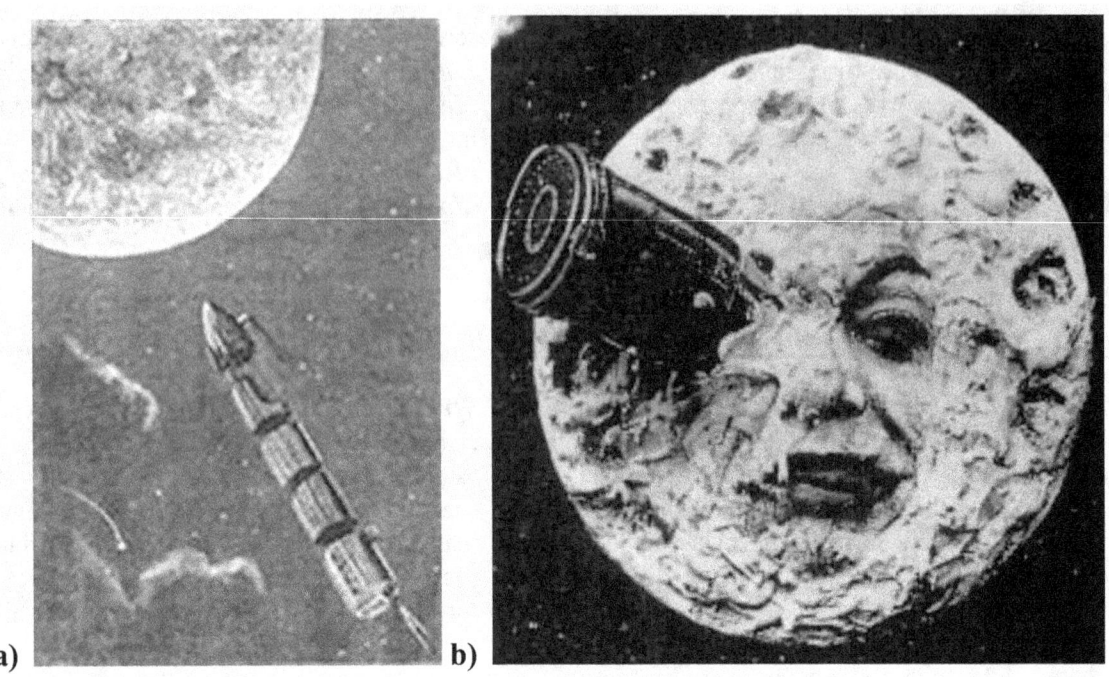

**Image 7.7
The Moon in the popular scope**

a) Illustration of the novel *From the Earth to the Moon* by Jules Verne, drawn by Henri de Montaut in 1868.
b) *Trip to the Moon* frame by *Georges Méliès* of 1902.

Bibliographic Citations

[1] Hervás and Panduro, Lorenzo. *Viaje estático al mundo planetario*. Tomo II Madrid: Imprenta de Aznar. 1.793
Internet Archive.
https://archive.org/details/viageestticoalm00pandgoog/page/n8

[2] Internet Archive: Cornell University Library. Jules Verne. *From Earth to the Moon*.
https://ia600206.us.archive.org/22/items/cu31924052535725/cu31924052535725.pdf

[3] Internet Archive. James Roy Newman. The World of Mathematics, Volume 1. New York: Simon and Schuster,
1956. Pag 257. https://archive.org/details/world1ofmathemati00newm

[4] Abetti, Giorgio. *The history of astronomy*. London: Sidgwick and Jackson. 1954
World Digital Library. Newton, Isaac, Sir. *Philosophiae naturalis principia mathematica*.
https://www.wdl.org/es/item/17842/
Internet Archive. Newton, Isaac, Sir. *Mathematical principles of natural philosophy*.
https://archive.org/search.php?query=Mathematical%20principles%20of%20natural%20philosophy

[5] MacTutor History of Mathematics archive. Simon Newcomb: http://www-groups.dcs.st-and.ac.uk/history/Biographies/Newcomb.html

Chapter 8

Twentieth Century: Wars, Space Race And Manned Missions to the Moon

Our satellite as it could be seen from Earth on Wednesday, February 14, 2018 at 16:00 UCT, 28 days and 13.5 hours after the New Moon, reaching to see 1,4% of its illuminated hemisphere. Now at the end of its monthly cycle, the visibility of the satellite has fallen to minimum levels, in the next hours we will not be able to see any illuminated part of the Moon; a lunar month will ends, but another will begins and the New Moon phase soon will be reached again.

Courtesy of NASA's Scientific Visualization Studio: https://svs.gsfc.nasa.gov/4604

"The first of December had arrived, the decisive day because if the projectile's departure did not take place that same night, at ten forty-six minutes and forty seconds p.m., more than eighteen years must roll by before the Moon would again present herself under the same conditions of zenith and perigee. The weather was magnificent. Despite the approach of winter, the Sun shone brightly and bathed with its radiant effluvia the Earth, which three of its denizens were going to leave in search of a new world."

From the Earth to the Moon, direct route in 97 hours. (1865)[1]
Jules Verne (1828 - 1905)

"However, about seven o'clock, the heavy silence was dissipated. The Moon rose above the horizon, millions of hurrahs hailed her appearance; she was punctual to the meeting. The clamors rose to heaven, the applauses came from all points; and meanwhile the white Phoebe, shining peacefully in an admirable sky, caressed the crowd with its most affectionate rays. At that time the intrepid travelers appeared."

From the Earth to the Moon, direct route in 97 hours. (1865)

"Immediately Murchison, pressing the switch of the device with his finger, set the current and threw the electric spark at the bottom of the Columbiad. An appalling, unprecedented, superhuman detonation, of which there is no rumble that can give the faintest idea, nor the bursts of lightning, nor the crash of eruptions, occurred instantaneously. An immense beam of fire came out of the bowels of the Earth like a crater. The earth heaved up, and with great difficulty some few spectators obtained a momentary glimpse of the projectile victoriously cleaving the air in the midst of the fiery vapors!"

From the Earth to the Moon, direct route in 97 hours. (1865)

The Bittersweet Twentieth Century

The first decades

The scientific-technological history of the twentieth century must be studied by going back a little towards the previous century; and it must be centralized in the development of fuels, which are one of its fundamental features, at least until the advent of the digital electronic age. Since prehistoric times solid fuels, vegetable and minerals coals, were those that provided the energy demanded by human societies. But this would change towards the second half of the 19th century when chemical processes began to develop for the large-scale production of liquid fuels derived from petroleum; which initially were mainly dedicated to both public and private lighting: one of the most striking icons of the nineteenth century is the oil lamp.

Also, in parallel during the second half of the nineteenth century there was a great boom in the theoretical foundation, the design and construction of the internal combustion engine based on liquid fuels; which would be incorporated, during the first decade of the next century, at two inventions that would become the maximum symbols of the twentieth century: the automobile and the airplane.

In the philosophical, physical and cosmological planes the entrance of this new XX century brought also very important innovations. In 1905 the young German Jewish physicist Albert Einstein published his *Special Theory of Relativity*; and ten years later he extends and expands it to give rise to his new *General Theory of relativity*. These theories were developed by the physicist to include the relative character between time and space, great absent in classical Newtonian physics; and in this way to account for some cosmological inconsistencies manifested in this last theory; particularly in the case of the celestial bodies' movement.

It would be unacceptable to make an exposition about the Moon and eclipses, and not referring to what is catalogued as the most famous in history: the solar eclipse of *Albert Einstein*. This German physicist published in 1915 the *General Theory of Relativity*, with which he inaugurated the modern era of cosmology and in which he completely reformulated the concept of *Gravity* previously introduced by Newton. Now *Einstein* explained the gravitational interaction in terms of a deformation or curvature of the continuum space-time, which is due to the presence of very massive bodies, such as stars or suns; and at the same time he states that light is subjected to gravitational effects in precisely the same way as ordinary matter. This means that, in the presence of a considerable mass, anything that passes through space-time, including light, must undergo a change of trajectory due to that deformation on space-time: the rays of light move always following such curvature. In conclusion, the light's rays coming from distant stars and that pass close to the Sun, would follow a curved trajectory so that, contemplated from the Earth, they seem to come from a different location to the real one. Quantifying the idea, Einstein's general theory of relativity predicts a deviation of 1,75 arc seconds, twice that predicted by Newton's gravitational theory.

An experimental verification of the aforesaid phenomenon would be impossible in broad daylight; but an exceptional opportunity to do so would be during a total solar eclipse on May 29, 1919. For this purpose, the Royal Astronomical Society delegates astronomers Frank W. Dyson and Arthur S. Eddington the organization of two expeditions of observation and data collection with the purpose of evaluating the relativistic theories of Einstein. The two selected sites were the town of *Sobral* in the northeast of Brazil and the island of Príncipe in the Gulf of Guinea, because they are located right on the line where the phenomenon would be seen in all its magnitude. The eclipse had duration of 6 min and 50 seconds, and during this time it was possible to measure the position of 12 stars with errors lower than 6%. The data obtained confirmed that the light rays, by passing very close to our Sun, followed a curved trajectory and therefore also confirmed the relativistic predictions of Einstein: these data were considered at that time as a strong proof of the validity of the Theory of general relativity versus Newtonian physics.[2] This result was corroborated repeatedly by different astronomical observatories through measurements made during the total solar eclipses of 1922, 1929, 1952 and finally on June 30, 1973. The bending of light around massive objects is now named as *Gravitational lenses*, and has become an important tool in astrophysics widely used for the detection of very distant objects, such as exoplanets, and even for the detection of dark matter.

With all aforementioned, the liquid fuels, the automobile, the airplane and the new scientific theories, we have that the first two decades of the new twentieth century were cheerful, sweet. But this happiness would rather last very little, because soon the sour character of the century would be manifested: the fateful wars.

The century of the world wars

The citizens of the world of the first half of this century had to witness two bloody and inhumane world wars. The World War I, also called Great War, occurred in Europe between July 28, 1914 and November 11, 1918. The World War II was also unleashed in Europe between September 1, 1939 and last until August 14, 1945; to date the latter is the greatest war in history, of truly universal scope. Together with the great powers most of the nations of the world were involved divided into two military alliances: the *Allied Countries*, with the United Kingdom, the United States, the Soviet Union and China at the head of a long list of countries; and the *Axis Powers* to which Germany, the Empire of Japan and the Kingdom of Italy mainly belonged.

The characteristics of the causes of these wars are very similar, so that some historians argue that it is the same war but with an interlude, a recess of twenty years. But, for the purposes of this text, which now interests us are some of the consequences of the World War II: the Cold War, the Nuclear Arms Race and the Space Race.

Immediately after the World War II it was clearly evident that the world now was divided into two great political-economic blocks: the Western, capitalist and led by the United States; and the Eastern bloc, communist and at the head of the Union of Soviet Socialist Republics, USSR. These two nations emerged as the two greatest powers of the post-war period, and among them a tenacious confrontation breaks out in the political, economic, social, scientific, technological, military and armament; and by which each of the great potencies sought to impose its power over the other and implement its political economic model of governance throughout the planet. Political economic confrontation that has been named *Cold War* and which had its beginnings in 1945 and lasts until the definitive dissolution of the USSR, that occurred between the years of 1989 with the fall of the Berlin Wall and 1991 with the coup of state in the USSR.

The most important component of that Cold War was the Arms race, by which each one of the powers sought to maximize the variety, quantity and quality of its military weapons inventory; involving both traditional and non-conventional combat weapons, including in the latter the mass destruction weapons such as chemical and nuclear ones. Although the process did not lead to a direct military confrontation between the two potencies, world peace was permanently seriously threatened; to the point that in some media the possibility of humanity self-destruction was commented.

Brief history of the rocket

It is a very widespread fact that gunpowder was an invention or development of the ancient Chinese alchemists between the seventh and ninth centuries AD, and that its most primitive applications were of a harmless nature, such as the animation of religious ceremonies, something similar to what we now call fireworks. But it would not be long before the Chinese generals found its great potential within the military arts; soon the *arrows of fire*, bombs and cannons based on gunpowder appeared.

Also soon, the new military weapons would reach importance throughout the Asian continent. In India, and during the Anglo-

Mysore Wars of the late eighteenth century, the troops of Sultan *Fateh Ali Sahab Tipu* of the Kingdom of Mysore, considered a pioneer in the use of rocket artillery, successfully used in 1782 metal rockets propelled with gunpowder against the British armies. These, impressed, showed great interest in such innovative technology, copied it, took it to Europe, and redesigned and improved it considerably; with which the use of rockets based on gunpowder and for military purposes rapidly spread throughout Europe in the coming times.

By the early twentieth century, the technology of rockets became of universal knowledge: During the 1920s, important technological researches and developments were carried out in Europe, Asia, Russia, and the United States. In this last country the engineer Robert H. Goddard launched the first liquid-fuel based rocket on March 16, 1926; so demonstrating that liquid propellants were a viable option. On the other hand, in Germany during the decade of 1930 the *Association for Space Navigation* made the first tests with rockets based on liquid fuels.

In Germany, during the Second World War, the famous Rocket V2 was designed and built under the direction of German mechanical and aerospace engineer Wernher von Braun. From September of 1944 to March of 1945 the German Armed Forces, *Wehrmacht*, launched several thousand of them against Allied territory, most of them against London in England and Antwerp in Belgium. This rocket is considered the first long-range ballistic missile in the world, and the first documented human device that made a suborbital flight; and, therefore, it is also the forerunner of all modern rockets, including those used in the space programs of both the Soviet Union and the United States.

The Space Race

During the course of the Arms race, the powers soon realized that the presence and military dominance in space was a crucial factor, inescapable to achieve a complete and lasting universal supremacy. For this reason they focused their attention and allocated great financial, human, scientific and technological resources with the purpose of achieving, each on their own, the supremacy and absolute control of the space with marked political and economic purposes.

Space Race means a series of events carried out by the two great powers in the aerospace field and during the period of time between 1957 and 1975; which is part of another major period known as the *Arms race*, which in turn has its origins in the *Cold War* caused by the denouement of the Second World War. This Space Race was characterized by a strong scientific and technological competitiveness between the United States and the Soviet Union, with a view to obtaining technological and military dominance in space, the primacy in space exploration and guaranteeing human presence in space.

During its development the first great successes were obtained by the Soviet Union in October 1957 with *Sputnik 1*, the first artificial satellite to orbit the Earth; fact that is normally considered as the provocateur of the space race. And then in 1959 with a series of unmanned space probes called *Luna*[3], whose their main achievements were: first vehicle to escape from Earth's gravity and enter the solar orbit, first impact of a space vehicle on the lunar surface, first telemetry communication to and from Earth, first successful lunar flyby and first pictures of the hidden side of the satellite. But the success that most fame gave to the Soviet Union was to place the first man in space on April 12, 1961: the cosmonaut Yuri Gagarin orbited the Earth at an altitude of 315 km and in about 108 minutes in the space capsule *Vostok* 1, to then land safely. Finally, the Russians achieved the first unmanned moon landing in 1966 with the *Luna 9* probe, which transmitted images of the satellite' surface to the Earth.

A simple definition of a *Space probe* is that it consists of an artifact with its own energy, propulsion, instrumentation and communication systems; which, after being launched into outer space with the help of a rocket, is capable of flying in a remote-controlled manner from an operations' center on Earth until it reaches its final goal in space. On the other hand, an *artificial satellite* is a space probe inserted intentionally and indefinitely in the orbit of some celestial body.

The first successes of the United States began on January 31, 1958 with its first artificial satellite, *Explorer 1*. Later, 23 days after Gagarin's first orbital flight, on May 5, 1961, the first American astronaut, Alan Shepard, was placed in space using a terrestrial suborbital flight in the *Mercury Redstone 3* capsule. Afterward in 1962 for the first time a space probe flied over another planet: the *Mariner 2* probe passed at 34800 km from Venus and scanned its surface; then, in 1964, the planet Mars is flown for the first time by the *Mariner 4* probe, which took close-up images of it. Continuing with their space program, in 1967 the Americans made with Surveyor 5 their first unmanned lunar landing, with transmission of images and analysis of the land; and the following year they carried out the first orbital manned mission to the Moon with the Apollo 8 mission. But the greatest success of the entire history of space exploration was obtained by the United States on July 20, 1969 with the Apollo 11 mission, the first manned moon landing and the first men to walk on the Moon: Neil Armstrong, seconded by Edwin E Aldrin. Until 1971 the Americans made a total of six successful manned landings for a total of 12 men who "have had their feet" not only on Earth, but also on the Moon.[4]

The space race is understood to be ended up in 1975, when the relations between the two powers are relaxed and the first joint mission between the American Space Agency, NASA, and the Soviet Space Agency is carried out: an American Apollo spaceship and a Soviet Soyuz one are coupled in the space and their occupants perform joint tasks. But out of inertia and enthusiasm, space missions and space exploration continued in a less tense environment.

Orbiting the Earth: First men in space

At first the rockets were designed specifically for military applications and are traditionally called Ballistic Missiles. But during the space race the Americans and the Russians putted on way their respective space programs that were based on rockets redesigned for applications on astronautics and space exploration, which are now recognized as Space Rockets or Launch vehicles too.

Of the series of rockets designed and built in the United States the Vanguard, Redstone, Atlas, and the series Delta, Titan and Saturn stand out; of the latter the Saturn V was the greatest rocket of all time and it made possible the American Apollo lunar program. On the other hand, in the Soviet Union the best-known rockets are those of the Proton and Vostok series.

The technological scientific advantage of the Soviet Union over the Americans is accentuated considerably on April 12, 1961 when it managed to place the first man in space through its *Vostok space program*: the cosmonaut *Yuri Gagarin* orbited the Earth in the space capsule *Vostok 1*, which climbed to a maximum altitude of 315 km and circled the planet on a flight that, incredibly for the time, lasted 108 minutes. Equally incredible for almost everyone, was that the cosmonaut performed such a feat and managed to return safely and healthy to Earth.

Two years after the achievement of Gagarin and within the Vostok program, another great feat of the Soviets was carried out by the cosmonaut *Valeri F. Bykovski*, who on June 14, 1963, orbited the Earth 81 times in the space capsule *Vostok 5*, in a lapse of 120 hours, or five days. Then, on June 16, 1963, Russian engineer and cosmonaut *Valentina Tereshkova* became the first woman to fly into space: manning the spaceship *Vostok 6*, she made a total of 48 laps around the planet in a three-day mission. Later, in March of 1965 the Soviet cosmonaut *Alekséi A. Leónov* made the first spacewalk passing 12 minutes outside the *Vosjod 2* spacecraft; an authentic feat for that time.

But the Americans soon equaled the Soviets in the development of space technology. In order to face the great Soviet challenge, on July 29, 1958, the *National Aeronautics and Space Act* was approved. Through this act the former

National Advisory Committee for Aeronautics, NACA, was closed, and at the same time the actual National Aeronautics and Space Administration, NASA, was created. This agency would be responsible for the design, planning and execution of all future American space exploration programs.

The Mercury Project

Designed and developed between 1961 and 1963 within the general frameworks of the Cold War and the Space Race, the first manned space program in the United States was named *Mercury Project*; which constituted the American response to the remarkable leadership of the Soviet Union in that space race, and that began a year after such nation put the first satellite in space, the *Sputnik* 1. The project began in August of 1959 with a series of test flights without human crew, and finished in November of 1961 with an orbital flight of a chimpanzee named *Enos*, with a duration of 1 hour and 28,5 minutes and that ended successfully. The manned trips began in May 5, 1961 with a series of six successful missions that took the first Americans into space, and which ended on May 16, 1963.

From this Mercury project, two missions acquired great historical importance: on May 5 of 1961 astronaut Alan Shepard became the first American to perform a suborbital flight, in his Freedom 7 space capsule propelled by the *Redstone III* rocket, which reached an altitude of 187 kilometers above the Earth's surface on a flight that lasted 15,5 minutes. Subsequently on February 20, 1962, Astronaut John H. Glenn Jr. was the first American to orbit the Earth in his Friendship 7 space capsule propelled by the Atlas VI rocket, which reached a maximum altitude of 260 km and made three orbits to Earth with an orbital period of 88,5 minutes, for a total duration of 4 hours 55 minutes and 23 seconds in the mission.

The primary objectives of the Mercury Project were the following: To place a manned spacecraft in orbital flight around the Earth, investigate man's performance capabilities to function in an unknown space environment, and to develop technologies for safely recovering both crew and spaceships. These objectives were met and for this reason the project is considered very successful.[5]

The Gemini Program

The second manned space program developed by the United States was the *Gemini* Program, which was officially announced to the public on January 3, 1962. Its name comes from its crew of two men and the third constellation of the Zodiac and its twin stars, *Castor* and *Pollux*. *Gemini* involved 12 missions, including two unmanned flights for testing spaceships and machines. Its main objectives were: to subject men and equipment to space flights of up to two weeks; to carry out a space meeting, to assemble two vehicles in orbit and maneuver the set using the propulsion system of the target vehicle; and finally, to perfect the methods of entering the atmosphere and landing at a preselected point on Earth.

The fundamental purpose of the Gemini program was to develop techniques for the encounter and spatial coupling between two spaceships; processes known as *Rendezvous* and that would be very important during future manned space missions to the Moon. The objective and reuniting capsule was an unmanned *Agena*, which consisted of the upper stage of some rockets, and that was launched ahead of the Gemini Capsule. The launch rockets for these missions were the Titan II.

The program consisted of a total of twelve missions, the first two unmanned; and it covered about thirty months and a thousand hours of flight. As the most notables of these missions it should be stressed: firstly, the *Gemini* 4 of June 3, 1965, put into orbit by a *Titan II* rocket and that had duration of 4 days 1 hour and 56 minutes in space, completing a total of 62 orbits at a maximum altitude of 282 km. Its crew was James McDivitt and Edward White; the latter made the first spacewalk of the United States with duration of 22 minutes. And then the *Gemini 8* mission of March 16, 1966, also launched by a Titan II rocket, with duration of

10 hours 41 minutes and 26 minutes, made a total of six orbits at a maximum height of 272 km. With its two crewmen Neil Armstrong and David Scott, in this mission it occurred the first coupling of a Gemini space capsule with another unmanned aircraft, an *Agena*; this mission also involved the first emergency landing of a manned American spaceship.

The importance of the Gemini missions lied in the fact that the number of crewmembers was increased to two and the duration of the missions was extended to two weeks; with which American astronauts were prepared to live, work and sleep in space in more extreme conditions. Likewise, in the course of these missions, the first American spacewalks were conducted with their respective *Extra Vehicular Activities*, EVA, which were indispensable for future lunar missions. As important legacies of the Gemini missions we appreciate that they made the operations of on fly encounter and coupling between spaceships routine and safe, and they also managed to allow the astronauts to carry a safe and without major complications life during appreciable periods of time in the space.

First Unmanned Lunar Missions

Soviet space program

In January 1959 the Soviet *Luna 1* space probe went down in history as the first space probe to escape Earth's gravity, to successfully fly over 6000 km from the lunar surface, and finally to enter in solar orbit. Likewise, the first ignition of a rocket in terrestrial orbit and the first telemetry communication to and from Earth were made. On the other hand, the *Luna 2* probe was the first spacecraft to reach the lunar surface and impacted near the *Mare Serenitatis* plain in September 1959. Finally, in January of 1966 the *Luna 9* probe carried out the first unmanned and properly controlled moon landing, descending on the *Oceanus Procellarum* plain and managing to take the first images of the Moon' surface from the same terrain.

North American Space Program

Ranger probes: Spies and lunar projectiles

The *Ranger* program was a series of unmanned space missions from the United States in the 1960s. It aimed to achieve a lunar impact trajectory, to obtain the first close-up images of the satellite surface and transmit them to Earth before the spacecraft was destroyed by being intentionally crashed against the lunar surface. However, a set of setbacks led to the failure of the first six flights.

Ranger 7 was launched on 28 July 1964, it was the first fully successful mission of the *Ranger* program and the first US spacecraft to successfully transmit to Earth close-up images of the lunar surface. In addition to transmitting some video sequences, it also transmitted more than 4300 photographs during the last 17 minutes prior to the impact. After 68,6 hours of flight, the spacecraft was intentionally crashed at 2,6 km per second against the satellite surface.

The other two successful missions of this program were the *Ranger 8* launched on February 17, 1965, and the *Ranger 9* launched on March 21 of the same year. The images and videos obtained by the Ranger program were broadcast on live television to millions of viewers throughout the United States; and they served to select the respective landing sites for future Apollo missions and also were used for other scientific studies.

Lunar Orbiter: The first satellites of our satellite

After the successes of those initial space missions, the new successful program of the Americans was called *Lunar Orbiter*; and it consisted of designing, constructing and operating space probes that could become the first artificial satellites of the Moon. The program had two main objectives: to take the first photographs from the lunar orbit trying to cover the entire area and to provide evidence for selecting the landing sites for future Apollo manned missions. The program comprised a series of five successful unmanned lunar missions launched from 1966 to 1967; for all of them launch vehicles, or multistage rockets, *Atlas-Agena-D* were used.

The first three missions had low-altitude orbits and were dedicated to taking images of 20 possible sites for future manned landings, previously selected from Earth-based observations. Some relevant facts of the first two missions are the following:

Launched on August 10, 1966, the *Lunar Orbiter 1* space probe was the first to become an artificial satellite of the Moon, orbiting it at a maximum altitude of 1867 km. The nominal duration of the mission was one year, but due to technical failures it only lasted 80 days, and it crashed onto the lunar surface on the hidden side on October 29, 1966 during its 577th orbit. Even so, the probe achieved to take a total of 42 high resolution and 187 medium resolution frames that covered more than 5 million square kilometers of the Moon's surface and were transmitted to the Earth, achieving approximately 75% of the planned mission.

The *Lunar Orbiter 2* was launched on November 6, 1966, it was active for 339 days giving a total of 2346 orbits to our satellite; it acquired photographic data from November 18 to 25, 1966; 609 high-resolution and 208 medium-resolution frames were returned to Earth. These include a spectacular image of the Copernic lunar crater, which was cataloged by the media as one of the best images of the century. The spacecraft orbited the Moon at a maximum altitude of 3598 kilometers until it impacted the surface on October 11, 1967.

The fourth and fifth missions were dedicated to broader scientific objectives and they flew in high altitude polar orbits. The Lunar Orbiter 4 photographed the entire near side and nine percent of the opposite side of the Moon; while the Lunar Orbiter 5 finished covering the far side, or *hidden side of the Moon*, and additionally it took medium and high resolution images of another 36 preselected areas. In summary, 99 percent of the satellite surface was mapped from photographs taken with very good resolution.

Finally, as if that were not enough, the first photographs of the entire Earth were taken from space during these missions: on August 8, 1967 the *Lunar Orbiter 5* achieved the first image covering an entire hemisphere of our planet from the lunar orbit.

Surveyor Program: First controlled moon landings and first lunar topographers

After the Lunar Orbiter, the *Surveyor Program* was the third and last North American program of unmanned automatic lunar probes. These missions had lunar landing capability, or controlled and safe descent on the satellite's surface; and also soil excavation capacity, collection of samples and the subsequent chemical analysis of the same; additionally they had filmic and photographic capacity.

The program consisted of seven missions of which five were successful, and was developed between May 31, 1966 and January 7, 1968. Its fundamental objectives were: to develop and evaluate technologies that made it possible to perform controlled lunar landings, in such a way that the probe could function as a *Robotic Lunar Base*; and that additionally made it possible to realize excavations and physical chemical analyzes of the lunar surface to obtain basic geologic information of the satellite that could be useful in the future *Apollo Program* missions. None of the missions included returning the probes to Earth; whereby the seven spacecraft, or their remains, are still on the Moon.

The Surveyor 1 soft lunar lander module was launched on May 30, 1966 and prudently landed in the region known as Ocean of Storms, on following June 2; becoming the first North American probe to make a soft and successful touchdown over any other extraterrestrial body; and only four months after the first moon landing of the Soviet Union Luna 9 probe. This lunar soft-landing module collected data on the lunar surface that would be necessary for the moon landings of the future Apollo missions of 1969: it transmitted 1237 photos of the lunar

surface to Earth using a television camera and a sophisticated radio-telemetry system. The most important achievement of this mission was to have established that the lunar surface can support both spacecraft and men.

After Surveyor 2 failed and crashed into the Moon on September 23, 1966; the Surveyor 3 was launched on April 17, 1967, which landed on the satellite on April 20 of the same year in the *Mare Cognitum* area of the *Ocean of Storms*; then it transmitted 6315 TV images to Earth. This mission was the first that carried a digger spoon to sample the lunar soil, which was mounted on a robotic arm driven by an electric motor. The spoon was used to dig four trenches in the lunar soil, up to 18 centimeters deep; this was the first space probe to do this in any extraterrestrial body. The soil samples from the trenches were placed in front of the television cameras of the probe to be photographed and to transmit the images to the Earth. When the first lunar night, equivalent to 14 Earth days or around 336 hours, arrived on May 3, 1967, Surveyor 3 was turned off because its solar panels no longer produced electricity; and at the next lunar dawn the probe could not be reactivated due to the extremely cold temperatures it had experienced. This was in contrast to Surveyor 1 that could be reactivated twice after the long, freezing lunar nights, but never again.

On November 19, 1969, the lunar module of the Apollo 12 mission landed about 180 meters from the *Surveyor 3* probe, and its crew took photographs and removed almost 10 kilograms of its components, including the television camera; with the purpose of return them to Earth and to be evaluated.

The *Surveyor 4* probe also failed and crashed into the satellite on July 17, 1967; but missions 5, 6 and 7 were successful. The *Surveyor 5* made the first chemical analysis *in situ* of the lunar surface, it discovered that the satellite rock has a basaltic composition similar to that of our planet, and made the first reboot of a rocket engine on the Moon. The lunar module Surveyor 6 performed the first ascent from the lunar terrain and the first flight with controlled movement above the Moon's surface. Finally, the Surveyor 7 carried out activities very similar to the predecessor probes, but in very different regions of the satellite surface.

The Manned Space Program Apollo

With that aforesaid we can see that the Soviet Union took the lead in much of the unmanned lunar missions: first orbiter, first lunar landing and first images of the satellite's surface. But this nation did not pass from there: neither the Soviet Union nor any other country, excepting the United States, has achieved until this year 2019 to carry out some successful manned mission to the Moon. So in this chapter it only remains to expose the US manned space program.

General objectives

The American *Apollo* space program was designed during the 1960s to land humans on the Moon, to make a short stay on the satellite surface and then return them safely to Earth; all this according to the national objective proposed by President *John F. Kennedy* in a speech before the Congress on May 25, 1961. Other general objectives of the program were: to develop the necessary technologies to satisfy other national interests in the space, to reach the pre-eminence on space for the United States, to develop the capacity of the man to work in the lunar environment and to carry out a program of scientific exploration of the Moon. The program was named after the multi-faceted Greco-Roman Olympic god of beauty, music, light and of the Sun.

Specific objectives

The specific objectives of the lunar exploration were established considering that the study of the geological process of its evolution is of great importance, that its peculiar environment free of atmosphere provides an observation of our Solar System without obstructions, and finally that the proximity of the Moon to Earth they make it the first logical step for the future manned exploration of our Solar System as a whole.

So the specific objectives of the human exploration of our satellite were established as: To obtain information for determining the environment, the composition and the physical and geological properties of the Moon; determining whether the unique characteristics of the satellite can be used for establishing observatories and long-term scientific research laboratories; and determining if the natural resources of our satellite could be used for extended lunar operations and for future interplanetary exploration. The results and knowledge gained from these missions allowed us to obtain a deeper understanding of the Moon; as well as providing good information on the history and evolutionary sequence of the respective processes involved in the formation of our Solar System.[6]

Components and General Structure of the *Apollo* Missions

Schemes or general concepts of operation and functioning

Once the objectives of the US space program were established, the Apollo missions planners faced the challenge of designing a spacecraft that could meet them and that at the same time it would minimize demands on astronaut capabilities and the risks for human life, as well as minimizing the technological requirements and the respective economic costs of the entire program.

Among four possible schemes or flight configurations for the missions, the *Lunar Orbit Rendezvous* turned out to be the selected one and which managed to accomplish the objectives in the Apollo 11 mission on July 24, 1969: a single rocket *Saturn V* launched a 44100 kilogram spacecraft that was composed of a 28900 kg *Mother ship*, which remained waiting in orbit around the Moon; while a *Landing module* of two stages and about 15200 kg descended two astronauts to the satellite surface, waited for them and later ascended and returned to assemble once again with the mother ship, to finally be discarded later. Landing merely a part of the spacecraft and returning an even smaller part of only 4600 kg to the lunar orbit minimized both the total mass and the economic cost for launching from Earth; but this was the last scheme considered due to the great risks involved during the encounter and spatial coupling phases.

In aeronautics the word of French origin *Rendezvous* is translated into English as a space encounter and coupling in plenary flight of two spacecraft, with the purpose of functioning as a one. Likewise, the flight scheme called *Lunar Orbit Rendezvous* is a procedure or maneuver to send a manned modular ship to the Moon; which, when reaching the satellite orbit, it must first perform a rendezvous or spatial encounter and coupling of two of its modules before effecting the touchdown itself. This procedure had the advantage of allowing the Lunar Module to be used as a possible lifeboat if there was any failure in the Command Module.

Spacecraft: Modules and Space Capsules[7]

Once President Kennedy's lunar landing proposal became official, it began the detailed design of a *Command and Service Module*, CSM, for the *Apollo* missions. The final choice of *Lunar Orbit Rendezvous* determined the role of the CSM as a translunar shuttle; whose function would be to transport a new spacecraft, the Lunar Module, LM, which would take two men to the lunar surface, would wait for them and then it would return them to the CSM in the satellite orbit.

The Apollo spacecraft were not designed or built in one piece, but the dominant scheme were the modular ships: a complex spacecraft that involved up to four modules that could work both autonomous and independently, as well as an assembly of the ones over the others, forming another spaceship with different characteristics and functions.

Command and Service Module, CSM

Command and Service Module functioned as a mothership carrying three independent systems but at the same time they could be assembled together: the service module, the command module and the space capsule or *Apollo Lunar Module*. The CSM transported the astronauts to the lunar orbit, waited for them there and then put them on the back path to Earth. It consisted of two units: The cylindrical-shaped service module contained storage systems for various consumable elements needed during a mission, and which provided the necessary propulsion and electrical energy. The conical shaped Command Module was a cabin that housed the crew of three astronauts and transported the necessary equipment for atmospheric reentry and the splashdown when returning to the planet. A series of umbilical connections transferred power and consumables between the two modules: during the return trip, and just before re-entry, these connections were cut and the Service Module was completely disconnected and allowed that, by friction and overheating, it would burn up in the Earth's atmosphere.

An engine of the propulsion service system was used during the flights to place the Apollo spacecraft inside and outside the lunar orbit, and also for the corrections midway between the Earth and the Moon.

The CSM made a total of nine manned flights to the Moon, six of them involved successful touchdowns. After the Apollo lunar program the CSM was used as a shuttle for the crew on the Skylab program, and also in the Apollo-Soyuz test project in 1975, in which a US spacecraft met and docked with a Soviet Soyuz one in earthly orbit; fact that basically marks the end of the Cold War.

Command Module, CM

The Command Module, CM, was the cabin with a cone's trunk shape for the crew, it was designed to transport three astronauts from launch until reach the lunar orbit, and then back to an Earth's ocean. The *Front compartment* contained two reaction control system to control its position and direct its atmospheric entry path; a coupling tunnel and the system's components for the splashdown; it also carried parachutes to cushion this last operation. The Central or *Crew compartment* was an internal pressure vessel and was its only habitable room; it had an internal volume of 6,2 m^3 and contained the accommodation for the crew: three astronaut couches were lined up facing forward in the center of the compartment. It also housed the control panels and main screens, equipment bays, the guidance and navigation systems, food and equipment cabinets, the waste management system and the docking tunnel, among other components. The last section, the *Aft compartment*, contained 10 reaction control engines and their related propellant tanks, freshwater tanks and umbilical cables for interconnection with the other modules.

In numbers, the Command Module had a diameter of 3,91 m at the base and 3,48 m high, for a total volume of 10,4 m^3, of which 6,2 m^3 were conditioned to keep the three astronauts alive; it weighed approximately 5560 kg.

An ablative thermal shield, or thermal insulator armour, located outside the CM protected the capsule from the scorching heat generated by friction with Earth's atmosphere during reentry to the planet, heat that was sufficient to melt most metals; this material absorbed and diverted the intense heat of the process, and was charred and almost melted.

Service Module, SM

A cylindrical Service Module, SM, contained a service propulsion motor, a reaction control system with their respective propellant compounds, as well as a power generation system based on fuel cells that were powered by liquids hydrogen and oxygen; and it also supported the Command Module. A high-gain S-band antenna was used for long-distance communications in lunar flights. The Service Module was discarded just before re-entering the Earth's atmosphere. The dimensions of this module were 7,5 m long and 3,91 m in diameter; the initial version for the lunar mission weighed approximately 23300 kg fully fuelled.

The SM was assembled with the CM using three tension lashings and six compression pads, and so remained together for most of the mission until it was released and discarded just before re-entering the Earth's atmosphere. After being released, the stern thrusters of the SM were continuously firing for alienating it from

the CM and so ensuring that it followed a different path than the MC, and thus avoiding a disastrous collision.

Lunar Module, LM

The Lunar Module, LM, was designed to descend from the lunar orbit until touchdown with two astronauts, stay in the satellite surface for a time and then take them back to the same orbit for finding and coupling again with the Command Module. It was not designed to fly through terrestrial atmosphere or return to Earth alone, its fuselage was designed completely without aerodynamic considerations, and it was extremely light in construction. It consisted of two independent but arrangeable stages: one for the descent to the lunar surface and another for the respective ascent and coupling with the CM in satellite orbit; each one with its own engines. The descent stage contained storage for the descent propellant, consumables for a stay of two astronauts for about 34 hours on the satellite's surface, and equipment for ground exploration. The ascent stage contained the crew cabin, the lift propellant and a jet control system. A forward hatch provided access to and from the lunar surface, while an upper hatch and a docking port provided access to and from the Command Module.

The *Apollo* missions required that the LM be coupled with the CSM at the beginning of the lunar descent and ending the ascend maneuvers; the coupling mechanism was a system that consisted of a probe located in the nose of the CSM, which was connected to an assembly funnel located in the Lunar Module. The initial LM model weighed approximately 15100 kg; but an expanded Lunar Module, for more durable missions, weighed more than 16400 kg and allowed stays of up to 3 days on the satellite surface.

Emergency Abortion and Escape System

The Apollo ships had an emergency Launch and Escape System, LES, whose objective was to abort the mission by removing the command module and the crew cabin of the launch vehicle in case of emergency, such as a fire on the platform before launch, failure in control systems or probable failure of the launch vehicle that would lead to an imminent explosion.

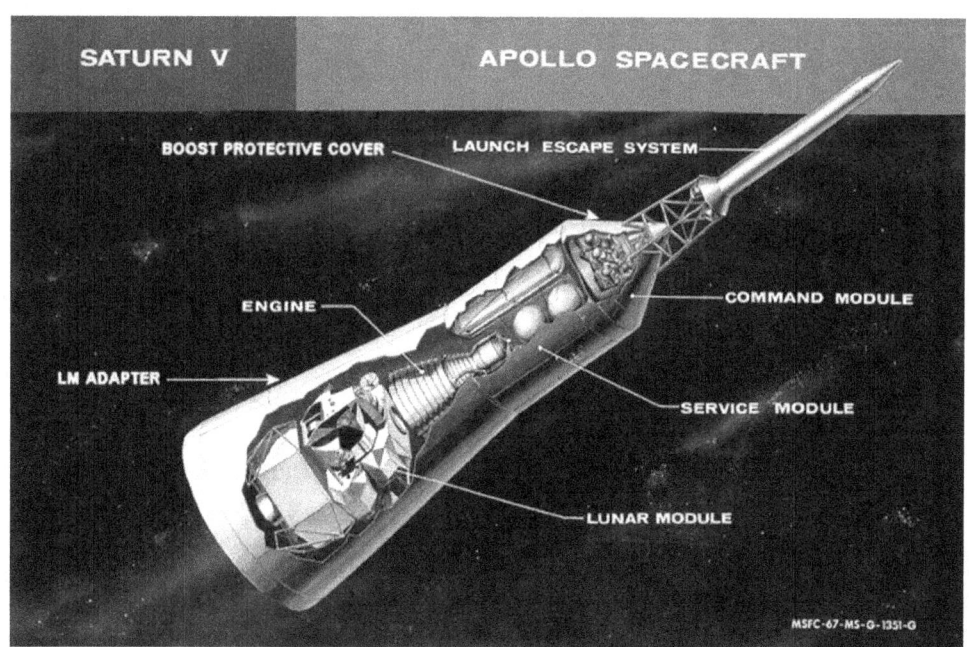

Figure 8.1
General diagram of an Apollo spacecraft

When the LES was activated, a small solid fuel escape rocket was fired and a system was activated to direct the Command Module out of the path of a major rocket in troubles. If the emergency occurred on the launch pad, the LES would raise the CM to a sufficiently safe height, and then the LES would be discarded to allow the recovery parachutes to safely be deployed before coming into contact with the ground.

In the absence of an emergency, the LES was routinely discarded about 20 or 30 seconds after the second stage of the main launch rocket was turned on. The LES was always carried but never used on four unmanned *Apollo* flights, nor on the other *Apollo* manned flights.

Communications system

The short-range communications between the CSM and the Lunar Module used two very high frequency antennas, VHF, installed in the SM.

A unified S-band high-gain addressable antenna for long-range communications with Earth was mounted on the aft bulkhead of the CSM. Four omnidirectional S-band antennas were used in the CM for when the position of the CSM would prevent the high gain antenna from pointing to the Earth.

Computers to the Moon

The *Apollo Guidance Computer*, AGC, was a digital computer designed and built by the Instrumentation Laboratory of the Massachusetts Institute of Technology, MIT, in the early 1960s especially for the Apollo space program; and which was installed onboard both Command Module and the Apollo Lunar Module. The device was one of the first computers based on integrated circuits, and it provided the electronic and computing interfaces to control and guide the spacecraft in mid-flight. Most software on the AGC was stored in a special read-only memory, although a small amount of read and write memory was available too. The crew communicated with the Apollo computer via a user interface which consisted of a calculator-style keyboard, numeric displays and of an array of indicator lights. Commands were entered numerically as two-digit numbers: standing for a Verb and a Noun.

Its role in the Apollo space program was to provide the computational ability demanded to control the orientation and navigation of both the Command Module and the Lunar Module. It was used for the first time in August 1966 and, with the exception of Apollo 8 that did not carry a lunar module, each Apollo mission carried two of these computers, one in the Command Module and another in the Lunar Module; constituting in each case the system of guidance, navigation and control. The AGC was in service until July of 1975.

Splashdown and Astronaut's recovery system

Splashdown, or water landing, is an aeronautical process to intentionally and carefully descend and park a spacecraft by parachute over a body of water, usually a sea; in a way very similar to a landing on the ground. A splashdown is achieved after having made a decrease in the altitude of the flight, have reduced the speed and achieved a certain inclination angle to then follow a gliding and approach pattern to the exact place of the touchdown, either on the surface of a river, on a lake or on the sea. A special design and the physical properties of water sufficiently cushion the impact of the spacecraft so that there is no need for a braking rocket. This operation should not be confused with an aeronautical accident or an uncontrolled landing on water.

The CM's Splashdown System consisted of three main parachutes, two drogue-type, braking or deceleration chutes; three pilot parachutes to deploy the electric network, three inflation bags to straighten the capsule in the sea if would be necessary, a sea recovery cable, an ink marker and an umbilical swimmer. At 7,3 km altitude the front heat shield was detached using four pressurized gas compression springs. Drogue parachutes were deployed to reduce the ship's speed to 201 kilometers per hour. At 3,3 km the drogue were discarded and the pilot parachutes were deployed, which reduced the speed of the CM to 35 kilometers per hour for the

splashdown. The portion of the capsule that first came into contact with the surface of the water contained four elastic ribs to further mitigate the impact's force.

After the landing, the astronauts were rescued by helicopter. The Apollo 11 mission was universally the first to make a manned moon landing and marked the first time that humans walked on the surface of another celestial body. The possibility of astronauts would bring dangerous lunar microorganisms or germs to Earth was remote, but not entirely impossible. To detect and contain any possible contaminants, the astronauts wore special biological isolation garments and when they returned were escorted to a mobile quarantine facility.

The *Mobile Quarantine Facility* was an appropriate Airstream type trailer used by NASA to quarantine for a period of 21 days the astronauts returning from the Apollo lunar missions. Its purpose was to prevent the spread of any contaminant of space or lunar origin, although the existence of such contagions was considered unlikely. It worked by keeping a lower pressure inside and filtering any ventilated air. It was made up of adequate spaces for living, eating and sleeping, as well as communication equipment that astronauts used to talk with their families. The trailers housed the three crew members, plus a doctor with their respective endowments, and an assistant to clean and cook.

Launch vehicles or Carrier rockets

By the end of 1944, it was obvious that Germany would not achieve victory in the war, so the German aerospace engineer Wernher Von Braun, thinking about his future, contacted the Allied countries and surrendered along with almost 500 scientists from his team before the American military forces, also delivering his aeronautical designs and several test rockets. Von Braun was transferred to the United States where he promptly obtained that nationality on April 14, 1955, and joined the United States Air Force. Later he was transferred from the Army to NASA and appointed director of the Marshall Space Flight Center; there he would be the chief designer of the Saturn rocket family for the Apollo space missions, which during the years of 1969 and 1972 would take Americans to the Moon.

The development of the Saturn rockets basically involved four functional series: Little Joe II, Saturn I, Saturn IB and Saturn V finally.

The Saturn V Rocket

The history of the lunar rocket Saturn-V is the history of the development of rockets; which began in Germany during the Second World War with the development of the V-2 missile; and continued in the United States with the development of the Redstone, Jupiter and Saturn-1 rockets. This was the work of the von Braun team at the Army Arsenal in Redstone Test Center, later the Marshall Space Flight Center, in Huntsville, Alabama.

NASA thought it would be easier to start producing the Saturn rockets since many of the components were designed to be transported by air; so only a new factory was required. The Saturn C-5, later called Saturn V and the most powerful of the configurations, was selected as the most suitable design for the Apollo space missions during the exploration of the Moon, because it was classified to transport human beings. The super-heavy three-stage launch vehicle powered by liquid fuel was used by NASA between 1967 and 1973. It was later used to launch the *Skylab*, the first US space station. The name of *Saturn* is due to a proposal by the engineer von Braun in 1958 as a logical successor to the previous series of *Jupiter* rockets, as well as a remembrance of the powerful position of that Roman god.

The three-stage *Saturn V* rocket was designed to send a CSM with its respective manned LM to our satellite, and the assembly fully provisioned. It was 10,1 m in diameter and 110,6 m high. Fully loaded, including fuel, the *Saturn V* weighed 2.950.000 kg, with a payload for the lunar mission of 48600 kg. The first stage, called S-IC, measured 42 meters high and 10 meters in diameter, weighed more than 2.300.000 kg and provided the necessary thrust to get the first 67 km of ascent; it worked with Rocket Propellant-1, or RP-1, and liquid oxygen. During the launch this first S-IC phase ignited its engines for 168 seconds, and at the time of separation the vehicle was at an altitude of approximately 67 km and was moving at a speed of about 2300 m/s.

The second phase, known as S-II, measured 24,5 meters in height and 10 meters in diameter, weighed about 500000 kgs; and its function was to accelerate the Apollo ship through the upper atmosphere. Finally, the third stage, S-IVB, measured 17,8 m high and 6,6 m diameter, and weighed about 119000 kgs fully supplied. For the lunar missions it was necessary to turn on this last stage twice: first for about 2,5 minutes for the insertion in Earth orbit after the separation of the second stage, and finally for approximately 6 minutes for the translunar injection. These two phases were based on liquids hydrogen and oxygen.

The Saturn V rockets were used in thirteen space missions launched from the Kennedy Space Center in Florida, without any loss of crew or payload. The Saturn V launch vehicles and their missions were designated with an AS-500 serial number, AS indicates Apollo-Saturn and 5 indicates Saturn V. Until the year 2018 it remains the tallest, heaviest, most powerful rocket, with a larger total boost achieved and with the highest payload that has launched manned spacecrafts into outer space.

Launching platforms for Space missions

Almost a month before President J.F. Kennedy's proposal, only one American had flown into space and none had made an earthly orbital flight. In some sectors there were doubts that NASA could meet the ambitious goal. But the approval by the Congress of the presidential proposal led to the development of the Apollo program, which required a massive expansion of the Agency's operation facilities.

NASA established the Marshall Space Flight Center in Huntsville, Alabama, on July 1, 1960, as the center for North American rocket and spacecraft research. The first task of the Marshall Center was to develop the heavy Saturn launch vehicles for the Apollo lunar program.

By that time it was clear that the Apollo program would exceed the capabilities of the Cape Canaveral launch facility in Florida: an even larger installation would be needed for the mammoth rocket required for manned lunar missions; therefore, a Launch Operations Center was designed, whose construction began in November of 1962. But after the tragic death of Kennedy on November 22, 1963, successor President Lyndon B. Johnson ordered on November 29 to baptize such center in honor to Kennedy. The Kennedy Space Center, located on the east coast of Florida, was so born; and since December 1968 it has been the main launching center for manned space flights of NASA.

When the launch complex for the Apollo lunar program was built at the Kennedy Space Center, it was designated as Launch Complex 39 and was designed to handle the launch of the giant Saturn V rocket; at that time the largest and most powerful rocket that had been designed and which was indispensable to launch the Apollo ships to the Moon. The launch operations of the Apollo, Skylab and Space Shuttle programs were carried out from the Launch Complex 39 belonging to the Kennedy Space Center.

It was also clear that NASA would soon surpass its ability to manage and control missions from its Cape Canaveral Air Force Station facilities in Florida; so a new Mission Control Center would be developed at the Marshall Center. For this purpose, in November 1961, the Manned Spacecraft Center was created, where research, development and personnel training for manned space flights would be carried out. Later this center was renamed in February of 1973 as the Johnson Space Center, in honor of the deceased American president Lyndon B. Johnson.

Unmanned Experimental Apollo Missions

Between 1961 and 1968 Saturn launch vehicles and the components of the Apollo spacecrafts were tested and evaluated on unmanned experimental flights.

There is some inconsistency in the denomination of the first three unmanned Apollo-Saturn flights, AS, or also in the Apollo flights; which is because the mission AS-204 was renamed posthumously to Apollo 1; although this manned flight must have followed the first three unmanned flights. However, after the fatal fire in which the entire crew of the AS-204 died during a test and training exercise, the unmanned Apollo flights that had already been terminated, resumed to again test both the Saturn V launch vehicle like the other components of the Apollo spacecraft. So these missions were designated as Apollo 4, 5 and 6, and the first three unmanned flights are known as AS-201, AS-202 and AS-203. With all this, the first manned and successful Apollo mission was Apollo 7.

Apollo 1

On January 27, 1967, the tragedy hit the Apollo program seriously, almost completely frustrating the United States' goal of moon landing by the end of the decade. The Apollo AS-204 was to be the first manned mission of the American Apollo space program; which was designed as the first low-Earth orbit test for the Command and Service Module and planned for February 21, 1967. But the mission was never able to take off because a fire in the cabin during a launch test on January 27 destroyed the Command Module and ended the lives of the three crew members: Command Pilot Virgil I. Grissom, Senior Pilot Ed White and Pilot Roger B. Chaffee. Subsequently, this mission AS-204 was renamed posthumously as Apollo 1.

Apollo 4

The Apollo 4 mission was the first unmanned test flight of the Saturn V launch vehicle, and it was also the first to be launched from the John F. Kennedy Space Center in Florida, from the facilities built specifically for that rocket. The mission placed a CSM in a high orbit around the Earth, and tested and rated the heat-sink armour of the CM at the re-entry speed into the Earth's atmosphere. It was launched on November 9, 1967; the flight lasted 8 hours and 37 minutes and gave a total of three complete orbits to our planet with a maximum altitude of 18090 kilometers. It was a complete test in the sense that all the stages of the rockets and the spacecraft were fully functional during the flight.

**Image 8.2
Saturn V Rocket in action**

Moment of engines' ignition and takeoff of the Saturn V Rocket during the launch of the Apollo 11 manned mission from launch complex 39 of the Kennedy Space Center on July 16, 1969 at 13:32 UCT.

Apollo 5

Apollo 5 was the first unmanned flight of the Apollo Lunar Module, which later would carry astronauts to the lunar surface. It was launched on January 22, 1968 with a Saturn IB rocket in an orbital flight over Earth; and lasted 11 hours and 10 minutes completing a total of seven orbits at a maximum height of 214 kilometers. It was the first unmanned test flight of the Lunar Module in Earth orbit, with a fundamental purpose of testing the LM in a space environment, particularly its descent and ascent systems.

Apollo 6

Subsequently, on April 4, 1968, the unmanned Apollo 6 mission was launched, which carried a CSM and a test replica of the LM. The fundamental purpose of this mission was to achieve the *Trans-lunar injection* stage, TLI, including abortion tests of such operation. In aeronautics it is called Trans-lunar injection to a propulsion maneuver used to place a spacecraft in a trajectory that will lead it directly to the Moon. The launch site was the Kennedy CL-39A, the mission lasted 9 hours 57 minutes and 20 seconds and a total of three terrestrial orbits were completed, in a regime of highly elliptical orbit with a maximum altitude of 22533 kilometers.

The Saturn V launch vehicle systems experienced some irregularities that could be compensated properly. But the damage on the engine of the third stage was more severe, preventing the translunar injection operation could be performed. Even so, the mission controllers were able to use the Service Module engine to basically repeat the same flight profile of Apollo 4. In any case, and based on the good performance of this mission and in the identification of satisfactory solutions for its problems, NASA declared that the Apollo missions were indeed ready to be launched, canceling then a fourth unmanned test.

Apollo Manned Missions

The Apollo missions were developed between 1961 and 1972; the first successful manned flight was made in 1968 and they achieved their fundamental objective of manned lunar landing on July 1969; very much in spite of the fatal fire in the Apollo 1's cabin on 1967 during a launching test, accident in which all its crew died. After the first successful moon landing, programs for space exploration and lunar geology studies were extended to complete six successful manned missions to the satellite.

General Profile and Development of a Manned Lunar Mission

The nominal planned Manned Moon Landing mission proceeded according to the following general flight profile:

0) Selection of astronauts

After having fulfilled all the respective requirements demanded by NASA in the calls for astronauts, a total of thirty-two men were chosen as candidates to fly in the manned missions of the Apollo program. Twenty-four of them actually became astronauts, abandoned gravity and Earth's orbit, and flew around the Moon between December 1968 and December 1972, three of them twice. Half of these 24 astronauts actually managed to walk on the satellite's surface. The astronauts for the Apollo missions were chosen primarily from the veterans of the previous Mercury and Gemini projects; and all the missions were commanded by veterans of the latter.

1) Launch

During launch the three stages of the Saturn V rocket worked sequentially for approximately eleven minutes in order to send the craft into a circular parking orbit at an approximate height of 190 km. The third stage burned only a small portion of its fuel to achieve said orbit.

2) Trans-lunar Injection

After one or two terrestrial orbits to verify the adequacy of the spacecraft systems, the third stage S-IVB turns on again for approximately 6 minutes to direct the spacecraft towards the Moon.

A trans-lunar injection, TLI, is an aeronautic propulsion maneuver used to place a spacecraft, which is initially in a low circular parking orbit around the Earth, in a trajectory that will make it reach the Moon directly.

As the spacecraft begins to navigate in the lunar transference arc, its trajectory approaches an elliptical orbit around the Earth with an apogee near the radius of the lunar orbit. The functioning of the propulsion engines is synchronized so that the spacecraft achieves its apogee as it approaches the Moon; finally, the spaceship will enter the sphere of satellite's gravitational influence. The lunar journey takes between 2 and 3 days. The mid-course corrections are made as necessary using the SM motor.

3) Transposition and coupling

The transposition and coupling was a maneuver made during the manned missions of the Apollo program. It involved the separation of the Command and Service Module from the adapter that tied it to the third stage of the Saturn V launch vehicle, then turning it 180° to connect its nose to the Lunar Module and finally separating the whole assembled array from that upper launching platform. The sequence of operations was like this:

The Lunar Module Adapter panels are separated to release the CSM and expose the LM. The Command Module Pilot moves the CSM to a safe distance and rotates it 180°; then the pilot couples the CSM with the LM and withdraws the entire spacecraft thus assembled from the third stage S-IVB, and the latter is then discarded by sending it to the solar orbit.

4) Lunar Orbit Insertion

The spacecraft passes approximately 110 km behind the Moon and the engine of the SM is turned on to brake it and put it in a lunar orbit of 110 by 310 km, which soon becomes circular to 110 km by means of a second motor action.

5) Insertion in the descending orbit

After a rest period, the Commander and the Lunar Module Pilot are transferred to this module, turn on their systems and deploy the landing gear. The CSM and LM are separated and then the crew directs the LM to a safe distance and starts the descent motor for the insertion in the descending orbit, which takes it to a distance of about 15 km from the satellite surface.

6) Descent and controlled landing

In the minimum distance to the lunar surface, the descent motor is turned on again to begin the final stage of landing and the Commander takes over the manual control for a properly controlled vertical landing.

7) Lunar Stay: Inspection and Sampling

The Commander and the Lunar Module Pilot descend to the lunar surface, deploy the respective equipment for the experiments, perform one or more extravehicular activities exploring the terrain, collecting samples and alternating with rest periods.

8) The ascent stage

The ascent stage contained the crew cabin with instrument panels and flight controls, its own upward propulsion engine system and two propellant tanks to return to lunar orbit and meet and couple with the Command and Service Module. It also contained a Reaction Control System for position and direction control. This stage ascends using the descent stage as a launch pad, which is abandoned on the satellite surface.

9) Coupling

Finally the LM ascent stage meets and merges with the CSM. The Commander and LM Pilot are transferred back to the CM with the samples of lunar material; then such ascent stage is discarded so that finally it goes out of orbit, falls and crashes on the lunar surface.

10) Trans-terrestrial injection

The SM motor is turned on to put the CSM on its back way to Earth. The Service Module was never conceived for returning to Earth; only the Command Module was aerodynamically designed to safely go down through the earthly atmosphere. All the flown service modules were left to burn up in the upper atmosphere, and unburned pieces fell into the Pacific Ocean. The SM was discarded just before the reentry into the earth's atmosphere, and then the CM rotated 180° to put its blunt end forward an so to enter in the terrestrial atmosphere.

11) Atmospheric reentry

All space flights are, including the very launch moment, highly risky. And, paradoxically, the return to home is probably the most critical, stressful and risky stage of these missions. This is due to the high speeds at which these spaceships travel, and the relatively large thickness and density of the Earth's atmosphere.

The atmospheric reentry is the income to the planet's atmosphere of spaceships coming either from outer space or from suborbital flights, which occurs about 100 km high in the terrestrial case. In aeronautics, it is understood that the process of reentry of space vehicles is intentional and controlled, and that it has the purpose of reaching the planet's surface securely and with their crewmember safe.

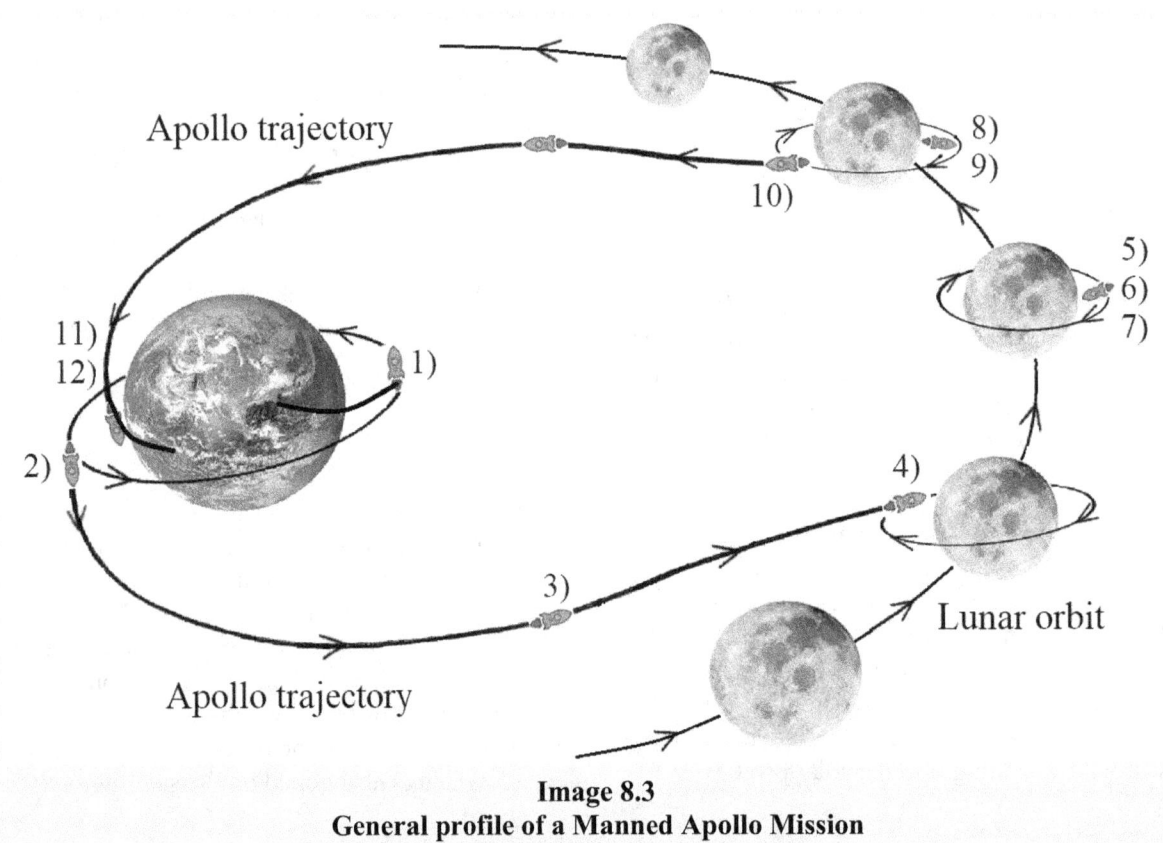

Image 8.3
General profile of a Manned Apollo Mission
Schematic diagram

Manned spacecraft must reduce their speeds to subsonic values before they can deploy its parachutes. The required amount of rocket fuel to decelerate the vehicle would be almost equal to the amount used initially to accelerate it during the launch, and therefore it is very impractical to use retro propulsion rockets for the whole procedure of re-entry to the planet. So the atmospheric heat dissipation is the only way to spend and reduce the high kinetic energy of spacecraft returning to Earth.

Given the special characteristics of the Earth's atmosphere and the high speeds of spacecraft, the phenomenon of friction between the air molecules and the construction materials of the ship considerably manifests itself. This leads to a drastic decrease in the speed of the vehicle and the consequent transformation of its kinetic energy into heat, which is known as aerodynamic heating and that if left unchecked could burn, melt and vaporize the spaceship and its crew.

From the aeronautical point of view, and to perform a safe re-entry, we have the concept of *Re-entry Corridor* which consists of a narrow corridor or atmospheric conduit, centralized at an angle of incidence of 6.2° and with a margin of error of only 0,7°, necessary to surely enter the terrestrial atmosphere so much for the ship as for its crew. If the vehicle entered with a lower angle, it would bounce off the upper layers of the atmosphere, it would take an escape path and could be permanently lost. If the incidence angle were higher, the physical shock due to atmospheric drag and aerodynamic heating due to friction would be very high, so as to melt most metals and the spacecraft could be burned up and killing its crew. In practice, and to partially absorb this heat and minimize risks, the spaceship is externally coated with a protective shield made of thermal insulation or ablative materials.

During atmospheric reentry in the Apollo missions, the atmospheric resistance slowed down the Command Module and the aerodynamic heating surrounded it with an envelope of supremely hot ionized air, which gave occasion to a blackout of communications for several minutes. Finally, when the speed and height were the indicated ones, the parachute systems were sequentially deployed, which made the CM acquires a sufficiently safe speed to get in the ocean.

12) Splashdown and rescue

The Splashdown method was used for the Mercury, Gemini and Apollo missions. In the first Mercury flights, a helicopter connected a cable to the capsule, lifted it from the water and took it to a nearby ship. Subsequently all the space capsules had a flotation collar, similar to a rubber life raft, attached to the spaceship to increase its buoyancy. After hooking the flotation collar, a hatch in the space capsule is usually opened and then the astronauts decide whether to be hoisted individually by a helicopter to take them to the recovery ship, or stay inside the spacecraft and thus be lifted aboard the ship by a crane. Subsequently, the astronauts were taken to the mobile quarantine station.

Experimental Apollo Manned Missions

The Apollo 7 and 9 missions were flights in earthly orbit to test the Command and Lunar modules. Apollo 8 and 10 tested various components and returned photographs of the lunar surface while orbited the Moon. The Apollo 13 mission did not land on the Moon due to a mechanical breakdown, but it also returned photographs. Six of the missions, the 11, 12, 14, 15, 16 and 17 ones achieved the primary objectives of these manned missions; which collected a large amount of scientific data and 381 kilograms of lunar soil samples. The experiments carried out there included studies on seismic and soil mechanics, on magnetic fields, solar wind experiments, heat flow, and studies on lunar meteorites, etc.

Apollo 7

Apollo 7 was the first manned space flight of the Apollo spacecraft and with terrestrial orbit. Its main objectives were demonstrating the operation of the Saturn launch vehicle and of the Command and Service Module, evaluating the performance of the crew and the personnel supporting the mission; and demonstrating the CSM's ability to meeting and coupling. The crew was formed by the astronauts Walter Schirra Jr, Donn Bisele and Walter Cunningham, who should also carry out two photographic experiments and three medical experiments.

Apollo 7, launched on October 11, 1968, was an Earth orbital flight with 10 days 20 hours and 9 minutes duration, completing a total of 163 orbits at a maximum altitude of 301 kilometers. During this mission, the first public and live television broadcast of a manned space mission was made.

Apollo 8

The Apollo 8 mission was launched on December 21, 1968, was the second manned mission of the Apollo Space Program of the United States and the first manned mission to leave Earth orbit, reach the Moon and orbiting it without landing for finally returning to the planet. The crew consisted of three astronauts: the Gemini 7 veteran Frank Borman was the Commander, the Command Module Pilot was James Lovell also a Gemini 7 veteran, and rookie astronaut William Anders as Lunar Module Pilot. They would be the first humans to leave Earth orbit, to see the complete figure of our planet from space, to see the Earth's dawn from the lunar orbit and, finally, the first to enter the gravitational influence of another celestial body and to see the hidden side of the Moon.

Originally planned as a second test of the Command Module and the Lunar Module in an elliptical orbit in early 1969, the initial mission profile was changed in August of 1968 to a more ambitious lunar orbital flight to test the CM alone, because the LM was not yet ready to make its first real flight.

The mission had a total duration of 6 days and 3 hours, took 2 days and 21 hours to enter the lunar orbit and in a span of 20 hours 10 minutes and 13 seconds it managed to complete ten revolutions to the satellite at an average height of 112,4 kilometers. Thus, during the Nativity celebrations on Earth, the crew was able to make a television broadcast of Christmas Eve from that lunar orbit, in which they read the first ten verses of the biblical Genesis' Book. This transmission would become the most watched in history until then. The crew of Apollo 8 returned home on December 27, 1968 and achieved splashdown in the northern part of the Pacific Ocean.[8]

The objectives of the Apollo 8 mission included establishing a coordinated performance between the crew, the Command and Service Module and the support facilities. Other purposes were to test and demonstrate the Trans-lunar injection, the navigation and communication systems of the CSM, as the evaluation of mid-course corrections, etc. All the main objectives of the mission and the detailed objectives of the test were achieved. All the launch vehicles and spacecraft systems worked according to the plan.

Apollo 9

Apollo 9 was the third manned flight of Apollo and the first that also included the Lunar Module. The crew consisted of Commander James McDivitt, CM Pilot David Scott, and LM Driver Russell Schweickart. The mission was launched from Kennedy Space Center Launch Complex 39 on March 3, 1969.

Apollo 9 was composed of a Command Module, a Service Module, a Lunar Module and an Instrumentation Unit. It was launched by a Saturn V rocket that had its respective three stages, S-IC, S-II and S-IVB. The CM served as a command, control and communications center; and complemented by the SM it provided all the life support elements for the three crew members. The Command Module allowed coupling with the LM and served as a floating ship in the sea. The CSM provided the main propulsion and maneuverability capability and was discarded just before reinsertion into the Earth's atmosphere of the CM. The Lunar Module was a two-stage vehicle that accommodated two men and could transport them to the lunar surface, had its own propulsion, communication and life support systems.

The main objective of Apollo 9 was an engineering test of the first manned Lunar Module and in Earth orbit, including its

218

performance as an independent and self-sufficient spacecraft for the execution of encounter and assembly maneuvers, simulating the operations that would be carried out in real future lunar missions. Other concurrent objectives included were the general verification of the launch vehicle and spacecraft systems, and the performance of the crew and the respective procedures. The mission was launched on March 3, 1969, lasted 10 days 1 hour and 54 seconds and gave 151 revolutions to Earth at an average altitude of 191 kilometers.[9]

Apollo 10

The Apollo 10 mission encompassed all aspects of a real manned lunar mission, except the moon landing; it was the first flight of a manned Apollo spacecraft and fully equipped to operate around the satellite in a circular orbit of about 130 kilometers. The fundamental objectives of this mission were to demonstrate the performance of the CSM and LM in the lunar gravitational field, and to evaluate the navigation systems of such modules there. For this purpose an eight-hour moon orbit of the decoupled LM was included, a descent to about 16,66 kilometers from the lunar surface and then the respective ascent for its reunion and coupling with the CSM were programmed. The relevant data gathered in this lunar landing trial was related to the lunar gravitational potential, which is essential to verify and refine the lunar flight control systems and the programmed LM trajectories. All the objectives of the mission were achieved, including twelve television transmissions to Earth.

The main crew of Apollo 10 consisted of the veteran astronauts of the Gemini missions: Commander Thomas P. Stafford, Command Module Pilot John W. Young, and Eugene A. Cernan as Lunar Module Pilot. The mission was launched from Kennedy Space Center Launch Complex 39 on May 18, 1969, to a circular orbit of land 207 km; after one and a half orbits the translunar injection occurred. The first live and color TV broadcasts to Earth began three hours after launch.

The insertion into lunar orbit occurred 76 hours after the start of the mission and when the service propulsion system was activated. About 4,5 hours later, and with a second engine actuation, the lunar orbit of Apollo 10 became circular at approximately 124 km; and then the first color images of the satellite surface were taken.

Approximately 100 hours after launch, on May 22, the LM disconnected from the CSM and flied briefly in a stable moon orbit of 120 by 129 kilometers. Next, and to make a simulation of the future Apollo 11 moon landing, the LM descent motor was operated for 27,4 seconds to take it to a new orbit of 17,5 by 120 km. About eight hours after being separated, and after some maneuvers to test the flight control systems, the ascent engines were activated and the LM returned for coupling with CSM and so start its journey back to home.

The mission had a total duration of 8 days 3 minutes and 23 seconds; the Command and Service Module was in lunar orbit for 61,6 hours and completed 31 revolutions; while the uncoupled Lunar Module made four orbits around our satellite and descended to a minimum height of 15,2 kilometers. Basically all the objectives of the mission were fulfilled.[10]

Moon Landing Apollo Manned Missions

After all those experimental missions, both unmanned and manned, the US space agency is now in a position to fulfill the proposal of the dead President Kennedy to bring humans to the satellite, land them and bring them back home safely.

Apollo 11: The first humans on the Moon

The Apollo 11 mission was the first in which humans descended to the lunar surface, walked over it, and then returned safely to Earth: On July 20, 1969, two astronauts, the mission's Commander Neil A. Armstrong and the Lunar Module Pilot Edwin E. "Buzz" Aldrin Jr., landed in this module in the *Mare Tranquilitatis*, the Sea of Tranquility or Tranquility Base, on the Moon; while the Command and Service Module with its Pilot, astronaut Michael Collins, waited for them in lunar orbit. During their stay on the Moon, the astronauts conducted scientific experiments, took photographs and collected samples from the satellite terrain.

Let's start first with the main protagonists, the astronauts with their exploits and their achievements; and let's leave for later the machines with their functioning and performance. Let's also start with what undoubtedly is the most famous phrase in history: *"That's one small step for a man, one giant leap for mankind."*[11], which was pronounced by the American astronaut Neil A. Armstrong on the lunar surface on July 21, 1969 at 02:56 UCT.

Neil Alden Armstrong[12]

Neil Alden Armstrong (August 5, 1930 - August 25, 2012), the first man to pose his feet and walk on the lunar surface, was an aeronautical engineer, test pilot, military aviator and American astronaut.

Armstrong was born on August 5, 1930 in Wapakoneta city, county of Auglaize, Ohio. The young man entered to study aeronautical engineering in 1947 at the public research university Purdue University in West Lafayette, Indiana. And very quickly, in January 1949, he was called by the United States Navy for training on flights at Naval Air Station Pensacola, in Florida. After a quick training, in August of 1950 Armstrong received notification that he was already a fully qualified naval aviator.

Once he carried out some minor missions, as of August 1951 Armstrong had to serve as an aviator in the Korean War, which had begun in June of the previous year; there the young pilot flew in a total of 78 missions totaling 121 flight hours, and was awarded several times. In September 1951, he was hit by anti-aircraft fire while making a low bombing run, and was forced to retreat.

After participating in the Korean War and retiring from the Navy, Neil Armstrong returned to Purdue University where he graduated with a bachelor's degree in Aeronautical Engineering in January of 1955, and he decided to become a test pilot. He applied at the High Speed Flight Station that the National Advisory Committee for Aeronautics, NACA, had at the Edwards Air Force Base, which he joined on July of 1955. After a long list of services as a test pilot that includes more than 200 different models of aircraft, Neil Armstrong became an employee of the National Aeronautics and Space Administration, NASA, when it was created in October 1958.

In April of 1962 NASA opened the competition for selecting the second group of aspiring astronauts for the Gemini Project, which duly qualified civilian test pilots could apply. Armstrong applied for the call and attended on the respective tests and exams. In a press conference on September 17, 1962, NASA publicly announced the selection of its second astronaut group; he was one of the two civilian pilots selected for this set, the other was Elliot McKay See Jr. From initially flying civil and war planes, now Armstrong would become space flights pilot.

After having been part of the alternate crew for the Gemini 5 mission on August 21, 1965, Neil Armstrong was assigned as Pilot Commander for the Gemini 8 space mission, which was launched on March 16, 1966 for a flight of six terrestrial orbits at a minimum altitude of 160 km. Then Armstrong became the first American civilian pilot for going into space; almost three years after the Soviet Valentina Tereshkova on board the Vostok 6 on June 16, 1963.

Later, once Armstrong served as backup commander for the Apollo 8 mission on December 21, 1968, the crew for Apollo 11 was officially announced on January 9, 1969: Armstrong as General Commander of the mission, the Lunar Module Pilot would be Buzz Aldrin and Michael Collins as Command Module Pilot.

Shortly after his successful mission to the Moon, Armstrong announced that he didn't plan to perform more space flights and resigned from NASA in 1971. He accepted a teaching position in the Department of Aerospace Engineering at the University of Cincinnati, where he was Professor of Aerospace Engineering. Over time, he completed his mastery by presenting a report on various aspects of Apollo missions, rather than a thesis on hypersonic flight simulation.

After the accident and failure of the Apollo 13 mission in 1970, Armstrong was part of the investigation team. On the other hand, in 1986 President Ronald Reagan requested that Armstrong join the Rogers Commission that would investigate the Challenger space shuttle disaster that occurred on January 28 of that year, and he was elected as vice president of the commission. Also President Reagan appointed him for a commission of fourteen members that would develop a plan for American civil space flights in the 21st century.

On August 7, 2012 astronaut Neil Armstrong underwent vascular derivation surgery, or vascular bypass, which had the purpose of solving circulatory problems due to some blocked coronary arteries. Although in principle he was recovering well, a few weeks later he developed complications in the hospital and died on August 25 in Cincinnati, Ohio, at age 82. Once the news was known, the expressions of sentiment were immediate, both from the official sector and from the private sector. The following phrases are included in a statement by the Armstrong family:

"Neil was our loving husband, father, grandfather, brother and friend. Neil Armstrong was also a reluctant American hero who always believed he was just doing his job. He served his Nation proudly, as a navy fighter pilot, test pilot, and astronaut. He remained an advocate of aviation and exploration throughout his life and never lost his boyhood wonder of these pursuits. For those who may ask what they can do to honor Neil, we have a simple request. Honor his example of service, accomplishment and modesty, and the next time you walk outside on a clear night and see the Moon smiling down at you, think of Neil Armstrong and give him a wink"[13]

Edwin Eugene Aldrin Jr: *Buzz Aldrin*

Edwin Eugene Aldrin Jr (January 20, 1930), *Buzz Aldrin*, is an American engineer, command pilot in the United States Air Force, and former astronaut of the Gemini 12 mission and in the Apollo 11 mission as the Lunar Module Pilot, in which he was the second man to walk on our satellite.

In 1947 Buzz graduated from Montclair High School, in New Jersey, and then entered the US Military Academy at West Point, where he graduated 1951 with a Bachelor of Science and Mechanical Engineering. His father, a man of military career and colonel of the US Air Force, encouraged his interest in aviation. Buzz officially joined the United States Air Force in 1951 and began combat training that same year. Initially Aldrin was a military combat pilot and, like Armstrong, participated in the Korean War; during which the air quadrille in which Aldrin belonged was responsible for breaking the record of shooting down enemy planes: during one month of combat they eliminated 61 enemy MIG aircrafts and punished another 57.

Once finished that war in 1953, Buzz Aldrin returned to his country and remained linked to the Air Force. Then, under the auspices of the Air Force Institute of Technology, he began working as a graduate student at the Massachusetts Institute of Technology in 1959 with the intention of obtaining a master's degree. So, in January of 1963 he obtained a doctorate in science and astronautics; with a thesis related to techniques for manned orbital rendezvous, and he was then assigned to the Gemini Target Office of the Air Force Space Systems Division in Los Angeles. There he was part of a third astronauts group selected by NASA on October 18, 1963, to carry out training for space flights. Then he was the first astronaut with a PhD and the one in charge of creating coupling and encounter techniques for spacecraft. He also pioneered underwater training techniques to simulate spacewalks, indispensable for future manned missions.

Astronauts Jim Lovell and Buzz Aldrin were designated on June 17, 1966, as prime crew for

Gemini 12 mission, with Gordon Cooper and Gene Cernan as their backups. That mission was ran on November 11, 1966 and during it the fourth docking with an Agena target vehicle was carried out; also, Aldrin performed three EVAs for a total of 5 hours 30 minutes on spacewalks, the longest and most successful record until that date.

Later Aldrin was assigned along with Neil Armstrong and Fred W. Haise Jr. to the reserve crew for the Apollo 8 space mission. Finally, he was assigned as Lunar Module Pilot for the historic Apollo 11 moon landing mission. On July 20, 1969, he went down in history as the second man to walk on the surface of our satellite, seconding the mission commander Neil Armstrong. Provisioned by the Lunar Module, these astronauts spent a total of 21 hours in the satellite environment, made hikes that were televised for the spectators on Earth and collected about 22 kilos of lunar rocks.

Subsequently on July of 1971 Aldrin left NASA and was assigned as Commander of the Test Pilot School at Edwards Air Force Base, California. After being briefly hospitalized, Aldrin retired from active duty after 21 years of aeronautical performance, and returned to the Air Force in a management role in March of 1972.

After retiring from NASA, Aldrin continued to be linked to space exploration, both from the public and private sectors; placing a special emphasis on the planet Mars: In June 2013 the New York Times published an article by Aldrin in which he supports a manned mission to Mars, and where he considers the Moon "not as a destination but as a starting point, which would place humanity on a path to the planet Mars and so would become a kind of two planets".[14]

He also founded Share Space Foundation, a non-profit organization dedicated to promoting space education, exploration and the experiences of affordable space flights. In August of 2015, he launched the Buzz Aldrin Space Institute in Florida to promote and develop his vision of a permanent human settlement on the planet Mars, according to his official website.[15]

Michael Collins

Michael Collins is the son of a US Army serviceman, who was serving in Rome at the same time Michael was born there on October 31, 1930.

After the United States entered World War II, the family moved to Washington DC, where Collins attended St. Albans School and graduated in 1948. He then entered the United States Military Academy at West Point where he graduated in 1952 with a Bachelor of Science degree. He then joined the Air Force that same year and completed flight training in Columbus, Mississippi; then he serves in the Nellis Air Force Base, in Nevada, and in George Air Force Base, California, where he learned to handle nuclear weapons. He also served as an experimental flights test officer at Edwards Air Force Base in California, testing combat aircraft.

On August 1960, when the Air Force of the EE. UU. began researching space flights, Michael Collins entered the School of Aerospace Research Pilots. Next NASA requested again astronaut aspirants, and in 1.963 he was chosen by the agency to be part of the third group of fourteen spacemen.

At the end of June 1965, Collins received his first crew assignment as the backup pilot for the Gemini 7 mission, which he successfully completed on January 24, 1966. Then he was assigned to the prime crew of Gemini 10 mission of July 18, 1966, with John Young as a partner; there they performed orbital rendezvous with two different spacecraft and undertook two extravehicular activities or spacewalks.

The second mission where Michael Collins participated was Apollo 11 of July 20, 1969, which made the first crewed moon landing in history. The astronaut remained in lunar orbit in the Command and Service Module while his companions Neil Armstrong and Buzz Aldrin descended in the Lunar Module and then walked on the surface of our satellite.

After that lunar mission, Collins left NASA in January of 1970 and a year later he joined the administrative staff of the Smithsonian

Institution in Washington. In 1980 he joined the private sector working as an aerospace consultant.

Elements of the Apollo 11 Mission

Command Module *Columbia*

The Apollo 11 Command Module, named Columbia, was a conical pressure vessel with a maximum diameter of 3,9 m at its base and an altitude of 3,65 m; made with aluminum structures in the shape of honeycomb, and packaged between sheets of aluminum alloy. Externally its base contained a stainless steel heat shield in the shape of honeycomb and filled with a phenolic epoxy resin as an ablative material. At the tip of the cone was a gate and a coupling assembly designed to be coupled with the Lunar Module.

The CM was divided into three compartments. The forward on the nose of the cone held a set of parachutes. The aft compartment was located around its base and contained jet control engines, propeller tanks, wiring and pipes. The crew compartment comprised most of the volume, some 6,17 cubic meters of space; at its center, lined up and facing forward there were three sofas for the astronauts, with a large access hatch located above the central sofa. The Apollo Navigation Computer, controls, screens, navigation equipment and other systems used by astronauts were located in this room. After the separation of the Service Module, the CM provided the ability to re-enter the Earth's atmosphere and realizing the splashdown at the end of the mission; for which five silver/zinc oxide batteries provided the power after the CM and SM separated.

Service Module

The Service Module, SM, was a cylinder 3,9 meters in diameter and 7,6 m long that was attached to the back of the CM. The exterior coating of the SM was formed by aluminum panels or compartments in the shape of honeycomb of 2,5 cm thick. The interior was divided into six sections around a central cylinder by means of aluminum radial beams. The six sections of the SM contained electrolytic cells or cells based on hydrogen and oxygen that provided electrical power to 28 volts, cryogenic tanks of oxygen and hydrogen and four tanks for fuel and the oxidant of the main propulsion engine. The radiators of the electric power system were on top of the cylinder, and the radiator panels of the environmental control system were spaced at the bottom. A liquid propulsion system was mounted on the rear of the SM. The position control was provided by four jet controllers, each spaced at 90 degrees around the front of the SM.

Command and Service Module

The Command Module housed the crew, the operations systems of the spacecraft and the reentry systems to the Earth's atmosphere. The Service Module transported the main propulsion system and most of the consumables: oxygen, water, helium, fuel cells and the respective fuels. Telecommunications included voice, television, data and tracking subsystems for communications between astronauts, the CM, LM and the Earth.

When these two modules were coupled they formed a structure with a total length of 11,0 meters and with a maximum diameter of 3,9 meters, known as the Command and Service Module, CSM. The initial launch mass of the CSM was 28801 kg, including fuels, of which the SM weighed 23244 kg while the CM had a mass of 5557 kg.

The *Eagle* Lunar Module

The Apollo 11 *Eagle* Lunar Module was the first crewed vehicle to land on our satellite. It carried two astronauts, mission commander Neil Armstrong and its pilot Edwin Buzz Aldrin, the first men to walk on the satellite surface. It was equipped with an Apollo Computer and also transported the first Apollo Surface Experiment Package, which consisted of several autonomous experiments to be deployed and left on the lunar surface; it also carried other scientific and sample collection devices.

The Lunar Module was a two-stage vehicle designed for space operations in either orbit or the lunar terrain. Its total dimensions were 6,98 m in height and 9,45 in width; and it had the capacity to transport two astronauts. The total mass of the assembled spacecraft was 15100 kg including its respective propellants, fuels and oxidants. The LM's ascent and descent stages operated as a unit until the ascent stage was turned on, separated and functioned as a single spacecraft for the ascent, encounter and coupling with the CSM waiting for it in lunar orbit.

The descent stage comprised the lower part of the spacecraft and was an octagonal prism of 4,2 meters in width and 3,23 in height. It contained the landing rocket, two fuel tanks, two tanks of oxidant, water, oxygen and helium tanks; as well as storage space for equipment and lunar experiments. Four landing legs with round cushions were installed on the sides of the descent platform which maintained the floor of this stage at 1,5 meters above the ground; one of the legs had a small exit platform and a stairway for the astronauts. This descent stage served as a launch pad for the ascent stage, and was left on the satellite.

The ascent stage was an irregularly shaped unit of approximately 3,76 m in height and up to 4,3 m in width, installed on the upper part of the descent platform. There the astronauts were housed in a pressurized compartment with a volume of 6,65 cubic meters that served as the base of operations for lunar activities; there were no seats in the LM but during the rest periods, while they were parked on the Moon, the crew slept in hammocks hanging in the cabin. It contained its own Ascent Propulsion System (APS) engine and two hypergolic propellant tanks for return to lunar orbit and rendezvous with the CSM. It also contained a Reaction Control System (RCS) for attitude and displacement control. It had an inlet and outlet hatch on one side, and a coupling hatch on the top to connect to the CSM. An Apollo Navigation Computer and a control console were mounted in the front of the crew compartment; various communication antennas were also mounted along the top; while at the base was the ascent engine and tanks for fuels and oxidants. The ascent stage was ignited and launched from the satellite at the end of lunar surface operations, and it returned the astronauts to CSM which awaited them in lunar orbit.

Image 8.4

Command and Service Module

The *Endeavor* Command and Service Module of the Apollo 15 mission photographed by the ascent stage of the *Falcon* Lunar Module. Both modules were in lunar orbit and were preparing to perform the respective Lunar Orbital Encounter, or Rendezvous, on August 2, 1971 for then start its return for home. The Service Module is the central cylindrical part, and the Command Module is the upper conical one.

The descent and ascent maneuvers were achieved by means of reaction control systems. The telemetry, television and voice communications with the Earth were made through S-band antennas, while the very high frequency, VHF, antennas were used for communications between the astronauts and the LM, and between the latter and the CSM in orbit. An environmental control system recycled oxygen and maintained the proper temperature in the crew cabin and the electronic equipment. Orientation and navigation's control were provided by a radar distance determination system, an inertial measurement unit consisting of gyroscopes and accelerometers, and controlled by the Apollo Guidance Compute. Power was provided by six batteries of silver and zinc.

Profile and development of the mission[16]

After launching with a Saturn V rocket on July 16, 1969 at 1:32 pm UCT and from Launch complex 39 at the Kennedy Space Center, the Apollo 11 mission entered Earth orbit twelve minutes later. After one and a half orbits the S-IVB stage was restarted at 16:16:16 UCT for the translunar injection maneuver, and it worked for 5 minutes and 48 seconds to put the spacecraft on course to the Moon.

After 25 minutes the CSM separated from the S-IVB launch stage that contained the Lunar Module inside, turned around and coupled with the LM at 16:56:03. Approximately fifty-three minutes later the S-IVB stage was definitive separated from the CSM and was discarded by launching it into a heliocentric orbit. Ten hours after the launch, and during the translunar flight, a color television broadcast was made from Apollo 11; and at 16:16:58 on July 17 a start of the main engine was made by about 3 seconds for course correction.

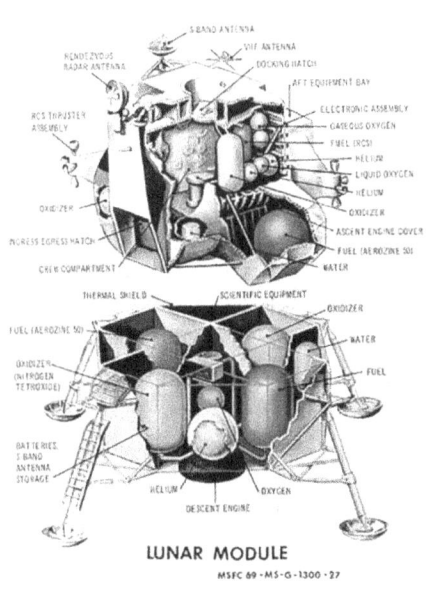

a)

b)

Image 8.5
The *Eagle* Lunar Module
a) Schematic diagram of the Lunar Module showing its two descent and ascent phases, and its main constituents. b) Apollo 11 *Eagle* Lunar Module in a moon landing configuration, it was photographed in lunar orbit from the *Columbia* Command and Service Module.

At 75 hours 49 minutes and 50 seconds on the mission, the insertion into lunar orbit was achieved on July 19 at 17:21:50 UCT by a retrograde ignition of the main engine for 3 minutes and 57 seconds, while the spacecraft was behind the Moon and out of contact with Earth. Almost four minutes later a 17-second ignition took the craft into a circular lunar orbit.

On July 20 at 12:52:00 Armstrong and Aldrin entered the LM for a final verification of the systems for the descent procedure. At 17:44:00 the LM and the CSM were uncoupled and after a visual inspection by Collins, at 18:11:53 a maneuver for distancing the two spacecraft was carried out.

The Lunar Module's descent stage engine was activated by 30 seconds at 19:08, putting the spaceship in a descent orbit with a maximum approach of 16,6 km over the Moon's surface. One hour later, at 20:05 the LM descent motor was activated again and the descent to the lunar surface began, which would last 12 minutes with 36 seconds.

After a journey of 102 hours 45 minutes and 39 seconds, equivalent to 4,3 days, the Eagle Lunar Module landed on the lunar surface at 20:17:40 UCT on July 20, 1969 at *Mare Tranquilitatis*, or the Sea of Tranquility. Commander Astronaut Neil Armstrong reported to the Apollo Mission Control Center in Houston pronouncing the phrase: *"Houston, Tranquility Base here - the Eagle has landed."*

After a brief period of rest, activities to perform the first Extra Vehicular Activity, EVA, of all human history on the Moon began. On July 21 Armstrong came down to the lunar surface at 02:56:15 UCT and said: *"That's one small step for man, one giant leap for mankind."* Buzz Aldrin followed him 19 minutes later and described the lunar surface as *Magnificent desolation*.

Then the astronauts uncovered the commemorative plaque mounted on a prop behind the descent module's ladder, and read the inscription aloud: *"Here men from the planet Earth first set foot on the Moon July 1969, AD; We came in peace for all mankind."*; it already was 03:24:40 UCT on our satellite. Following, at 03:41:43 they installed an American flag on the moon floor; they also placed a television camera on a tripod about nine meters from the Lunar Module, and seven minutes later they had telephone communication with President Richard Nixon.

Working on the Moon

Subsequently, the astronauts deployed their research equipment and instruments, conducted the respective experiments, made walks on the Moon's floor crossing a total distance of about 250 meters, took multiple photographs and collected about 21,55 kg of rock and soil from the satellite surface. The first experiments that were carried out on the Moon were:

Apollo Lunar Surface Experiments Package

Apollo 11 carried a small package of equipment called the Early Apollo Scientific Experiments Package, or EASEP. Later, the *Apollo Lunar Surface Experiments Package*, ALSEP, comprised a more complete set of scientific instruments placed by the spacemen at the landing site of each of the five Apollo missions following Apollo 11.

The first pack of surface experiments of the Apollo missions consisted of a set of scientific instruments deployed by the astronauts at the landing site. The Apollo Lunar Surface Experiments Package could only operate during the lunar day when enough sunlight was available, and it consisted of two solar panels to provide power, an antenna and a communications system to send data to terrestrial stations and to receive operating commands. It also contained a seismometer designed to measure the seismic activity and physical properties of the lunar crust and its interior; and a lunar dust detector to measure the accumulation of dust and radiation damage over the solar cells.

The Apollo 11 Experiment Package was deployed approximately 17 meters south of the LM and was turned on by ground command at 04:40:39 on July 21, 1969, while the astronauts were still on the surface. Although the operating temperatures exceeded the maximum planned at 30° C, the Package worked normally. The unit had a total mass of 48 kg, and received link commands and transmitted telemetry data to Earth.

Laser Ranging Retro-Reflector Experiment

The experiment called *Laser Ranging Retro-Reflector* was also considered part of the Apollo Experiment Package even though it was not connected to the unit and did not require power. Its objective was to detect and reflect a beam of laser light towards Earth, which was also fired from our planet and with the purpose of measuring with high precision the distance at which the Moon is at a certain moment. The experiment was included in three of the Apollo missions: 11, 14 and 15.

a)

b)

Image 8.6
Tranquility Base
a) At Tranquility Base the commander of the Apollo 11 mission, Neil Armstrong, unpacks some equipment from the Eagle Lunar Module. b) The Eagle Lunar Module at the Tranquility Base was photographed by Neil Armstrong during the Apollo 11 mission, from the edge of the Little West Crater on the satellite surface.

The experiment consists of a series of reflecting mirrors with the property of reflecting an incoming beam of light always in the same direction in which it came. These reflectors can be illuminated with laser beams directed through large telescopes installed on Earth; then the reflected rays are also observed with a terrestrial telescope, which provides a measure of the distance traveled by the light on its round trip between the Earth and the Moon. Between 1969 and 1985 the experiments were done part-time using the 107-inch telescope of the McDonald Observatory; and since 1985 the determinations have been made using an exclusive 30-inch telescope.

The distance to the Moon is calculated as half the product between the speed of light and the time it takes for the laser beam to reach the mirrors on the lunar globe, reflecting on them and return to the station on Earth.

In reality, the round trip time of approximately 2,5 seconds is affected by multiple factors: the location of the Moon in the sky, the relative movement between Earth and the satellite, the Earth's rotation, the lunar libration, the weather; and also by the movement of the observation station due to the movement of the crust and the terrestrial tides, the different speed of light in various parts of the atmosphere and the relativistic effects. However, the Earth-Moon distance has been measured with very good precision for more than 35 years; and although it changes continuously for a number of reasons, its average value is 385000 km.

Lasers are used because they stay much focused at great distances. However there is enough dispersion of the ray, which is about 7 kilometers in diameter when it reaches the moon, and scientists compare the task of directing the ray with the use of a rifle to hit a dime located 3 miles away.

Image 8.7

Working on the Moon

Astronaut Buzz Aldrin has just deployed the Apollo Scientific Experiment Package. The Passive Seismic Experiment Package is next to him; the solar panels have been deployed to the East and West, and the antenna points towards the Earth. The retro reflector laser is beyond the antenna, and at the bottom is the television camera against the sky's black background.

On the other hand, when the laser beam returns to Earth it has a diameter of 20 kilometers, and it is too weak to be seen with the human eye: of 1017 photons directed to the reflector, only one is detected on Earth every few seconds, even in good conditions. Even so, the Moon's distance can be measured with an accuracy of approximately 3 centimeters.

The laser experiment has produced many important measurements and which lead to a better understanding of the lunar orbit and the speed at which the satellite is retreating from Earth, which currently is 3,8 centimeters per year. They also provide a better understanding of the variations in the lunar globe's rotation movement, which are related to the distribution of the mass within the satellite and involve the existence of a small liquid core, with a radius of approximately 350 kilometers. These measurements have also improved our knowledge of the changes in the Earth's rotation rate and the precession movement of its rotation axis; and they have been used to prove that the universal gravity force is very stable, and that Einstein's general theory of relativity predicts the satellite orbit within the precision of the experiments results.

Ascent and lunar orbit abandonment

Once all those activities on the Moon were done according to the established schedule, Buzz Aldrin returned first to the LM after having spent 1 hour and 45 minutes outside it and walking on the lunar surface; Armstrong followed him about 8 minutes later at 05:09:32 UCT and after having walked 2 hours and 14 minutes over the lunar globe. The EVA ended at 05:11:13 when the LM hatch was closed and Armstrong and Aldrin spent the next 7 hours resting and checking systems.

The propulsion engines were activated and the LM's ascent stage took off from the Moon at 17:54:01 UCT on July 21, after 21 hours with 36 minutes on the lunar surface and leaving there the descent stage. After reaching the CSM in lunar orbit and piloted by Michael Collins, the ascent stage docked with it at 9:35:00 and once Aldrin and Armstrong were transferred to the CSM, the LM's ascent phase was abandoned in the lunar orbit at 00:02:08 UTC on July 22. The actual destination of the ascent stage is unknown, but it is assumed that it crashed on the satellite surface at some point within the next 1 to 4 months.

In addition to the descent stage, commemorative medallions were left on the satellite surface with the three astronauts' names of Apollo 1 mission whom lost their lives during a fire on the launch pad, and of two Russian cosmonauts who also died in space accidents. An inch and a half silicon disc was also left there, containing micro-miniaturized friendly messages from 73 countries; along with the names of the leaders of Congress and NASA.

Trans-terrestrial injection and returning home

The Trans-terrestrial injection started at 04:55:42 on July 22 with a 2,5 minute ignition of the CSM main engine; the first course correction was made the same day at 20:01:57. The following day, July 23 at 01:00 UCT, a television broadcast that lasted 18 minutes was made, this was the first of a series of three broadcasts. After fifty-nine and a half hours of having performed the Trans-terrestrial injection, the SM definitively separated from the CM at 16:21:13 of July 24 and this one took a course away from the SM and prepared for reentry the planet. After separation, the Service Module will enter the Earth's atmosphere and eventually burn up.

Re-entry and splashdown

The reentry procedures to the Earth's atmosphere began on July 24 at 16:23:00 UCT, about sixty hours after leaving the lunar orbit; the CM entered the Earth's atmosphere twelve minutes later, reoriented to a position of thermal shield forward and about nine minutes later the deployment of the parachutes began. After an 8-day 3-hour and 18-minute flight, approximately 36 minutes longer than planned, the Columbia Command Module of the Apollo 11 mission, with its full crew safeguarded, achieved a splashdown in the Pacific Ocean on July 24 at

16:50:35. The drop point was about 24 km from the recovery ship USS Hornet. Of the 2950000 kgs thrown into space, only slightly less than 5000 Kgs represented in the CM and its astronauts returned to the Earth, this is only 0,17% of that initial weight.

The Apollo 11 mission in July 1969 was totally integrated by a team of Gemini veterans: Neil Armstrong, Michael Collins and Buzz Aldrin. The astronauts Armstrong and Aldrin would be the happy ones in charge to realize the first human lunar landing in the Sea of the Tranquility on the Moon on July 20, 1969. They spent a total of 21 hours and 36 minutes on the satellite surface, and 2 hours and 31 minutes outside the space capsule: they were walking on the lunar terrain, taking photographs, collecting samples of materials and deploying automated scientific instruments. In addition, they continuously sent black and white television signals to the Earth. The astronauts returned safely at home on July 24, eight days after their departure.

The spacecraft's performance was excellent throughout the mission. The mission's main objective of landing astronauts on the Moon and return them alive to Earth was achieved. The Apollo 11 command module is on display at the National Air and Space Museum in Washington, DC.

Collins, Armstrong and Aldrin were awarded by Richard Nixon with the Presidential Medal of Freedom. However, Aldrin and Armstrong ended up receiving the most public credit for the historic event, although Collins was also on the flight. Neil Armstrong and Michael Collins also received the Gold Medal of Congress in 2011, and were honored with four stars on the Hollywood Walk of Fame in California.

Subsequent Apollo Missions

Apollo 12

The main objectives of the second manned moon landing mission were to deploy a full version of the Apollo Surface Experiment Package, which was to be installed and left on the satellite surface to collect scientific, seismic and engineering data for an indefinite period of time. Additionally, astronauts had to perform an extensive series of lunar exploration and research tasks.

During the Apollo 12 mission of November 1969, Gemini veteran Charles Conrad Jr. as Commander and rookie Alan L. Bean as Lunar Module Pilot, made a precision moon landing in the *Oceanus Procellarum*, or Ocean of Storms, at short distance from the unmanned lunar probe Surveyor 3 which had landed on April 1967. They carried the first color television camera to the satellite surface, but it was damaged when it was accidentally directed to the Sun. They also performed two EVAs for a total of 7 hours and 45 minutes, during which they collected 34,3 kg of lunar samples. In one of the activities they walked towards the Surveyor 3, they photographed it and took some of its parts to return them to Earth. The Command Module Pilot was Gemini veteran Richard F. Gordon.

Apollo 13 a *Successful failure*

The success of the first two lunar landings made it possible for future missions to take as a crew a single veteran as Commander, along with two rookie astronauts. The Apollo 13 mission launched on April 11, 1970, carried astronauts James A. Lovell Jr. as Commander, John L. Swigert Jr. as CM Pilot and Fred W. Haise Jr. as LM Pilot; and it was destined for the lunar formation Fra Mauro. But at mid-way, two days after the launch, some imperfect components in a liquid oxygen storage tank made it to blow up, disabling the SM which had to be abandoned. Which forced the crew to abort the moon landing mission although even in lunar orbit, and to return home on the Command Module, but using the Lunar Module as an improvised *lifeboat* and power source before it was also abandoned. Finally, the CM with the safe astronauts re-entered Earth and managed to splashdown 5 days 22 hours and 54 minutes after its launch. Even so, the mission was classified as a *successful failure* due to the experience gained in the salvage and rescue of the crew.

Apollo 14

After conducting the respective investigations on the Apollo 13 mission's accident, NASA published a preliminary list of eight additional projected moon landing sites; along with plans for increasing the cargo capacity of the Saturn V rocket, the CSM and of the LM for the next five missions; which would allow the astronauts to remain on the Moon for more than three days. Additionally, in order to increase the exploration area and to broadcast the LM lift off from the satellite surface on television, these missions would also carry a *Lunar Exploration Vehicle*, better known as *Rover*. The Block II space suit would also be reevaluated for extended missions with the aim of enabling greater flexibility and visibility when driving the Rover.

Due to the failure of Apollo 13, the mission to Fra Mauro was reassigned to Apollo 14 which was launched on February 31, 1971, and was commanded by Mercury veteran astronaut Alan B. Shepard Jr., with Stuart A. Roosa as the CM Pilot and Edgar D. Mitchell as LM pilot. Shepard and Mitchell spent 33 hours and 31 minutes on the lunar surface, and performed two EVAs over a period of 9 hours and 24 minutes, setting a record for that time. During their stay in the satellite they collected a total of 42,80 kg of lunar soil and rock samples.

Apollo 15

Apollo 15 was the first mission with the greatest sojourn time capacity over the Moon, and the first to carry a Lunar Exploration Vehicle for a greater displacement capacity on the satellite surface. The objectives of the mission were to explore the Hadley Rille-Apennine Mountains region, configure and activate scientific experiments on the lunar surface, perform engineering assessments of new Apollo equipment, conduct lunar orbital experiments and take photographs of the terrain.

Apollo 15 was launched on July 26, 1971, and the astronauts David R. Scott, Alfred M. Worden and James B. Irwin were its crew. Scott and Irwin landed in the LM on July 30 on *Palus Putredinis* in *Mare Imbrium*, or Sea of Showers, near the Hadley Rille-Apennine Mountains region. They spent little less than two days 20 hours on the lunar surface: in about 18 hours on two EVAs, they collected approximately 77 kilograms of lunar material.

The Apollo 15 *Falcon* Lunar Module was the fourth manned spacecraft to successfully land on the Moon. It carried two astronauts, Commander David R. Scott and its Pilot James B. Irwin, the seventh and eighth man to walk on the Moon respectively. In addition to traditional equipment, this LM also carried the first Lunar Exploration Vehicle.

The terrain exploration and the geological investigations at Hadley Rille-Apennine Mountains region, near the landing site, were greatly improved with the incorporation of the Lunar Exploration Vehicle. The configuration of the Apollo Experiments Package on the surface was the third of a trio of complete and operational packages incorporated in missions 12, 14 and 15. Experiments on orbital theories focused on a set of instruments and cameras in the Module of Scientific instruments.

Another important objective of the mission consisted on the launch from the CSM of a sub satellite of particles and fields to the lunar orbit, shortly before beginning the phase of returning to Earth. The subsatélite was designed to investigate the mass of the Moon and the respective gravitational variations, the composition of the particles of the space near the lunar globe, and the interaction of the lunar magnetic field with that of the Earth.

Driving and exploring the Moon

The Lunar Exploration Vehicle[17], better known as Rover, was an electric vehicle designed to operate in the vacuum of low lunar gravity and to be able to roll on the difficult satellite soil, which allowed the Apollo astronauts to extend the scope of their Extravehicular Activities in the satellite surface. Three Rovers were driven on the Moon, one on Apollo 15 by astronauts David Scott and Jim Irwin, another on Apollo 16 by John Young and Charles Duke, and finally that one of Apollo 17 driven by Gene Cernan and Harrison Schmitt. Each Rover was used in three voyages, one per day during the three days duration on each mission. During Apollo 15, the Lunar Exploration Vehicle was driven a total of 27,8 km in 3 hours 2 minutes of time, the longest single crossing was 12,5 km and the maximum distance from the LM was 5,0 km. In the Apollo 16 mission the vehicle crossed 26,7 km in 3 hours and 26 minutes of driving; the longest crossing was 11,6 km and the Rover reached a distance of 4,5 km from the LM. During the Apollo 17 the vehicle traveled 35,9 km in 4 hours and 26 minutes in total driving time; the longest crossing was 20,1 km and the greatest distance from the LM was 7,6 km.

The rovers had a mass of 210 kg and were designed to contain an additional load of 490 kg on the lunar surface. The chassis was 3.1 meters long with a wheelbase of 2.3 meters, while the maximum height was 1.14 meters. The structure was made of aluminum alloy assemblies and consisted of a 3-part chassis that could be folded and stored in a compartment of the Lunar Module.

It had two folding seats made of tubular aluminum side by side, and aluminum panels as floor. A large mesh dish antenna was mounted on a mast in the front center of the Rover. Fully loaded, the rover had a ground clearance of 36 cm.

Apollo 16

The Apollo 16 mission was launched on April 16, 1972, and landed in Moon's Descartes Highlands on April 20, 1972. The crew was commanded by John Young, whereas Ken Mattingly was the Command Module Pilot and Charles Duke the Lunar Module Pilot. The astronauts Young and Duke spent little less than three days on the satellite surface, performing extravehicular activities for about 20 hours, and during which they collected 94,30 kg of satellite samples. The mission returned to Earth on April 27 and had a total duration of eleven days one hour and fifty-one minutes.

Image 8.8

Driving and Exploring the Moon

a) Astronaut Commander Eugene A. Cernan tests the Lunar Exploration Vehicle during the first Extra Vehicular Activity of Apollo 17 at the *Taurus-Littrow* landing site.

Apollo 17

Apollo 17 was the last manned mission of the Apollo space program and landed on the Moon in *Taurus-Littrow* region on Dec 11, 1972. Astronaut Eugene A. Cernan commanded the mission; his companions were the Command Module Pilot Ronald E. Evans and NASA's first scientist astronaut, the geologist Dr. Harrison H. Schmitt as the LM Pilot, who was originally scheduled for Apollo 18 mission, but due to the cancellation of such mission, the lunar geological community pushed for including him in the last lunar landing. Cernan and Schmitt remained on the surface for a little more than three days and spent a total of 22 hours 3 minutes and 57 seconds in EVAs, during which they collected 110,52 kg of lunar soil samples.

Achievements and Legacies of the Apollo Missions

The Apollo program established several important landmarks of human space flights. Until now it is the only one to send crewed missions beyond Earth's low orbit. Apollo 8 was the first manned spacecraft to orbit another celestial body, while the final mission Apollo 17 was the ninth manned mission beyond low Earth orbit and the successful sixth manned moon landing. In general, the program brought 382 kg of soil and lunar rocks to the Earth, which largely served to study and understand both the chemical composition of the Moon and its geological history and structure. The program laid the groundwork for the further development of human spaceflight capability and promoted the construction of the Johnson Space Center and the Kennedy Space Center.

b)

Image 8.8

Driving and Exploring the Moon

b) Close-up view of the astronaut Buzz Aldrin's boot footprint on the lunar floor, photographed with the 70 mm lunar surface camera during the Moon sojourn of the Apollo 11 mission.

The Apollo program and its lunar landings are usually considered to be the greatest scientific and technological achievement in the history of humanity, including over nuclear technology and its bombs. Apollo program also favored developments in many other areas of technology related to rocketry, aeronautics and manned space flights, such as computers, telecommunications and avionics. The design of the flight computer used in this program was the major driving force for the research and development of integrated circuits, great protagonists of the modern electronic and digital era: by 1963 the Apollo program was using something like 60 percent of the integrated circuits total production of the United States.

The Apollo program included a large number of unmanned test missions and 12 manned ones: three missions in orbit around the Earth, Apollo 7, 9 and the Apollo-Soyuz mission; two voyages in lunar orbit, the Apollo 8 and 10; an aborted lunar mission, Apollo 13; and finally six successful landing missions: Apollo 11, 12, 14, 15, 16 and 17. Two astronauts from each of these six voyages walked on the satellite: Neil Armstrong, Edwin Aldrin, Charles Conrad, Alan Bean, Alan Shepard, Edgar Mitchell, David Scott, James Irwin, John Young, Charles Duke, Gene Cernan and Harrison Schmitt, becoming the only humans who have set foot in another body of the Solar System.

Collected Lunar Samples

The Apollo program collected and brought some 381 kg of lunar soil and rocks to our planet, which are now stored in nitrogen ambient to keep them free of moisture and are handled only indirectly, using special tools. The vast majority are conserved in the Lunar Samples Laboratory Facility at the Lyndon B. Johnson Space Center in Houston, Texas. There is also a smaller collection stored at the White Sands Test Facility in Las Cruces, New Mexico.

A very important sample collected during the Apollo 15 mission by astronauts David Scott and James Irwin is called the *Genesis Rock*; it is an anorthosite rock and is composed mostly of a type of plagioclase feldspar known as anorthite, a calcium-rich feldspar mineral; and it is believed to be representative of the highlands' crust. Initially it was thought that it was a sample of the Moon's primitive cortex, but a later analysis showed that it was formed after the satellite crust had been solidified. The rock is an extremely ancient sample that was formed in the early stages of the Solar System, at least 4 billion years ago and it corresponds to the pre-Nectarian period of the Moon's geologic history.

Cultural Impact

The crew of the Apollo 8 mission sent back to Earth the first live televised images of our planet and its nearby and faithful satellite; and on the day before Christmas 1968 they read the Story of the Universal Creation contained in the biblical Book of Genesis: Up to a quarter of the world's population could have seen, either live or delayed, the Christmas Eve broadcast during the ninth lunar orbit. On the other hand, about a fifth of the world's population saw the live broadcast of the lunar walks by the crew of Apollo 11.

During the next 1970s decade, the Apollo program also intensified the level of environmental awareness and positively influenced environmental activism, all this due to the multiple photos of our planet taken by astronauts. The most famous of all probably is the one realized by the Apollo 17's crew: *The Blue Marble*, as it is known from that moment, is considered as a representation of the isolation, fragility and the vulnerability of our planet in middle of the astronomical immensity, darkness and solitude of space; and it became an authentic icon of the Environmental Movement of those dates.

According to some newspapers of the time, the Apollo program managed to fulfill the objective established by President Kennedy to confront the Soviet Union in the space race and defeat it, conquering a monumental and universally significant achievement, demonstrating the great superiority of the capitalist free market system represented by the North American country. Even so, and in spite of the multiple and varied evidence for the contrary, there were and

continue to be sectors of the population that deny the truthfulness of such lunar landings, and rather are inclined to avail themselves of supposed *conspiracy theories* with which people are being deceived.

In Table 8.1 some very important numerical data are included about the set of six successful Apollo landing missions. The Landing Date refers to the day on which the respective Lunar Module landed on the satellite surface. The Total Duration of the mission is counted in days, hours and minutes from the launch day until the Command Module returns and gets splashdown. The astronauts' Sojourn Time on the surface is counted in days, hours and minutes from the moment the Lunar Module landed until it finally lifted off from the lunar surface. EVAs Time is the sum of the times in hours, minutes and seconds used by astronauts in the respective activities outside the Lunar Module. The Collected Samples is the sum of the weight in kg of all those samples collected from the satellite surface, packed and carried to the Earth in each of the missions.[18]

Table 8.1 The Apollo Missions in Numbers					
Mission	Landing Date d/m/a	Total Duration d:h:m	Sojourn Time d:h:m	EVAs Time h:m:s	Collected Samples Kg
Apollo 11	20/Jul/1.969	8:3:18	0:21:30	2:31:40	21,55
Apollo 12	19/Nov/1.969	10:4:30	1:7:30	7:24:00	34,35
Apollo 14	5/Feb/1.971	9:00:01	1:9:30	9:28:10	42,80
Apollo 15	30/Jul/1.971	12:7:12	2:18:57	18:22:48	76,70
Apollo 16	21/Abr/1.972	11:1:55	2:23:03	20:13:12	95,21
Apollo 17	11/Dic/1.972	12:13:55	3:3:07	21:39:00	110,42
Totals		63:6:51	12:11:37	79:38:50	381,03

a)

Image 8.9

The Moon and the Earth seen by astronauts

a) Exceptional photograph of the Full Moon taken from the Apollo 11 spacecraft during its journey back to Earth and at a distance of about 18500 kilometers.

Image 8.9

The Moon and the Earth seen by astronauts

b) The Blue Marble, one of the most famous Earth's photographs, was taken by the Apollo 17 mission crew as they travelled to the Moon.

Bibliographic Citations

[1] Internet Archive: Jules Verne. *From the Earth to the Moon; and Round the moon.* https://archive.org/details/cu31924052535725/page/n8

[2] Royal Astronomical Society. "Celebrating the 20th century's most important experiment". http://www.ras.org.uk/news-and-press/68-news2009/1627-ras-pn-0942-celebrating-the-20th-centurys-most-important-experiment

[3] ZARYA. Soviet, Russian and International Space Flight. http://www.zarya.info/Diaries/Luna/Luna.php

[4] NASA Solar System Exploration Page: http://solarsystem.nasa.gov/
Philip's Astronomy Encyclopedia, (London : Philip's, 2002). www.philips-maps.co.uk.

[5] NASA. Project Mercury. https://www.nasa.gov/mission_pages/mercury/index.html

[6] NASA. Apollo Lunar Surface Experiments Package. https://www.lpi.usra.edu/lunar/documents/NASA-CR-115109.pdf

[7] NASA. Apollo Operations Handbook, Block II Spacecraft, Volume 1, Spacecraft Description, SM2A-03-Block II-(1). https://history.nasa.gov/alsj/alsj-CSMdocs.html

[8] NASA. Apollo 8. https://www.nasa.gov/mission_pages/apollo/missions/apollo8.html

[9] NASA. Apollo 9. https://www.nasa.gov/mission_pages/apollo/missions/apollo9.html

[10] NASA. Apollo 10. https://www.nasa.gov/mission_pages/apollo/missions/apollo10.html

[11] NASA. July 20, 1969: One Giant Leap for Mankind. https://www.nasa.gov/mission_pages/apollo/apollo11.html

[12] NASA. Biography of Neil Armstrong. https://www.nasa.gov/centers/glenn/about/bios/neilabio.html
NASA. Armstrong Fact Sheet. https://www.nasa.gov/centers/armstrong/news/FactSheets/FS-111-AFRC.html

[13] NASA. News & Features. Family statement regarding the death of Neil Armstrong https://www.nasa.gov/home/hqnews/2012/aug/HQ_12_600_armstrong_family.html#.XKO3wvdKhhE

[14] The New York Times. The Call of Mars . https://www.nytimes.com/2013/06/14/opinion/global/buzz-aldrin-the-call-of-mars.html

[15] Buzz Aldrin Space Institute. https://aldrin.fit.edu/#!page_id=24875a285079e65e2

[16] NASA. Apollo 11 Timeline. https://history.nasa.gov/SP-4029/Apollo_11i_Timeline.htm

[17] NASA. The Apollo Lunar Roving Vehicle. https://nssdc.gsfc.nasa.gov/planetary/lunar/apollo_lrv.html

[18] Fricke. Robert W. J r. Apollo by the numbers: A Statistical Reference. https://history.nasa.gov/SP-4029/Apollo_00g_Table_of_Contents.htm

Image Catalog and Their Licenses for Use

I wish to express my deep gratitude to all those people and institutions that have put the following images in the field of the Public Domain; which have been of great value to illustrate the respective contents in this text.

Special thanks to the National Aeronautics and Space Administration of the United States, NASA, for having their images within the Public Domain; and in particular those contained in its website NASA's Scientific Visualization Studio: https://svs.gsfc.nasa.gov/4604

Cover and Back Cover

#	Image Title	Author	License Type	Link to the Image
1	Half Moon of the 24 of January of 2.018	NASA's Scientific Visualization Studio	Public Domain	https://svs.gsfc.nasa.gov/4604
2	Astronaut Buzz Aldrin's bootprint in the lunar soil	NASA	Public Domain	https://www.nasa.gov/mission_pages/apollo/40th/images/apollo_image_11a.html
3	Saturn V Rocket at the Launch of Apollo 11	NASA	Public Domain	https://history.nasa.gov/ap11ann/kippsphotos/saturn5.html
4	Apollo 11 Lunar Module (LM) "*Eagle*"	NASA	Public Domain	https://www.nasa.gov/multimedia/imagegallery/image_feature_1161.html
5	Buzz Aldrin deploys the Apollo 11 Scientific Experiments Package	NASA	Public Domain	https://www.hq.nasa.gov/alsj/HamishALSEP.html
6	Eugene A. Cernan tests the Apollo 17 Lunar Exploration Vehicle	NASA	Public Domain	https://en.wikipedia.org/wiki/File:NASA_Apollo_17_Lunar_Roving_Vehicle.jpg
7	Full Moon Photographed From Apollo 11 Spacecraft	NASA	Public Domain	https://www.nasa.gov/mission_pages/apollo/40th/images/apollo_image_25.html

Chapter 1

#	Image Title	Author	License Type	Link to the Image
1	Sunset at Stonehenge	Bkamprath	Public Domain	https://commons.wikimedia.org/wiki/File:Stonehenge_sunset.jpg
2	Cuneiform tablet		Public Domain Creative Commons	https://www.metmuseum.org/art/collection/search/321987
3	Egyptian papyrus	Anagoria	GNU_Free_Documentation_License	https://commons.wikimedia.org/wiki/File:0042_Planetentafel_anagoria.JPG#
4	Selene and Endymion	Sebastiano Ricci	Public Domain	https://commons.wikimedia.org/wiki/File:Sebastiano_Ricci_015.jpg
5	Dragons and eclipses	The author		
6	Ixchel, the Mayan lunar goddess	Unknown Maya artist	Public Domain	https://commons.wikimedia.org/wiki/File:Goddess_O_Ixchel.jpg
7	Archaeoastronomy. The Nebra Sky Disc	Dbachmann, Theway	Creativecommons.org/license 4.0	https://commons.wikimedia.org/wiki/File:Nebra_sky_disk.png
8	Archaeoastronomy. The Antikythera Mechanism		GNU_Free_Documentation_License	https://commons.wikimedia.org/wiki/File:NAMA_Machine_d%27Anticyth%C3%A8re_1.jpg
9	The different Cycles of the Moon	The author		
10	The synodic month and the lunar phases	The author		

Chapter 2

#	Image Title	Author	License Type	Link to the Image
1	Anaximander Cosmology	The author		
2	Pythagorean Cosmology	The author		
3	Geometry of the Sun-Earth-Moon system according to Aristarchus	The author	Public Domain	https://commons.wikimedia.org/wiki/File:Aristarchus_working.jpg
4	General geometry of a Solar Eclipse.	The author		
5	Half Moon: The Moon just half illuminated	The author		
6	General geometry for a Lunar Eclipse.	The author		
7	Celestial Uniform Circular Movement	The author		
8	The Eratosthenes's experiment.	The author		

Chapter 3

#	Image Title	Author	License Type	Link to the Image
1	Some geographical areas of the ancient Greek-Roman world	Modified	Creativecommons.org/licenses3.0	https://commons.wikimedia.org/wiki/File:Extent_of_the_Roman_Republic_and_the_Roman_Empire_between_218_BC_and_117_AD.png
2	The Parallax	The author		
3	Movement of the Sun: The Ecliptic, the Equinoxes and Solstices	The author		
4	The Eccentric Deferens and the Apses	The author		
5	The Hipparchus' lunar model	The author		
6	The Ptolemy's lunar model	The author		
7	The Ptolemaic Geocentric Universe.	The author		

Chapter 4

#	Image Title	Author	License Type	Link to the Image
1	Medieval astrological celestial map	Pintores del Sultán Murad III	Public Domain	https://commons.wikimedia.org/wiki/File:Celestial_map,_signs_of_the_Zodiac_and_lunar_mansions..JPG?uselang=fr
2a	Kitāb al-Majisṭī: The medieval Arab Almagest	Qatar Digital Library's digital archive	Public Domain/mark/1.0/	https://www.qdl.qa/en/archive/81055/vdc_100023514339.0x00000e
2b	Kitāb al-Majisṭī: The medieval Arab Almagest	Qatar Digital Library's digital archive	Public Domain/mark/1.0/	https://www.qdl.qa/en/archive/81055/vdc_100023514339.0x00000e
3	The Lunar phases according to medieval astronomy	Al-Biruni	Public Domain	https://commons.wikimedia.org/wiki/File:Lunar_eclipse_al-Biruni.jpg
4	A medieval look for the Moon	Istanbul University Library	Creative Commons 2.0	https://en.wikipedia.org/wiki/File:Astronomes_-_miniature_ottomane_XVIIe.jpg
5	Medieval diagram for a lunar eclipse	British Library	Granted Permission	http://www.bl.uk/catalogues/illuminatedmanuscripts/ILLUMIN.ASP?Size=mid&IllID=10144
6	Medieval outline of the geocentric Universe	Sacro Bosco	Public Domain	https://digitalcollections.nypl.org/search/index?filters%5BnamePart_mtxt_s%5D%5B%5D=Sacro%20Bosco%2C%20Joannes%20de%20%28fl.%201230%29&keywords=&layout=false
7	The Lunar phases according to medieval astronomy	British Library	Granted Permission	http://www.bl.uk/catalogues/illuminatedmanuscripts/ILLUMIN.ASP?Size=mid&IllID=10159
8	The eclipses in medieval astronomy	metmuseum.org/art/	creativecommons.org/publicdomain	https://www.metmuseum.org/art/collection/search/356589

Chapter 5

#	Image Title	Author	License Type	Link to the Image
1	Lunar Eclipse of Christopher Columbus of 1,504	Nicolas Camille_Flammarion	Public Domain	https://en.wikipedia.org/wiki/March_1504_lunar_eclipse
2	The Copernican Heliocentric System	Nicolaus Copernicus	Public Domain	https://www.wdl.org/es/item/3164/
3	Copernican Planisphere	Andreas Cellarius (1596–1665)	Public Domain	https://commons.wikimedia.org/wiki/File:Cellarius_Harmonia_Macrocosmica_-_Planisphaerium_Copernicanum.jpg
4	Copernican Lunar Model	El autor		
5	Tychonian model of the Universe	Fastfission.	Public Domain	https://commons.wikimedia.org/wiki/File:Tychonian_system.svg

Chapter 6

#	Image Title	Author	License Type	Link to the Image
1	Elliptical geometry	The author		
2	Elliptical Lunar orbit	The author		
3	A Voyage to the Moon	Gustave Dore	Public domain	https://www.wikiart.org/en/gustave-dore/a-voyage-to-the-moon
4	Lunar Day	Henry White Warren	Public Domain	https://en.wikipedia.org/wiki/File:Old_view_moon.jpg
5	Earth and moon in space	Henry White Warren	Public Domain	https://archive.org/details/recreationsinast00warriala
6	Galileo teaching the Doge of Venice the use of the telescope	Giuseppe_Bertini	Public Domain	https://commons.wikimedia.org/wiki/File:Bertini_fresco_of_Galileo_Galilei_and_Doge_of_Venice.jpg
7	Cover of the Sidereus nuncius, Venice edition of 1610	Galileo Galilei	Public Domain	https://en.wikipedia.org/wiki/File:Houghton_IC6.G1333.610s_-_Sidereus_nuncius.jpg
8	The Moon as Galileo saw it	Galileo	Public Domain	https://en.wikipedia.org/wiki/File:Galileo%27s_sketches_of_the_moon.png
9	Calculation of the height of the lunar mountains	The author		
10	The Moon photographed by the Apollo 17 mission	NASA	Public Domain	https://en.wikipedia.org/wiki/File:Mare_Imbrium-Apollo17.jpg

Chapter 7

#	Image Title	Author	License Type	Link to the Image
1	Observatory of Paris	Wolf, Charles J. E.	Public Domain	https://en.wikipedia.org/wiki/File:Paris_Observatory_XVIII_century.png
2	Carte de la Lune by Giovanni D. Cassini	Giovanni_Cassini	Public Domain	https://commons.wikimedia.org/wiki/File:Carte_de_la_Lune_de_Giovanni_Domenico_Cassini.jpg
3	The Three Bodies Problem	The author		
4	The 40-foot telescope		Public Domain	https://commons.wikimedia.org/wiki/File:Herschel_40_foot.jpg
5	Luna by William Parsons	Richard Adams Locke, Joseph Nicolas Nicollet	Public Domain	https://commons.wikimedia.org/wiki/File:Moon_Rosse_Telescope_1856.png
6	Earth - Moon System: Tides and Torque	The author		
7a	From Earth to the Moon by Jules Verne	Henri de Montaut	Public Domain	https://commons.wikimedia.org/wiki/File:%27From_the_Earth_to_the_Moon%27_by_Henri_de_Montaut_31.jpg
7b	Voyage to the Moon by Georges Méliès		Public Domain	https://commons.wikimedia.org/wiki/File:Le_Voyage_dans_la_lune.jpg

Chapter 8

#	Image Title	Author	License Type	Link to the Image
1	General diagram of an Apollo Spacecraft	NASA	Public Domain	https://en.wikipedia.org/wiki/File:Apollo_Spacecraft_diagram.jpg
2	Rocket Saturn V	NASA	Public Domain	https://history.nasa.gov/ap11ann/kippsphotos/39525.jpg
3	General profile of a Manned Apollo Mission	The author		
4	Command and Service Module	NASA	Public Domain	https://en.wikipedia.org/wiki/File:Apollo_CSM_lunar_orbit.jpg
5a	The Lunar Module Eagle	NASA	Public Domain	https://en.wikipedia.org/wiki/File:LM_illustration_02.jpg
5b	The Lunar Module Eagle	NASA	Public Domain	https://www.nasa.gov/multimedia/imagegallery/image_feature_1161.html
6a	Neil Armstrong at the Tranquility Base	NASA	Public Domain	https://www.nasa.gov/mission_pages/apollo/apollo11.html
6b	The Lunar Module at the Tranquility Base	NASA	Public Domain	https://www.nasa.gov/image-feature/lunar-module-at-tranquility-base
7	Working on the Moon	NASA	Public Domain	https://www.hq.nasa.gov/alsj/HamishALSEP.html
8a	Eugene A. Cernan tests the Lunar Exploration Vehicle	NASA	Public Domain	https://en.wikipedia.org/wiki/File:NASA_Apollo_17_Lunar_Roving_Vehicle.jpg
8b	Astronaut bootprint Buzz Aldrin	NASA	Public Domain	https://www.nasa.gov/mission_pages/apollo/40th/images/apollo_image_11a.html
9a	Full moon photographed from the Apollo 11 spacecraft	NASA	Public Domain	https://www.nasa.gov/mission_pages/apollo/40th/images/apollo_image_25.html
9b	The Blue Marble by Apollo 17 astronauts	NASA	Public Domain	https://www.nasa.gov/image-feature/apollo-17-blue-marble

www.ingramcontent.com/pod-product-compliance
Lightning Source LLC
Chambersburg PA
CBHW080452220526
45465CB00006B/2249